Ben Stacy Jerrik (Ed.)

Evolution Of Reptiles

AF693906

Ben Stacy Jerrik (Ed.)

Evolution Of Reptiles

Dinosaur, Dinosauriformes, Theropoda

Part Press

Imprint

Permission is granted to copy, distribute and/or modify this document under the terms of the GNU Free Documentation License, Version 1.2 or any later version published by the Free Software Foundation; with no Invariant Sections, with the Front-Cover Texts, and with the Back- Cover Texts. A copy of the license is included in the section entitled "GNU Free Documentation License".

All parts of this book are extracted from Wikipedia, the free encyclopedia (www.wikipedia.org).

You can get detailed informations about the authors of this collection of articles at the end of this book. The editors (Ed.) of this book are no authors. They have not modified or extended the original texts.

Pictures published in this book can be under different licences than the GNU Free Documentation License. You can get detailed informations about the authors and licences of pictures at the end of this book.

The content of this book was generated collaboratively by volunteers. Please be advised that nothing found here has necessarily been reviewed by people with the expertise required to provide you with complete, accurate or reliable information. Some information in this book maybe misleading or wrong. The Publisher does not guarantee the validity of the information found here. If you need specific advice (f.e. in fields of medical, legal, financial, or risk management questions) please contact a professional who is licensed or knowledgeable in that area.

Any brand names and product names mentioned in this book are subject to trademark, brand or patent protection and are trademarks or registered trademarks of their respective holders. The use of brand names, product names, common names, trade names, product descriptions etc. even without a particular marking in this works is in no way to be construed to mean that such names may be regarded as unrestricted in respect of trademark and brand protection legislation and could thus be used by anyone.

Cover image: www.ingimage.com
Concerning the licence of the cover image please contact ingimage.

Publisher:
Part Press is a trademark of
International Book Market Service Ltd., 17 Rue Meldrum, Beau Bassin, 1713-01 Mauritius
Email: info@bookmarketservice.com
Website: www.bookmarketservice.com

Published in 2013

Printed in: U.S.A., U.K., Germany. This book was not produced in Mauritius.

ISBN: 978-613-9-16048-8

Contents

Articles

References

Peter_Brownback

Peter E. Brownback III	
Allegiance	United States of America
Service/branch	United States Army
Rank	Colonel
Unit	173rd Brigade
Battles/wars	Vietnam War
Other work	Presiding Officer Lawyer

Peter E. Brownback III is a retired military officer and lawyer.[1] He was appointed to be a Presiding Officer on the Guantanamo military commissions, by retired general John D. Altenburg.

Brownback is also a combat veteran who served two tours in Vietnam. The first as part of the 173rd Brigade and later as a member of the US Special Forces. He was the honour graduate of his special forces course that he completed in 1972. The Washington Post reported: *"...that Brownback and Altenburg have known each other since 1977, that Brownback's wife worked for Altenburg, and that Altenburg hosted Brownback's retirement party in 1999."*[2]

Defense attorneys challenged the bias of five of the six presiding officers, including Brownback.[3] Three officers were removed by Altenburg, but Brownback was allowed to remain. Brownback was the only officer appointed to the first commission who had any legal training.

Dismissed charges against Omar Khadr

On June 4, 2007, in a move the *BBC* described as a "stunning blow" to the Bush Presidency's detainee policy, Brownback dismissed all charges against Canadian youth Omar Khadr.[4]

In the Summer of 2006 the United States Supreme Court overturned the then current version of the Guantanamo military commission on constitutional grounds.[4] The Supreme Court had ruled that the Bush Presidency lacked the constitutional authority to institute military commissions. The Supreme Court however, did rule that the United States Congress did have the constitutional authority to set up military commissions.

In the fall of 2006 the Congress passed the Military Commissions Act, which authorized military commissions similar to those the Supreme Court overturned, to try "unlawful enemy combatants".[4]

However, Brownback and Captain Keith J. Allred, ruled that since Khadr and Salim Ahmed Hamdan's Combatant Status Review Tribunals had not determined that they were "unlawful enemy combatants", but merely "enemy combatants", the commissions lacked jurisdiction to try them.[4] [5]

None of the remaining Guantanamo captives have had a "competent tribunal", like the AR-190-8 Tribunal, make a determination as to whether they broke any of the laws of war that would strip them of the protections of Geneva Conventions Prisoner of War status.[4]

November 2007 Guantanamo Military Commission hearings

An appeal court determined that the Presidents of Military Commissions themselves had the authority to determine whether the captives were lawful combatants.[6] [7] Brownback reconvened Khadr's proceedings on Thursday November 8, 2007. Before Brownback made his determination as to whether Khadr was an unlawful combatant lawyers were allowed to challenge Brownback over whether he was an appropriate person to sit as Khadr's judge. When one of Khadr's lawyers had asked his whether he had said he had "taken a lot of heat" for his June ruling he acknowledged he had. But he ruled that he was sufficiently independent to serve as a neutral judge.

Brownback postponed making a ruling on whether Khadr was a lawful or unlawful combatant, giving attorneys more time for preparation.[8] The next two sessions were scheduled for December 7, 2007 and January 13, 2008.

Replaced

On May 29, 2008 a new officer was appointed to replace Brownback.[9] [9] [10] [11] [12] [13] [14] [15]

References

[1] Toni Locy (August 24, 2004). "Guantanamo hearings start today" (http://www.usatoday.com/news/washington/2004-08-23-gitmo-hearings_x.htm). USA Today. . Retrieved 2007-06-07.

[2] John Mintz (September 16, 2004). "Presiding Officer At Guantanamo Faces Questions" (http://www.washingtonpost.com/ac2/wp-dyn/A24538-2004Sep15?language=printer). Washington Post. . Retrieved 2007-06-07.

[3] Three Guantanamo panelists "dismissed over bias allegations" (http://www.usatoday.com/news/nation/2004-10-21-gitmo-panelists_x.htm). USA Today. August 21, 2004. Three Guantanamo panelists. Retrieved 2007-06-07.

[4] sergeant Sara Wood (June 4, 2007). "Charges Dismissed Against Canadian at Guantanamo" (http://www.defenselink.mil/news/newsarticle.aspx?id=46281). Department of Defense. . Retrieved 2007-06-07.

[5] sergeant Sara Wood (June 4, 2007). "Judge Dismisses Charges Against Second Guantanamo Detainee" (http://www.defenselink.mil/news/newsarticle.aspx?id=46288). Department of Defense. . Retrieved 2007-06-07.

[6] Jane Sutton (November 8, 2007). "Guantanamo judge says he took "heat" from Pentagon" (http://africa.reuters.com/world/news/usnN08228199.html). Reuters. . Retrieved 2007-11-08.

[7] "Trial of Canadian Guantanamo detainee gets go-ahead" (http://afp.google.com/article/ALeqM5hSnhIKDk4rmaMKFST9yG6SskIbmw). Armed Forces Press. November 8, 2007. . Retrieved 2007-11-08.

[8] "Guantanamo judge postpones question of trial eligibility for Canadian detainee" (http://www.pr-inside.com/guantanamo-judge-postpones-question-of-r291337.htm). PR Inside. November 8, 2007. . Retrieved 2007-11-18.

[9] Jane Sutton (May 29, 2008). "Guantanamo judge dismissed in Canadian's case" (http://ca.reuters.com/article/domesticNews/idCANASU5290220080530?sp=true). Reuters. . Retrieved 2008-05-29.

[10] Michael Melia (May 29, 2008). "Gitmo judge removed from Canadian's case" (http://www.miamiherald.com/news/world/AP/story/551526.html). Miami Herald. . Retrieved 2008-05-29.

[11] Omar El Akkad (May 29, 2008). "Khadr judge at Guantanamo Bay relieved of duties" (http://www.theglobeandmail.com/servlet/story/RTGAM.20080529.wkhadr0529/BNStory/National/?page=rss&id=RTGAM.20080529.wkhadr0529). Globe and Mail. . Retrieved 2008-05-29.

[12] Steven Edwards (May 29, 2008). "Pentagon fires Khadr judge: Col. Peter Brownback replaced after chastising prosecution" (http://www.nationalpost.com/news/world/story.html?id=549763). National Post. . Retrieved 2008-05-29.

[13] Michael Melia (May 29, 2008). "Gitmo judge removed from Canadian's case" (http://www.kansascity.com/451/story/641581.html). Kansas City Star. . Retrieved 2008-05-29.

[14] "Khadr judge fired, says his military lawyer" (http://www.cbc.ca/world/story/2008/05/29/judge-khadr.html). CBC News. May 29, 2008. . Retrieved 2008-05-29.

[15] "U.S. military dumps judge in Khadr terrorism case in surprise move" (http://canadianpress.google.com/article/ALeqM5gMFtTJrQcHknWUFeUfvS9Lo9_7zA). Canadian Press. May 29, 2008. . Retrieved 2008-05-29.

Ramzi_bin_al-Shibh

Ramzi bin al-Shibh	
 FBI photo of al-Shibh	
Born	May 1, 1972[1] Ghayl Bawazir, Eastern Yemen
Detained at	CIA black sites, Guantanamo
ISN	10013
Charge(s)	Charged before a military commission in 2008, no trial date set yet.

Ramzi bin al-Shibh (Arabic: رمزي بن الشيبة, *Ramzī bin ash-Shībah*; also transliterated as **bin al-Shaibah**) (born May 1, 1972[2]) is a detainee at Guantanamo Bay and stands accused of being a "key facilitator for the September 11 attacks."[3]

In the mid-1990s, bin al-Shibh moved as a student to Hamburg, where he soon became close friends with Mohamed Atta, Ziad Jarrah and Marwan al-Shehhi. Together, they became the masterminds of the September 11 attacks, forming the Hamburg cell. He was the only one out of the four who failed to obtain a US Visa and therefore acted as an intermediary for the hijackers in the United States, by wiring money and passing on information from key al-Qaeda figures. After the attacks, bin al-Shibh was the first to be publicly identified as the "20th hijacker", of whom there were several more likely candidates thought to have been tasked to fill out the single missing slot among the United 93 hijacking.

Bin al-Shibh has been in custody since he was captured on September 11, 2002, in Karachi, Pakistan.[4]

Early life

Ramzi bin al-Shibh was born in the Yemeni province of Hadhramaut.[3] [5] When he was young, his family moved to a working-class neighborhood in the capital, Sana'a.[6] In 1987, his father died and he was then cared for by his older brother, Ahmed, and his mother.[6] In 1987, while still in high school, bin al-Shibh worked part-time as a clerk for the International Bank of Yemen.[7] He continued working there until 1995.[1]

Bin al-Shibh applied for a U.S. visa in 1995, but his request was denied.[1] He then went to Germany, where he requested political asylum, claiming that he was a political refugee from Sudan. He lived in Hamburg until 1997 when the judge refused his asylum request.[1] Bin al-Shibh returned to the Hadramaut region of Yemen, but a short while later bin al-Shibh received a German visa under his real name.[1] While he was in Germany, bin al-Shibh used the name **Ramzi Omar**. In 1997, bin al-Shibh met Mohamed Atta, the leader of the Hamburg cell, at a mosque.[8] For two years, Atta and bin al-Shibh were roommates in Germany.[8]

Al Qaeda training

In late 1999, bin al-Shibh traveled to Kandahar in Afghanistan, where he received training at Al Qaeda camps, and met others involved in planning the September 11 attacks.[8]

Attempts to come to the United States

Original plans for the 9/11 attacks called for bin al-Shibh to be one of the hijacker pilots, along with three other members of the Hamburg cell, including Mohamed Atta, Marwan al-Shehhi, and Ziad Jarrah.[9] From Hamburg, bin al-Shibh applied to take flight training in the United States. At that time, he also applied to Aviation Language Services, which provides language training for student pilots.[10] Bin al-Shibh applied for an entry visa to the United States, four times, and was refused each time. He made visa applications in Germany on May 17, 2000, and again in June, on September 16, and October 25, 2000.[2] [10] According to the 9/11 Commission, this refusal of a visa was motivated by general concern by U.S. officials that people from Yemen would illegally overstay their visit and seek work in the United States. His friend, Zakariyah Essabar, was also denied visas. After he failed to enter the United States, bin al-Shibh took on more of a "coordinator" role in the plot, and a link between Atta in the United States and Khalid Sheikh Mohammed in Afghanistan.[11] [12]

Connections to 9/11 attacks

Saeed al-Ghamdi

According to al-Jazeera reporter Yosri Fouda's documentary *Top Secret: The Road to September 11*, three weeks prior to the attacks Saeed al-Ghamdi is believed to have used the name "Abdul Rahman" to message bin al-Shibh (who was posing as a girlfriend) online, where he messaged

> "The first semester commences in three weeks. Two high schools and two universities. ... This summer will surely be hot ...19 certificates for private education and four exams. Regards to the professor. Goodbye."

This was said to be a reference to two military targets and two civilian, 19 hijackers.[13]

Mohamed Atta

Similarly, bin al-Shibh later said that Mohamed Atta had phoned him on the morning of August 29.

> "He said, 'A friend of mine gave me a puzzle and I want you to help me out.' I said to him, 'Is this the time for puzzles, Mohamed?' He said, 'Yes, I know, but no one else but you could help me.' He said, 'Two sticks, a dash and cake with a stick down. What is it?' I said, 'Did you wake me up just to tell me this?' As it turns out, two sticks is the number 11. A dash is a dash. And cake with a stick down is the number nine. And that was September 11.[14]

Ziad Jarrah

In August 2000, Ziad Jarrah attempted to enroll bin al-Shibh in a Florida flight school.[2]

Marwan al-Shehhi

Bin al-Shibh sent money via wire transfer on September 25, 2000 to Marwan al-Shehhi in Florida.[2]

Zacarias Moussaoui

In August 2001, bin al-Shibh sent approximately $14,000 to Zacarias Moussaoui, using the alias **Ahad Sabet**,[15] a few days after receiving transfer of $15,000 from Hashim Abdulrahman in the United Arab Emirates.[2]

20th hijacker

Bin al-Shibh was the first to be publicly identified as the "20th hijacker," someone who was thought to have been tasked to fill out the single missing slot among the four terrorist teams. This spot was never filled, and consequently United Flight 93 had four hijackers, not five, which is believed to have led in part to the success of the passenger revolt. The passengers likely caused the crash of the plane near Shanksville, Pennsylvania.

Videos

After January 14, 2002 bin al-Shibh was found to be among five suspected al-Qaeda members delivering what United States Attorney General John Ashcroft described as "martyrdom messages from suicide terrorists" on five discovered videos. NBC News said that the five videos had been recorded after the September 11 attacks.

On September 8, 2006, al-Qaeda released a video that shows Osama bin Laden and some of the 9/11 hijackers. The tape identifies bin al-Shibh as the "coordinator of the 9/11 attacks" in its own English subtitles.[16] The video goes on to show bin al-Shibh and other hijackers training in kickboxing as well as disarming and concealing weapons at a terrorist training camp in or near Kandahar, Afghanistan.[16] [17]

FBI Most Wanted Terrorist List

On January 17, 2002 the FBI published the first **Most Wanted Terrorists Seeking Information** list (now known as the FBI's "Seeking Information - War on Terrorism" list), in order to profile the five wanted terrorists about whom very little was known, but who were suspected of plotting additional terrorist attacks in martyrdom operations.[18] (see current version displaying photos of five terrorists on the remaining martyrdom videos FBI list, as of June 2006)[19]

Ramzi bin al-Shibh was one of the four known names among the five. The other three are still featured in compiled video clips, in order of appearance, Muhammad Sa'id Ali Hasan, Abd al-Rahim, and Khalid Ibn Muhammad al-Juhani.[20] [21] [22] The fifth was identified a week later as *Abderraouf Jdey*, alias: *Al-Rauf bin al-Habib bin Yousef al-Jiddi*.

Ashcroft said the five videotapes, shown by the FBI without sound, had been recovered from the rubble of the home of Mohammad Atef outside of Kabul, Afghanistan. Ashcroft called upon people worldwide to help "identify, locate and incapacitate terrorists who are suspected of planning additional attacks against innocent civilians." "These men could be anywhere in the world", he said. The sound was left out to guard against the possibility that the messages contained signals for other terrorists. Ashcroft added that an analysis of the audio suggested "the men may be trained and prepared to commit future suicide terrorist acts." Ashcroft said not much was known about any of them except bin al-Shibh.

Other attacks

Ramzi bin al-Shibh is also suspected to be involved in the 2000 USS Cole bombing, and the 2002 Ghriba synagogue bombing in Tunisia.[8]

Capture and detention

Bin al-Shibh was captured in Pakistan on September 11, 2002, after a gun battle in Karachi with the Pakistani ISI and the CIA's Special Activities Division.[23] On September 14, 2002 he was subsequently turned over to the United States, which transferred him to a black site in Morocco for interrogation. The CIA admitted in August 2010 that it has video tapes of these interrogations.[24][25] [26] His profile was removed from the FBI Seeking Information wanted list by October 17, 2002.[27]

Ramzi bin al-Shibh at Guantanamo Bay, Cuba (2010)

Bin al-Shibh remained a prisoner of the U.S., at an undisclosed CIA-led location, until September 2006. On September 6, 2006 U.S. President George W. Bush announced that bin al-Shibh and other CIA held prisoners had been transferred to Guantanamo Bay, Cuba.

Bin al-Shibh is also wanted by German courts where he shared a Hamburg apartment with Mohamed Atta, the suspected ringleader of the September 11 hijackers. In 2005 the USA denied a German request for bin al-Shibh's extradition, which resulted in 9/11 suspect Abdelghani Mzoudi being acquitted of the charges presented at that time.[28]

Combatant Status Review Tribunal

A three page long Summary of Evidence memo was prepared for Ramzi bin al-Shibh on February 8, 2007.[29] The transcript of his hearing was eight pages long. It said that he choose not to attend to his Tribunal that was held on March 9, 2007. The first two pages of the transcript were consumed with the Tribunal's officers swearing oaths, and the reading out of the Tribunal mandate and authority.[30]

Personal Representative interviews with the captive

The Tribunal's President then called on the captive's Personal Representative to explain his efforts to explain the captive's right to be present at his Tribunal.

Personal Representative interviews with the captive

February 9, 2007	• Purpose of the meeting was to present the official notification that a Combatant Status Review Tribunal was being scheduled. • Captive stated he would not attend the Tribunal. • Captive stated he would not meet again with the Personal Representative or the translator.
February 13, 2007	• Purpose of the meeting was to present the Unclassified Summary, containing the summary of the unclassified allegations, to the captive. • Captive chose not to leave his cell to attend the interview.
February 16, 2007	• Purpose of the meeting was to present the Unclassified Summary to the captive. • Captive chose not to leave his cell to attend the interview.
March 5, 2007	• Purpose of the meeting was to present the Unclassified Summary to the captive. • Captive chose not to leave his cell to attend the interview.

Allegations

The allegations prepared for the first 558 captives whose status was examined by Combatant Status Review Tribunals, between August 2004 and January 2005, were all broken into two sections — the allegations that established a connection to terrorism, and the allegations that established hostile activity. The allegations were always numbered, and were generally only one or two sentences in length.

The allegations that Ramzi bin al-Shibh would have faced, during his Tribunal, were:

> "On the morning of 11 September 2001, four airliners traveling over the United States were hijacked. The flights hijacked were: American Airlines Flight 11, United Airlines Flight 175, American Airlines Flight 77, and United Airlines Flight 93. At approximately 8:46 a.m., American Airlines Flight 11 crashed into the North Tower of the World Trade Center, resulting in the collapse of the tower at approximately 10:25 a.m. At approximately 9:03 a.m., United Airlines Flight 175 crashed into the South Tower of the World Trade Center, resulting in the collapse of the tower at approximately 9:55 a.m. At approximately 9:37., American Airlines Flight 77 crashed into the southwest side of the Pentagon in Arlington Virginia. At approximately 10:03 a.m. United Airlines Flight 93 crashed in Stoney Creek Township Pennsylvania. These crashed and subsequent damage to the World Trade Center and the Pentagon resulted in the deaths of 2972 persons in New York, Virginia, and Pennsylvania.
>
> "According to court transcripts and evidence from *United States v. Zacarias Moussaoui*, the detainee was closely associated with three of the hijackers responsible for the "9/11" attacks, Mohamed Atta, Marwan al-Shehhi and Ziad Jarrah while they lived in Hamburg during the late 1990s and early 2000. The detainee, Atta and Shehhi are know to have lived at or frequented on particular address during the same time period, 54 Marienstrasse, Hamburg, Germany 21073.
>
> "Airline and immigration records indicate that from November 1999 through February 2000, the detainee, Mohamed Atta, Marwan al-Shehhi and Ziad Jarrah all traveled from Germany to Pakistan.
>
> "Sayf al-Adl is a senior al Qaida military commander with a long-term relationship with Usama bin Laden, Sayf al-Adl's role in the organization has been as a trainer, military leader, and key member of Usama bin Laden's security detail.
>
> "The diary of Sayf al-Adl was recovered during a raid in Saudi Arabia in 2004. The diary details the detainee's involvement in the 11 September 2001 terrorist plot and subsequent attack. The detainee is listed as a "highly professional jihadist" along with "9/11 hijackers", Mohamed Atta and Ziad Jarrah. The diary states that the three were briefed on an operation involving aircraft by Abu Hafs, a senior al Qaida planner. The detainee, Mohamed Atta, and Ziad Jarrah subsequently met with Usama bin Laden about the plan. Following the meeting, al Qaida began arrangements for the detainee, Mohamed Atta and Ziad Jarrah to receive pilot

training. The detainee handled administrative details for the "9/11 hijackers" while they were in the United States and the detainee served as an al Qaida Europe based liaison.

"The detainee was identified in a video tape of potential suicide operatives.

"The detainee attempted to obtain a United States visa on four occasions from May 2000 to November 2000 for the purpose of attending flight school in the United States. Each application was rejected by *[sic]* United States Department of State.

"The detainee attempted to enroll in the Florida Flight Training School, where 9/11 hijacker Ziad Jarrah was a student. The detainee put down a 2,350 United States dollars *[sic]* deposit for flight training.

"Ziad Jarrah repeatedly attempted to assist the detainee's travel to the United States and enrollment in the Florida flight training center.

"The detainee attempted to enroll at the *[sic]* Florida-based aviation language school.

"The detainee, while in Germany, wired 9/11 hijacker Marwan al-Shehi, who was in the United States 2708.33 *[sic]* United States dollars *[sic]* on 13 June 2000 via moneygram.

"The detainee, while in Germany, wired 9/11 hijacker Marwan al-Shehhi, who was in the United States) *[sic]* 1760.61 *[sic]* United States dollars on 26 July 2000 via Western Union.

"The detainee, while in Germany, wired 9/11 hijacker Marwan al-Shehhi who was in the United States) *[sic]* 4,118.13 *[sic]* United States dollars on 25 September 2000 via Western Union.

"In June 2002, the detainee was personally interviewed by Yosri Fouda, an investigative journalist for Al-Jazeera television. The interview took place over the course of 48 hours in Karachi, Pakistan. Also present at the meeting was Khalid Sheikh Mohammed, a senior al Qaida planner. Fouda conducted the interview in person with both the detainee and KSM. The detainee and KSM detailed how the 9/11 attacks were planned and executed during the course of the interview. KSM identified the detainee as the coordinator of the 9/11 attacks. The detainee displayed items he claimed were "souvenirs" of the 9/11 attacks. The items included: an air navigation map of the American eastern seaboard, flight simulator CD-Roms and Boeing manuals and a flight instruction book the detainee claimed had 9/11 hijacker Mohamed Atta's handwritten notes. The detainee stated Mohamed Atta left them in the Hamburg, Germany, apartment he shared with the detainee. The detainee stated that he later met with Atta in July, 2001 in Madrid, Spain, to finalize the operational details of the 9/11 plot. The detainee stated he received a phone call on 29 August 2001 from Atta that gave the date fro the 9/11 attacks. After learning this, the detainee ordered active al Qaida cells in Europe and elsewhere to evacuate and then he fled to Pakistan.

"An unsigned letter found at the detainee's point of capture, and addressed to the detainee, asks follow on questions related to the detainee's Al-Jazeera interview detailing the 9/11 attacks.

"An article from the London *Sunday Times* published on 8 September 2002 listed excerpts from a 112 page document entitled "The Reality of the New Crusaders' War". The detainee passed the document to Al-Jazeera Yosri Fouda with a request for the document to be translated into English and entered into the Library of Congress. According to the London *Sunday Times* the document is al Qaida's written attempt to justify the 9/11 attacks through Islamic teaching.

"The London *Sunday Times* article published on 8 September 2002 listed excerpts from "The Reality of the New Crusaders' War" which contained statements from Taliban leader Mullah Mohammed Omar and Usama bin Laden which encourages jihad in service of the ousted Taliban regime.

"The detainee was captured in a safe house. Items also recovered at the safe house at the time of the detainee's capture were high explosives, sheet explosives, a large quantity of improvised detonation devices, passports for Usama bin Laden's family members, a handwritten note to a senior al Qaida operative, identification cards for a senior al Qaida operative, identification cards for Ahmed al-Haznawi, a 9/11 hijacker, and contact information for several known al Qaida operatives.

"Documents captured in a raid of a separate al Qaida safe house were identical to documents captured alnong with the detainee. The documents included training manuals, security information and combat related subjects.

"Letters and personal effect of a senior al Qaida operative were discovered in the safe house where the detainee was arrested.

"Letters found at the detainee's point of capture detailed a plan to egress Pakistan with forged identification. This plan was in conjunction with a senior al Qaida operative.

"A letter captured on an al Qaida courier detailed a senior al Qaida operative's instructions to the detainee to identify operatives to send to the United States or United Kingdom.

"The detainee wired approximately 15,000 United States dollars to Zacharias Moussaoui while Moussaoui was enrolled in pilot training."

The Department of Defense announced on August 9, 2007 that all fourteen of the "high-value detainees" who had been transferred to Guantanamo from the CIA's black sites, had been officially classified as "enemy combatants".[31] Although judges Peter Brownback and Keith J. Allred had ruled two months earlier that only "*illegal* enemy combatants" could face military commissions, the Department of Defense waived the qualifier and said that all fourteen men could now face charges before Guantanamo military commissions.[32] [33]

Habeas petition

On June 12, 2008 the United States Supreme Court ruled, in *Boumediene v. Bush*, that the Military Commissions Act could not remove the right for Guantanamo captives to access the US Federal Court system.

The first 22 captives had their habeas petitions re-initiated in August 2008.[34] Ramzi Bin AlShibh's habeas corpus petition was filed August 29, 2008.

Guantanamo military commission

Bin al-Shibh and four other captives classified as high value detainees (Khalid Sheikh Mohammed, Mustafa al-Hawsawi, Ammar al-Baluchi and Walid Bin Attash) were charged in Guantanamo military commissions in Spring 2008. The men triggered controversy when they announced that they did not want US-appointed attorneys and they planned to boycott their commissions. The military commissions, authorized by President George W. Bush, did not permit suspects to forgo legal representation, to act as their own attorneys, or to boycott their commissions. The later commissions, authorized by the Military Commissions Act of 2006, did authorized suspects to serve as their own attorneys.

The other four men did eventually agree to attend their commissions. Bin al-Shibh, however, has continued to try refuse to attend. His appointed attorneys had expressed concern that he had refused to meet with them, and since the location of camp 7, where the high value detainees are held is so secret even military attorneys are not allowed to travel there to visit him. The men are hooded when they travel from the camp to their commission, and his attorney offered to allow herself to be hooded. This offer was turned down.

The judge presiding over the commission's pre-trial motions ordered bin al-Shibh and Hawsawi to undergo mental competency hearings. On December 8, 2008, Mohammed told the judge that he and the other four indictees wished to confess and plead guilty; however, the plea would be delayed until after the competency hearings for bin al-Shibh and Hawsawi, so that all five men could make their plea together.[35]

On May 17, 2010, *Saba News* reports that *'"Ramzi Al-Shaibah"*, and four other Yemenis would face charges in the summer of 2010.[36]

Two of the other Yemenis 'Saba News' reported would face charges were: Walid Bin Atash and Abdul Rahim Al-Nasheri. *Saba News* did not name the fourth and fifth individuals. Lawyers of Bin al-Shibh are argue that bin al-Shibh may be unfit to stand trial and have asked that the proceedings against him and his four co-accused be stayed until his mental state is determined. They say he has been prescribed psychotropic drugs of the sort that are used to treat Schizophrenia. Bin al-Shibh claims that he is mentally fit, has denounced his lawyers, and says that he wants to represent himself before the commissions.[37]

See also

- Shaker Aamer

References

[1] "9/11 Commission Report, Chapter 5" (http://www.9-11commission.gov/report/911Report_Ch5.htm). National Commission on Terrorist Attacks Upon the United States. 2004. .

[2] Indictment of Zacarias Moussaoui (http://www.usdoj.gov/ag/moussaouiindictment.htm), with supporting conspirators, Ramzi bin al-Shibh and Mustafa al-Hawsawi. Filed in the United States District Court for the Eastern District of Virginia.

[3] "Detainee Biographies" (http://www.webcitation.org/query?url=http://www.odni.gov/announcements/content/DetaineeBiographies.pdf+&date=2009-08-31) (PDF). Office of the Director of National Intelligence. Archived from the original (http://www.odni.gov/announcements/content/DetaineeBiographies.pdf) on date=2009-08-31. .

[4] He was captured after a gun battle in Karachi with the Pakistani ISI and the CIA's Special Activities Division after a year later of the September 11 attacks.

[5] McDermott, Terry (2005). *Perfect Soldiers*. Harper. pp. 39.

[6] McDermott, Terry (2005). *Perfect Soldiers*. Harper. pp. 41.

[7] McDermott, Terry (2005). *Perfect Soldiers*. Harper. pp. 42.

[8] "Ramzi Binalshibh: al-Qaeda suspect" (http://news.bbc.co.uk/2/hi/south_asia/2257456.stm). BBC. September 14, 2002. .

[9] McDermott, Terry (2005). *Perfect Soldiers*. Harper.

[10] Zacarias Moussauoi v. the United States (http://cryptome.org/usa-v-zm-030706-02.htm), trial testimony on March 7, 2006.

[11] "Al-Jazeera reporter speaks on terrorist plans" (http://www.abc.net.au/lateline/stories/s689774.htm). Lateline / ABC (Australia). September 30, 2002. .

[12] "The Mastermind" (http://www.cbsnews.com/stories/2002/10/09/60II/main524947.shtml). CBS News. March 5, 2003. .

[13] (http://archives.cnn.com/2002/WORLD/meast/09/12/alqaeda.911.claim/index.html)

[14] (http://transcripts.cnn.com/TRANSCRIPTS/0209/14/smn.09.html)

[15] "Motion: 9/11 conspiracy suspect may have used ID of Arizona doctor" (http://archives.cnn.com/2002/LAW/08/07/inv.moussaoui.stolen.id/index.html). CNN. August 7, 2002. .

[16] "Video shows Osama and killers" (http://www.nydailynews.com/front/story/450616p-379142c.html). New York Daily News. 09-08-06. .

[17] "Al-Qa'ida releases film showing Bin Laden with the hijackers" (http://www.belfasttelegraph.co.uk/news/story.jsp?story=705570). Belfast Telegraph. 09-08-06. .

[18] Most Wanted Terrorists Seeking Information (http://www.fbi.gov/mostwant/seekinfo/seek.htm), January 17, 2002, (dead link)

[19] Martyrdom Messages/video, Seeking Information Alert (http://www.fbi.gov/terrorinfo/terseekinfomartyr.htm) video clips published by the FBI January 17, 2002, and photos of remaining 5 terrorists, FBI archival after September 2002

[20] FBI Seeking Information - War on Terrorism, Martyrdom Messages/video Seeking Information Alert (http://www.fbi.gov/mpg/persons.mpg), VIDEO 2 minutes 11 seconds, mpg (29.1 mb)

[21] FBI Seeking Information - War on Terrorism, Martyrdom Messages/video Seeking Information Alert (http://mfile.akamai.com/6066/rm/www.fbi.gov/mpg/persons.rm), VIDEO 2 minutes 11 seconds, rm (229 kb - stream)

[22] FBI Seeking Information - War on Terrorism, Martyrdom Messages/video Seeking Information Alert (http://www.fbi.gov/mpg/persons_asf.asf), VIDEO 2 minutes 11 seconds, asf (371 kb - stream)

[23] ^ http://www.washingtonpost.com/wp-dyn/content/article/2007/12/17/AR2007121702151.html

[24] "CIA tapes prove Morocco rendition" (http://www.theaustralian.com.au/news/world/cia-tapes-prove-morocco-rendition/story-e6frg6so-1225907071320). August 19, 2010. .

[25] Gutteridge, Clara (October 18, 2010). "New CIA Interrogation Tapes Hint at Legal 'Loophole' Allowing the US to Outsource Torture" (http://www.huffingtonpost.com/clara-gutteridge/new-cia-interrogation-tap_b_686897.html). *Huffington Post*. .

[26] "Binalshibh to go to third country for questioning" (http://edition.cnn.com/2002/WORLD/asiapcf/south/09/16/alqaeda.pakistan/). CNN. September 17, 2002. .

[27] FBI Seeking Information archive (http://web.archive.org/web/20021017112003/http://www.fbi.gov/terrorinfo/terseekinfo.htm), Internet Archive Wayback Machine, October 17, 2002

[28] CIA Red Cell Memorandum - 2 Feb 2010 (http://file.wikileaks.org/file/us-cia-redcell-exporter-of-terrorism-2010.pdf) CIA Red Cell Memorandum on United States "exporting terrorism"

[29] Margot Williams (2008-11-03). "Guantanamo Docket: Ramzi Bin al Shibh" (http://projects.nytimes.com/guantanamo/detainees/10013-ramzi-bin-al-shibh/documents/5/pages/800). New York Times. .

[30] OARDEC (2007-03-09). "Verbatim Transcript of Open Session CSRT Hearing for ISN 10013" (http://www.defenselink.mil/news/transcript_ISN10013.pdf). United States Department of Defense. . Retrieved 2010-04-05.
[31] Lolita C. Baldur (Thursday, August 9, 2007). "Pentagon: 14 Guantanamo Suspects Are Now Combatants" (http://www.time.com/time/nation/article/0,8599,1651680,00.html). Time magazine. . mirror (http://www.webcitation.org/query?url=http://findarticles.com/p/articles/mi_qn4176/is_20070324/ai_n18757735&date=2008-05-19)
[32] Sergeant Sara Wood (June 4, 2007). "Charges Dismissed Against Canadian at Guantanamo" (http://www.defenselink.mil/news/newsarticle.aspx?id=46281). Department of Defense. . Retrieved 2007-06-07.
[33] Sergeant Sara Wood (June 4, 2007). "Judge Dismisses Charges Against Second Guantanamo Detainee" (http://www.defenselink.mil/news/newsarticle.aspx?id=46288). Department of Defense. . Retrieved 2007-06-07.
[34] "Amended and Other Factual Returns Filed -- August 2008" (http://www.scotusblog.com/wp/wp-content/uploads/2008/08/table-on-returns-8-29-08.pdf) (PDF). United States Department of Justice. 2008-08-29. . Retrieved 2008-09-11.
[35] "Top 9/11 suspects to plead guilty" (http://news.bbc.co.uk/1/hi/world/americas/7770856.stm). BBC News. December 8, 2008. . Retrieved December 8, 2008.
[36] "U.S. to try five Yemeni Gitmo detainees" (http://www.sabanews.net/en/news214477.htm). Saba News. 2010-05-17. . Retrieved 2010-05-17. "The U.S. will start in this summer trying five Yemeni detainees at the U.S. Guantanamo Bay in Cuba including Ramzi Al-Shaibah, Walid Bin Atash and Abdul Rahim Al-Nasheri, the September 26 website has reported." mirror (http://www.webcitation.org/query?url=http://www.sabanews.net/en/news214477.htm&date=2010-05-17)
[37] http://www.hrw.org/news/2011/05/31/ramzi-bin-al-shibh

External links

- Pentagon charges 6 in 9-11 attacks (http://www.msnbc.msn.com/id/23108871)
- 'Clean team' interrogated 9-11 suspects (http://www.msnbc.msn.com/id/23120362)
- Shane, Scott (June 22, 2008). "Inside a 9/11 Mastermind's Interrogation" (http://www.nytimes.com/2008/06/22/washington/22ksm.html?partner=rssnyt&emc=rss). *New York Times*. Retrieved 2008-06-23.

Unlawful_combatant

An **unlawful combatant, illegal combatant** or **unprivileged combatant/belligerent** is a civilian who directly engages in armed conflict in violation of the laws of war. An unlawful combatant may be detained or prosecuted under the domestic law of the detaining state for such action.[1]

Capture of a Franc-Tireur, by Carl Johann Lasch.

The Geneva Conventions apply in wars between two or more sovereign states. Article 5 of the Third Geneva Convention states that the status of a detainee may be determined by a "competent tribunal." Until such time, he is to be treated as a prisoner of war.[2] After a "competent tribunal" has determined that an individual detainee is an unlawful combatant, the "detaining power" may choose to accord the detained unlawful combatant the rights and privileges of a prisoner of war as described in the Third Geneva Convention, but is not required to do so. An unlawful combatant who is not a national of a neutral State, and who is not a national of a co-belligerent State, retains rights and privileges under the Fourth Geneva Convention so that he must be "treated with humanity and, in case of trial, shall not be deprived of the rights of fair and regular trial."[3]

While the concept of an unlawful combatant is included in the Third Geneva Convention, the phrase itself does not appear in the document.[1] Article 4 of Third Geneva Convention does describe categories under which a person may be entitled to POW status; and there are other international treaties that deny lawful combatant status for mercenaries and children. In the United States, the Military Commissions Act of 2006 codified the legal definition of this term and invested the U.S. President with broad discretion to determine whether a person may be designated an unlawful enemy combatant under United States law. The assumption that such a category as unlawful combatant exists is not contradicted by the findings of the International Criminal Tribunal for the Former Yugoslavia in the Celebici

Judgment. The judgment quoted the 1958 ICRC commentary on the Fourth Geneva Convention: Every person in enemy hands must be either a prisoner of war and, as such, be covered by the Third Convention; or a civilian covered by the Fourth Convention. Furthermore, "*There is no* intermediate status; nobody in enemy hands can be outside the law,"[4] because in the opinion of the ICRC, "If civilians directly engage in hostilities, they are considered 'unlawful' or 'unprivileged' combatants or belligerents (the treaties of humanitarian law do not expressly contain these terms). They may be prosecuted under the domestic law of the detaining state for such action".[1] [5]

The Geneva Conventions do not recognize any lawful status for combatants in conflicts not involving two or more nation states. A state in such a conflict is legally bound only to observe Article 3 of the Geneva Conventions and may ignore all the other Articles. But each one of them is completely free—and should be encouraged—to apply all or part of the remaining Articles of the Convention.[6]

Camp X-ray, Guantánamo.

International law and practice

The term "unlawful combatant" has been used for the past century in legal literature, military manuals, and case law.[7] However, unlike the terms "combatant", "prisoner of war", and "civilian", the term "unlawful combatant" is not mentioned in either the Hague or the Geneva Conventions. So while the former terms are well understood and clear under international law, the term "unlawful combatant" is not.[3] [8]

At the First Hague Conference, which opened on 6 May 1899, there was a disagreement between the Great Powers—which considered *francs-tireurs* to be unlawful combatants subject to execution on capture—and a group of small countries headed by Belgium—which opposed the very principle of the rights and duties of armies of occupation and demanded an unlimited right of resistance for the population of occupied territories. As a compromise, the Russian delegate, F. F. Martens, proposed the Martens Clause, which is included in the preamble to the *1899 Hague Convention II – Laws and Customs of War on Land*. Similar wording has been incorporated into many subsequent treaties that cover extensions to humanitarian law.[9] [10] [11]

Prisoners of war

The Geneva Convention relative to the Treatment of Prisoners of War, 12 August 1949 (GCIII) of 1949 defines the requirements for a captive to be eligible for treatment as a POW. A lawful combatant is a person who commits belligerent acts, and, when captured, is treated as a POW. An *unlawful combatant* is someone who commits belligerent acts but does not qualify for POW status under GCIII Articles 4 and 5.

> Article 4
>
> A. Prisoners of war, in the sense of the present Convention, are persons belonging to one of the following categories, who have fallen into the power of the enemy:
>
> > 1. Members of the armed forces of a Party to the conflict as well as members of militias or volunteer corps forming part of such armed forces.
> >
> > 2. Members of other militias and members of other volunteer corps, including those of organized resistance movements, belonging to a Party to the conflict and operating in or outside their own territory, even if this territory is occupied, provided that such militias or volunteer corps, including such organized resistance movements, fulfill the following conditions:
> >
> > > (a) That of being commanded by a person responsible for his subordinates;
> > >
> > > (b) That of having a fixed distinctive sign recognizable at a distance;

(c) That of carrying arms openly;

(d) That of conducting their operations in accordance with the laws and customs of war.

3. Members of regular armed forces who profess allegiance to a government or an authority not recognized by the Detaining Power.

4. Persons who accompany the armed forces without actually being members thereof, such as civilian members of military aircraft crews, war correspondents, supply contractors, members of labour units or of services responsible for the welfare of the armed forces, provided that they have received authorization from the armed forces which they accompany, who shall provide them for that purpose with an identity card similar to the annexed model.

5. Members of crews [of civil ships and aircraft], who do not benefit by more favourable treatment under any other provisions of international law.

6. Inhabitants of a non-occupied territory, who on the approach of the enemy spontaneously take up arms to resist the invading forces, without having had time to form themselves into regular armed units, provided they carry arms openly and respect the laws and customs of war.

B. The following shall likewise be treated as prisoners of war under the present Convention:

1. Persons belonging, or having belonged, to the armed forces of the occupied country...

...

Article 5

...

Should any doubt arise as to whether persons, having committed a belligerent act and having fallen into the hands of the enemy, belong to any of the categories enumerated in Article 4, such persons shall enjoy the protection of the present Convention until such time as their status has been determined by a competent tribunal.

These terms thus divide combatants in a war zone into two classes: those in armies and organised militias and the like (**lawful combatants**), and those who are not. The critical distinction is that a "lawful combatant" (defined above) cannot be held personally responsible for violations of civilian laws that are permissible under the laws and customs of war; and if captured, a lawful combatant has to be treated as a prisoner of war by the enemy Power under the conditions laid down in the Third Geneva Convention.

If there is any doubt about whether a detained alleged combatant is a "lawful combatant" then the combatant must be held as a prisoner of war until his or her status has been determined by "a competent tribunal".[12] If that tribunal rules that a combatant is an "unlawful combatant" then the person's status changes to that of a civilian which may give them some rights under the Fourth Geneva Convention.[13]

Persons who are not prisoners of war in an international conflict

A non-combatant civilian "in the hands" of an enemy or an Occupying Power often gains rights through the *Geneva Convention Relative to the Protection of Civilian Persons in Time of War*, 12 August 1949 (GCIV), if they qualify as a "protected person".

Article 4. Persons protected by the Convention are those who, at a given moment and in any manner whatsoever, find themselves, in case of a conflict or occupation, in the hands of a Party to the conflict or Occupying Power of which they are not nationals.
Nationals of a State which is not bound by the Convention are not protected by it. Nationals of a neutral State who find themselves in the territory of a belligerent State, and nationals of a co-belligerent State, shall not be regarded as protected persons while the State of which they are nationals has normal diplomatic representation in the State in whose hands they are.

If the individual fulfills the criteria as a protected person, they are entitled to all the protections mentioned in GCIV. It should be emphasised that, in a war zone, a national of a neutral state, with normal diplomatic representation, is not a protected person under GCIV.

If a combatant does not qualify as a POW, then, if they qualify as a protected person, they receive all the rights which a non-combatant civilian receives under GCIV, but the party to the conflict may invoke Articles of GCIV to curtail those rights. The relevant Articles are 5 and 42.

> Part I. General Provisions
>
> ...
>
> Art. 5 Where in the territory of a Party to the conflict, the latter is satisfied that an individual protected person is definitely suspected of or engaged in activities hostile to the security of the State, such individual person shall not be entitled to claim such rights and privileges under the present Convention as would, if exercised in the favour of such individual person, be prejudicial to the security of such State.
> Where in occupied territory an individual protected person is detained as a spy or saboteur, or as a person under definite suspicion of activity hostile to the security of the Occupying Power, such person shall, in those cases where absolute military security so requires, be regarded as having forfeited rights of communication under the present Convention.
> In each case, such persons shall nevertheless be treated with humanity and, in case of trial, shall not be deprived of the rights of fair and regular trial prescribed by the present Convention. They shall also be granted the full rights and privileges of a protected person under the present Convention at the earliest date consistent with the security of the State or Occupying Power, as the case may be. Section II. Aliens in the territory of a party to the conflict Art. 42. The internment or placing in assigned residence of protected persons may be ordered only if the security of the Detaining Power makes it absolutely necessary.

It is likely that if they are found to be an "unlawful combatant" by "a competent tribunal" under GCIII Article 5, and if they are a protected person under GCIV, that the Party to the conflict will invoke GCIV Article 5. In which case, the "unlawful combatant" does not have the "rights and privileges under the present Convention as would, if exercised in the favour of such individual person, be prejudicial to the security of such State." They do, however, retain the right "to be treated with humanity and, in case of trial, shall not be deprived of the rights of fair and regular trial prescribed by the present Convention."

If, after a "fair and regular trial", the individual in question is found guilty of a crime, then the "unlawful combatant" can be punished by whatever lawful methods are available to the party to the conflict.

If the party does not use Article 5 of GCIV, the party may invoke Article 42 of GCIV and use "internment" to detain the "unlawful combatant".

For those nations that have ratified Protocol I of the Geneva Conventions, are also bound by Article 45.3 of that protocol which curtails GCIV Article 5.[3]

> Any person who has taken part in hostilities, who is not entitled to prisoner-of-war status and who does not benefit from more favourable treatment in accordance with the Fourth Convention shall have the right at all times to the protection of Article 75 of this Protocol. In occupied territory, any such person, unless he is held as a spy, shall also be entitled, notwithstanding Article 5 of the Fourth Convention, to his rights of communication under that Convention.

Persons who are not prisoners of war in an internal conflict

Civilians are covered by GCIV Article 3:

> Article 3
>
> > 1) Persons taking no active part in the hostilities, including members of armed forces who have laid down their arms and those placed *hors de combat* by sickness, wounds, detention, or any other cause, shall in all circumstances be treated humanely, without any adverse distinction founded on race, colour, religion or faith, sex, birth or wealth, or any other similar criteria.
> >
> > ...
> >
> > (d) the passing of sentences and the carrying out of executions without previous judgement pronounced by a regularly constituted court, affording all the judicial guarantees which are recognized as indispensable by civilized peoples.
> >
> > ...
> >
> > The Parties to the conflict should further endeavour to bring into force, by means of special agreements, all or part of the other provisions of the present Convention.
> >
> > ...

Combatants who do not qualify for prisoner of war status

If the combatant is engaged in "armed conflict not of an international character" then under the Article 3 of the general provisions of the Geneva Conventions they should be "treated humanely", and if tried "sentences must ... be pronounced by a regularly constituted court"[14]

The last time that American and British unlawful combatants were executed after "a regularly constituted court" was the Luanda Trial as mercenaries.[15]

Parole violation

A combatant who is a POW, and who is subsequently paroled on the condition that he will not take up arms against the belligerent power (or co-belligerent powers) that had held him as a prisoner, is considered a parole violator if he breaks said condition. He is regarded as guilty of a breach in the laws and customs of war, unless there are mitigating circumstances such as coercion by his state to break his parole. As with other combatants, he is still protected by the Third Geneva Convention (GCIII), until a competent tribunal finds him to be in violation of his parole.

The Geneva Convention (1929) made no mention of parole, but as it was supplemental to the Hague conventions, it relied on the wording of Hague to address this issue.[16] The authors of GCIII, 1949, decided to include a reference with some modification to parole, because during the Second World War, some belligerent countries did permit such release to some extent.[1]

Article 21 of GCIII (1949) reproduces the Articles 10 and 11 of the *Have IV: Regulations Respecting the Laws and Customs of War on Land*, 18 October 1907, but did not include Article 12, which provides: "Prisoners of war liberated on parole and recaptured bearing arms against the Government to whom they had pledged their honour, or against the allies of that Government, forfeit their right to be treated as prisoners of war, and can be brought before the courts."[17] Nevertheless, contained in the commentary on GCIII: The only safeguard available to a parole violator—who has been coerced into fighting, and who has been recaptured by the Power that detained him previously—is contained in the procedural guarantees to which he is entitled, pursuant to Article 85 of GCIII.[1]

In the opinion of Major Gary D. Brown, United States Air Force (USAF), this means that "[T]he Hague Convention specified that parole breakers would forfeit their right to be treated as prisoners of war if recaptured. The 1949 Geneva Convention is less direct on the issue. A recaptured parole violator under the Convention would be afforded the opportunity to defend himself against charges of parole breaking. In the interim, the accused violator would be entitled to P[o]W status." [18]

Mercenaries

Under Article 47 of Protocol I (Additional to the Geneva Conventions of 12 August 1949, and relating to the Protection of Victims of International Armed Conflicts) it is stated in the first sentence "A mercenary shall not have the right to be a combatant or a prisoner of war."

On 4 December 1989 the United Nations passed resolution 44/34 the *International Convention against the Recruitment, Use, Financing and Training of Mercenaries*. It entered into force on 20 October 2001 and is usually known as the UN Mercenary Convention.[19] Article 2 makes it an offence to employ a mercenary and Article 3.1 states that "A mercenary, as defined in article 1 of the present Convention, who participates directly in hostilities or in a concerted act of violence, as the case may be, commits an offence for the purposes of the Convention."[20]

Child soldiers

The United Nations Convention on the Rights of the Child, Article 38, (1989) proclaimed: "State parties shall take all feasible measures to ensure that persons who have not attained the age of 15 years do not take a direct part in hostilities".

In a 2003 briefing[21] for the 4th UN Security Council open debate on children and armed conflict by Human Rights Watch they state in their introduction that:

> In recent years progress has been made in developing a legal and policy framework for protecting children involved in armed conflict. The Optional Protocol to the Convention on the Rights of the Child on children in armed conflict, which came into force in February 2002, prohibits the direct use of any child under the age of 18 in armed conflict and prohibits all use of under-18s by non-state armed groups. By mid-December 2003, 67 states had ratified the Optional Protocol, including seven mentioned in this report (The seven are: Afghanistan, Democratic Republic of Congo, Philippines, Rwanda, Sierra Leone, Sri Lanka and Uganda). The UN Committee on the Rights of the Child had begun examining governments' reports on steps taken to implement the Protocol. [Articles 8(2)(b)(xxvi), (e)(vii)[22] of] the Rome Statute of the International Criminal Court (1998) defines the recruitment of children under the age of 15 as a war crime.[23]

On 26 July 2005, the United Nations Security Council unanimously passed UN Security Council Resolution 1612, the sixth in a series of resolutions about children and armed conflict.[24] Resolution 1612 established the first comprehensive monitoring and reporting system for enforcing compliance among those groups using child soldiers in armed conflict.[25]

Municipal law

United States

There are two separate issues to be determined in evaluating the category "unlawful combatant" as applied by the government of the United States. One issue is whether such a category *could* exist without violating the Geneva Conventions, and if such a category does exist, what steps the United States executive branch needs to take to comply with municipal laws as interpreted by the judicial branch of the United States government.

1942 Quirin case

The term *unlawful combatant* has been used for the past century in legal literature, military manuals and case law.[3] The term "unlawful combatants" was first used in US municipal law in a 1942 United States Supreme Court decision in the case *ex parte Quirin*.[26] In this case, the Supreme Court upheld the jurisdiction of a U.S. military tribunal over the trial of several German saboteurs in the US. This decision states (emphasis added and footnotes removed):

> By universal agreement and practice, the law of war draws a distinction between the armed forces and the peaceful populations of belligerent nations and also between those who are lawful and unlawful combatants.

> Lawful combatants are subject to capture and detention as prisoners of war by opposing military forces. **Unlawful combatants** are likewise subject to capture and detention, but in addition they are subject to trial and punishment by military tribunals for acts which render their belligerency unlawful. The spy who secretly and without uniform passes the military lines of a belligerent in time of war, seeking to gather military information and communicate it to the enemy, or an enemy combatant who without uniform comes secretly through the lines for the purpose of waging war by destruction of life or property, are familiar examples of belligerents who are generally deemed not to be entitled to the status of prisoners of war, but to be offenders against the law of war subject to trial and punishment by military tribunals.

The validity of this case, as basis for denying prisoners in the War on Terrorism protection by the Geneva Conventions, has been disputed.[27] [28] [29] A report by the American Bar Association commenting on this case, states:

> The Quirin case, however, does not stand for the proposition that detainees may be held incommunicado and denied access to counsel; the defendants in Quirin were able to seek review and they were represented by counsel. In Quirin, "The question for decision is whether the detention of petitioners for trial by Military Commission ... is in conformity with the laws and Constitution of the United States. " Quirin, 317 U.S. at 18. Since the Supreme Court has decided that even enemy aliens not lawfully within the United States are entitled to review under the circumstances of Quirin,11 that right could hardly be denied to U. S. citizens and other persons lawfully present in the United States, especially when held without any charges at all.
>
> —American Bar Association[30]

Since the 1942 Quirin case, the US signed and ratified the 1949 Geneva Conventions, which are, therefore, considered to be a part of US municipal law, in accordance with the Supremacy Clause in the Constitution of the United States.[31] In addition the US Supreme Court invalidated this premise, in Hamdan v. Rumsfeld, by ruling that Common Article Three of the Geneva Conventions applies to detainees in the War on Terror, and that the Military Tribunals used to try these suspects were in violation of US and international law.[32]

Congress addressed these issues in the Military Commissions Act of 2006, so that enemy combatants and unlawful enemy combatants might be tried under military commissions; however, on 12 June 2008, the Supreme Court ruled, in Boumediene v. Bush, that Guantanamo Bay captives were entitled to access the US justice system, and that the military commissions as constituted under the Military Commissions Act of 2006 fell short of what was required of a court under the United States constitution (see the section below for more details).

2001 Presidential military order

The suicide hijackers of the September 11, 2001 attacks have been widely referred to in the USA as "unlawful combatants" due to the nature of the devastating attacks of hijacking commercial civilian airlines, with passengers on board, and using them to crash into buildings with no warning, demands or negotiations. In the wake of the 9/11 attacks, the United States Congress passed a resolution known as the *Authorization for Use of Military Force* (AUMF) on 18 September 2001. In this, Congress invoked the War Powers Resolution and stated:

> That the President is authorized to use all necessary and appropriate force against those nations, organizations, or persons he determines planned, authorized, committed, or aided the terrorist attacks that occurred on September 11, 2001, or harbored such organizations or persons, in order to prevent any future acts of international terrorism against the United States by such nations, organizations or persons.[33]

Using the authorization granted to him by Congress, on 13 November 2001, President Bush issued a Presidential Military Order: "Detention, Treatment, and Trial of Certain Non-Citizens in the War Against Terrorism"[34] which allowed "individuals ... to be detained, and, when tried, to be tried for violations of the laws of war and other applicable laws by military tribunals", where such individuals are a member of the organization known as al Qa'ida; or has conspired or committed acts of international terrorism, or have as their aim to cause, injury to or adverse effects on the United States, its citizens, national security, foreign policy, or economy. The order also specifies that the detainees are to be treated humanely.

The length of time for which a detention of such individuals can continue before being tried by a military tribunal is not specified in the military order. The military order uses the term "detainees" to describe the individuals detained under the military order. The U.S. administration chooses to describe the detainees held under the military order as "**illegal enemy combatants**".

With the U.S. invasion of Afghanistan some lawyers in the Justice Department's Office of Legal Counsel and in the office of White House counsel Alberto Gonzales advised President Bush that he did not have to comply with the Geneva Conventions in handling detainees in the War on Terrorism. This applied not only to members of al Qa'ida but the entire Taliban, because, they argued, Afghanistan was a "failed state."[35]

Despite opposition from the U.S. State Department, which warned against ignoring the Geneva Conventions, the Bush administration thenceforth began holding such individuals captured in Afghanistan under the military order and not under the usual conditions of Prisoners of War.[36] For those U.S. citizens detained under the military order, US officials, such as Vice President Dick Cheney, argue that the urgency of the post-9/11 environment called for such tactics in administration's war against terrorism.

Most of the individuals detained by the U.S. military on the orders of the U.S. administration were initially captured in Afghanistan. The foreign detainees are held in the Guantanamo Bay detention camp established for the purpose at the Guantanamo Bay Naval Base, Cuba. Guantanamo was chosen because, although it is under the *de facto* control of the United States administration, it is not a sovereign territory of the United States, and a previous Supreme Court ruling *Johnson v. Eisentrager* in 1950 had ruled that U.S. courts had no jurisdiction over enemy aliens held outside the USA.

In *Rasul v. Bush* the Supreme Court ruled that "the US Guantanamo Bay, Cuba, Naval Base, which the United States occupies under a lease and treaty recognizing Cuba's ultimate sovereignty, but giving this country complete jurisdiction and control for so long as it does not abandon the leased areas."[37] and that as the United States had complete jurisdiction, the federal courts has the authority under the federal habeas corpus statute to decide whether foreign nationals (non-U.S. citizens) held in Guantanamo Bay were rightfully imprisoned. This ruling largely overturned the judicial advantage for the U.S. administration of using the Naval Base, that *Johnson v. Eisentrager* seemed to have conferred.

Legal challenges

There have been a number of legal challenges made on behalf of the detainees held in Guantanamo Bay detention camp and in other places. These include:

- On 30 July 2002, the U.S. District Court for the District of Columbia ruled in Rasul v. Bush, that it did not have jurisdiction because Guantanamo Bay Naval Base is not a sovereign territory of the United States. This decision was appealed to the D.C. Circuit Court of Appeals, which upheld the decision, (along with a related case in March 2003 — see Al-Odah v. United States). *Rasul v. Bush* was appealed to the United States Supreme Court on 2 September 2003.
- On 10 November 2003, the United States Supreme Court announced that it would decide on appeals by Afghan war detainees who challenge their continued incarceration at Guantanamo Bay Naval Base as being unlawful, (See Rasul v. Bush).
- On 10 January 2004, 175 members of both houses of Parliament in the UK filed an *amici curiae* brief to support the detainees' access to US jurisdiction.
- On 28 June 2004, the Supreme Court ruled in *Rasul v. Bush* that detainees in Guantanamo Bay Naval Base could turn to U.S. courts to challenge their confinement, but can also be held without charges or trial.
- On 7 July 2004, In response to the Supreme Court ruling, the Pentagon announced that cases would be reviewed by military tribunals, in compliance with Article 5 of the Third Geneva Convention.[38] [39]

- On 8 November 2004, a federal court halted the proceeding of Salim Ahmed Hamdan, 34, of Yemen. Hamdan was to be the first Guantanamo detainee tried before a military commission. Judge James Robertson of the U.S. District Court for the District of Columbia ruled in *Hamdan v. Rumsfeld*[40] that no competent tribunal had found that Hamdan was not a prisoner of war under the Geneva Conventions.
- By 29 March 2005, all detainees at the Guantanamo Bay Naval Base had received hearings before Combatant Status Review Tribunals. The hearings resulted in the release of 38 detainees, and confirmed the *enemy combatant* status of 520 detainees.[39] Reuters reported on 15 June 2005 only four detainees had been charged and that Joseph Margulies, one of the lawyers for the detainees said "The (reviews) are a sham ... They mock this nation's commitment to due process, and it is past time for this mockery to end".[41]

Yaser Hamdi was captured in Afghanistan in November 2001. He was taken to Guantanamo Bay Naval Base, but was transferred to jails in Virginia and South Carolina after it became known that he was a U.S. citizen. On 23 September 2004, the United States Justice Department agreed to release Hamdi to Saudi Arabia, where he is also a citizen, on the condition that he gave up his U.S. citizenship. The deal also bars Hamdi from visiting certain countries and to inform Saudi officials if he plans to leave the kingdom. He was a party to a Supreme Court decision *Hamdi v. Rumsfeld* which issued a decision on 28 June 2004, repudiating the U.S. government's unilateral assertion of executive authority to suspend the constitutional protections of individual liberty of a U.S. citizen. The Court recognized the power of the government to detain unlawful combatants, but ruled that detainees must have the ability to challenge their detention before an impartial judge. Though no single opinion of the Court commanded a majority, eight of the nine justices of the Court agreed that the Executive Branch does not have the power to hold indefinitely a U.S. citizen without basic due process protections enforceable through judicial review.

On 8 May 2002, José Padilla, also known as Abdullah al-Muhajir, was arrested by FBI agents at Chicago's O'Hare International Airport and held as material witness on the warrant issued in New York State about the 2001 9/11 attacks. On 9 June 2002 President Bush issued an order to Secretary Rumsfeld to detain Padilla as an "enemy combatant". The order legally justified the detention by leaning on the AUMF which authorized the President to "use all necessary force against those nations, organizations, or *persons*" and in the opinion of the administration a U.S. citizen can be an enemy combatant (this was decided by the United States Supreme Court in the case of *Ex parte Quirin*).[42] Padilla is being detained in Miami and is accused of providing material support for terrorism.

- The 13 November 2001, Military Order, mentioned above, exempts U.S. citizens from trial by military tribunals to determine if they are "unlawful combatants", which indicates that Padilla and *Yaser Hamdi* would end up in the civilian criminal justice system, as happened with John Walker Lindh.
- On 18 December 2003, the Second Circuit Court of Appeals declared that the Bush Administration lacked the authority to detain a U.S. citizen arrested on U.S. soil as an "illegal enemy combatant" without clear congressional authorization (per 18 U.S.C. § 4001 [43](a)); it consequently ordered the government to release Padilla from military custody within thirty days.[44] But agreed that he could be held until an appeal was heard.
- On 20 February 2004, the Supreme Court agreed to hear the government's appeal.
- The Supreme Court heard the case, *Rumsfeld v. Padilla*, in April 2004, but on 28 June it was thrown out on a technicality. The court declared that New York State, where the case was originally filed, was an improper venue and that the case should have been filed in South Carolina, where Padilla was being held.
- On 28 February 2005, in Spartanburg, South Carolina, U.S. District Judge Henry Floyd ordered the Bush administration to either charge Padilla or release him.[45] He relied on the Supreme Court's ruling in the parallel enemy combatant case of Yaser Hamdi (*Hamdi v. Rumsfeld*), in which the majority decision declared a "state of war is not a blank check for the president when it comes to the rights of the nation's citizens."
- On 19 July 2005, in Richmond, Virginia, the Fourth Circuit Court of Appeals began hearing the government's appeal of the lower court (the District of South Carolina, at Charleston) ruling by Henry F. Floyd, District Judge, (CA-04-2221-26AJ). Their ruling, decided 9 September 2005, was that "the President does possess such authority pursuant to the Authorization for Use of Military Force Joint Resolution enacted by Congress in the wake of the attacks on the United States of September 11, 2001. Accordingly, the judgment of the district court is

reversed."[46]

- In Hamdan v. Rumsfeld (29 June 2006) the US Supreme Court did not rule on the subject of unlawful combatant status but did reaffirm that the US is bound by the Geneva Conventions. Most notably it said that Common Article 3 of the Geneva Convention, regarding the treatment of detainees, applies to all prisoners in the War on Terror.

Combatant Status Review Tribunal

Following the *Hamdan v. Rumsfeld*-ruling (November 2004) the Bush administration has begun using Combatant Status Review Tribunals to determine the status of detainees. By doing so the obligation under Article 5 of the GCIII was to be addressed.

However, critics maintain these CSRTs are inadequate to warrant acceptance as "competent tribunal." Their principal arguments are:

- The CSRT conducted rudimentary proceedings
- The CSRT afforded detainees few basic protections
- Many detainees lacked counsel
- The CSRT also informed detainees only of general charges against them, while the details on which the CSRT premised enemy combatant status decisions were classified.
- Detainees had no right to present witnesses or to cross-examine government witnesses.

Notable cases pointed to by critics as demonstrating the flawed nature of the procedure include: Mustafa Ait Idir, Moazzam Begg, Murat Kurnaz, Feroz Abbasi, and Martin Mubanga. A comment by legal experts states:

> It appears ... that the procedures of the Combatant Status Review Tribunals do not qualify as status determination under the Third Geneva Convention. <......> The fact that no status determination had taken place according to the Third Geneva Convention was sufficient reason for a judge from the District Court of Columbia dealing with a habeas petition, to stay proceedings before a military commission. Judge Robertson in Hamdan v. Rumsfeld held that the Third Geneva Convention, which he considered selfexecuting, had not been complied with since a Combatant Status Review Tribunal could not be considered a 'competent tribunal' pursuant to article 5 of the Third Geneva Convention.[47]

James Crisfield, the legal advisor to the Tribunals, offered his legal opinion, that CSRT "do not have the discretion to determine that a detainee should be classified as a prisoner of war — only whether the detainee satisfies the definition of 'enemy combatant'".[48] Determining whether a captive should be classified as a prisoner of war is the sole purpose of a competent tribunal.

Analysis of these Tribunals by two lawyers for Guananamo detainees, Professor Mark P. Denbeaux of the Seton Hall University School of Law, his son Joshua Denbeaux, and some of his law students resulted in a report called *No-hearing hearings*. In essence it supports the criticism voiced above.[49] [50]

Military commissions

As of 17 October 2006, When President Bush signed the Military Commissions Act of 2006 into law, Title 10 of the United States Code was amended to include a definition of an "unlawful enemy combatant" as

> a person who has engaged in hostilities or who has purposefully and materially supported hostilities against the United States or its co-belligerents who is not a lawful enemy combatant (including a person who is part of the Taliban, al-Qaida, or associated forces); or a person who, before, on, or after the date of the enactment of the Military Commissions Act of 2006, has been determined to be an unlawful enemy combatant by a Combatant Status Review Tribunal or another competent tribunal established under the authority of the President or the Secretary of Defense.

The definition of a lawful enemy combatant is also given, and much of the rest of the law sets out the specific procedures for determining whether a given detainee of the U.S. armed forces is an unlawful enemy combatant and how such combatants may or may not be treated in general and tried for their crimes in particular. Among its more controversial provisions, the law stipulates that a non United States citizen held as an enemy combatant or is awaiting such determination may not seek habeas corpus relief. Such detainees must simply wait until the military convene a detainee status review tribunal (under the procedures described in the Detainee Treatment Act of 2005).

Immediately after Bush signed the Act into law, the U.S. Justice Department notified the U.S. Court of Appeals for the District of Columbia that the Court no longer had jurisdiction over a combined *habeas* case that it had been considering since 2004. A notice dated the following day listed 196 other pending habeas cases for which it made the same claim.[51]

Of the first three war crimes cases brought against Guantanamo Bay detainees under the Military Commissions Act, one resulted in a plea bargain and the two others were dismissed on jurisdictional grounds.

On 4 June 2007, in two separate cases, military tribunals dismissed charges against detainees who had been designated as "enemy combatants" but not as "unlawful enemy combatants". The first case was that of Omar Khadr, a Canadian who had been designated as an "enemy combatant" in 2004. Khadr was accused of throwing a grenade during a firefight in Afghanistan in 2002. Colonel Peter Brownback ruled that the military tribunals, created to deal with "unlawful enemy combatants", had no jurisdiction over detainees who had been designated only as "enemy combatants". He dismissed without prejudice all charges against Khadr.[52] Also on 4 June, Captain Keith J. Allred reached the same conclusion in the case of Salim Ahmed Hamdan.[53] [54] [55]

The United States Department of Defense responded by stating: "We believe that Congress intended to grant jurisdiction under the Military Commissions Act to individuals, like Mr. Khadr, who are being held as enemy combatants under existing C.S.R.T. procedures." That position was called "dead wrong" by Specter.[53]

Supreme Court ruling on Military Commissions Act of 2006

On 12 June 2008, the Supreme Court ruled, in Boumediene v. Bush, 5-4 that Guantanamo captives were entitled to access the US justice system.[56] [57] [58] Justice Anthony Kennedy wrote in the majority opinion:

> The laws and Constitution are designed to survive, and remain in force, in extraordinary times.

The Court also ruled that the Combatant Status Review Tribunals were "inadequate".[56] Ruth Bader Ginsburg, Stephen Breyer, David Souter and John Paul Stevens joined Kennedy in the majority.

Chief Justice John Roberts, in the minority opinion, called the CSR Tribunals[56]

> the most generous set of procedural protections ever afforded aliens detained by this country as enemy combatants.

Samuel Alito, Clarence Thomas and Antonin Scalia joined Roberts in the dissent.[57]

Vincent Warren, the executive director of the Center for Constitutional Rights, the organization that initiated the action that triggered the Supreme Court ruling responded:[58]

> The Supreme Court has finally brought an end to one of our nation's most egregious injustices. It has finally given the men held at Guantanamo the justice that they have long deserved. By granting the writ of habeas corpus, the Supreme Court recognizes a rule of law established hundreds of years ago and essential to American jurisprudence since our nation's founding. This six-year-long nightmare is a lesson in how fragile our constitutional protections truly are in the hands of an overzealous executive.

2009

In January and February 2009, President Barack Obama's nominees for Attorney General and Solicitor General, Eric Holder and Elena Kagan, both testified they agreed the U.S. government may detain combatants in accordance with the laws of war until the end of the war, (this sidesteps the issue of deciding whether the combatant is a lawful or unlawful combatant and the need to try them). When asked by Senator Lindsey Graham "If our intelligence agencies should capture someone in the Philippines that is suspected of financing Al Qaeda worldwide, would you consider that person part of the battlefield?" Both Holder and Kagan said that they would.[59] [60] [61]

Other countries

Israel, since the 2002 "Imprisonment of Illegal Combatants Law", makes theoretical distinctions between lawful and unlawful combatants and the legal status thereof.[62] [63] [64] [65]

The United Kingdom Crown Prosecution Service (CPS) makes the distinction. The CPS conducted a "through review of the evidence concerning the deaths of Sergeant Steven Roberts of the 2nd Royal Tank Regiment and Mr Zaher Zaher, an Iraqi national, at Az Zubayr, Iraq on 24 March 2003":[66] [67]

> In reviewing the case, the CPS lawyer considered the possible view that, because of his behaviour, Mr Zaher had become an unlawful combatant and therefore under the Rules of Engagement, under which the [British] soldiers were required to operate, they would have been entitled to take offensive action against him. Under the Rules of Engagement and the Geneva Convention, unless a person is positively identified as being a combatant, they should be considered a civilian and treated accordingly.
> As the alternative view would be that Mr Zaher was not an unlawful combatant but a civilian, the reviewing lawyer also considered whether the soldiers could rely on self defence. ...
> —Crown Prosecution Service.[66]

International criticism of unlawful combatant status

The designation of some prisoners as "unlawful combatants", has been the subject of criticism by international human rights institutions; including Amnesty International, Human Rights Watch and the International Committee of the Red Cross.

In response to the US-led military campaign in Afghanistan, a legal advisor at the Legal Division of the ICRC, published a paper on the subject,[3] in which it states:

> Whereas the terms "combatant" "prisoner of war" and "civilian" are generally used and defined in the treaties of international humanitarian law, the terms "unlawful combatant", "unprivileged combatants/belligerents" do not appear in them. They have, however, been frequently used at least since the beginning of the last century in legal literature, military manuals and case law. The connotations given to these terms and their consequences for the applicable protection regime are not always very clear.

Human Rights Watch have pointed out that in a judgement, the International Criminal Tribunal for the Former Yugoslavia interpreted the International Committee of the Red Cross, *Commentary: IV Geneva Convention Relative to the Protection of Civilian Persons in Time of War* (Geneva: 1958) to mean that:

> there is no gap between the Third and Fourth Geneva Conventions. If an individual is not entitled to the protection of the Third Convention as a prisoner of war ... he or she necessarily falls within the ambit of [the Fourth Convention], provided that its article 4 requirements [defining a protected person] are satisfied.[13]

This does not mean that the status of unlawful combatant does not exist because in the opinion of the ICRC "If civilians directly engage in hostilities, they are considered 'unlawful' or 'unprivileged' combatants or belligerents ... [and] They may be prosecuted under the domestic law of the detaining state for such action".[1] [68]

Many governments and human rights organizations worry that the introduction of the *unlawful combatant* status sets a dangerous precedent for other regimes to follow. When the government of Liberia detained American activist Hassan Bility in 2002, Liberian authorities dismissed the complaints[69] of the United States, responding that he had been detained as an unlawful combatant.

See also

- Irregular military
- Targeted killing

USA specific

- Criticisms of the War on Terrorism
- Enemy Combatant and No longer enemy combatant
- Military Police: Enemy Prisoners of War, Retained Personnel, Civilian Internees and Other Detainees
- Seton Hall study
- Department of Defense Directive 2310

Footnotes

[1] ICRC official statement: *The relevance of IHL in the context of terrorism* (http://www.icrc.org/web/eng/siteeng0.nsf/html/terrorism-ihl-210705), 21 July 2005

[2] "Unlawful Combatants" in the United States: Drawing the Fine Line Between Law and War (http://www.abanet.org/irr/hr/winter03/unlawful.html) Human Rights Magazine Winter 2003, published by the American Bar Association

[3] The legal situation of unlawful/unprivileged combatants (http://www.icrc.org/eng/assets/files/other/irrc_849_dorman.pdf) (IRRC March 2003 Vol.85 No 849)

[4] The International Criminal Tribunal for the Former Yugoslavia "Celebici Judgment: Prosecutor v. Delalic, Mucic, Delic, and Landzo, Case No." IT-96-21-T (http://www.un.org/icty/cases-e/index-e.htm) seems to return the Appeal Judgment instead of the Trial Judgment. However, the relevant section of the Judgment is available from the University of the West of England Delalic et al. (I.T-96-21) "Celebici" 16 November 1998 Part III B, Applicable law 2. Status of the Victims as "Protected Persons." See: Para. 271 (http://www.ess.uwe.ac.uk/documents/part3B.htm):

> In addition, the evidence provided to the Trial Chamber does not indicate that the Bosnian Serbs who were detained were, as a group, at all times carrying their arms openly and observing the laws and customs of war. Article 4(A)(6) undoubtedly places a somewhat high burden on local populations to behave as if they were professional soldiers and the Trial Chamber, therefore, considers it more appropriate to treat all such persons in the present case as civilians.
>
> It is important, however, to note that this finding is predicated on the view that there is no gap between the Third and the Fourth Geneva Conventions. If an individual is not entitled to the protections of the Third Convention as a prisoner of war (or of the First or Second Conventions) he or she necessarily falls within the ambit of Convention IV, provided that its article 4 requirements are satisfied. The Commentary to the Fourth Geneva Convention asserts that;
>
>> [e]very person in enemy hands must have some status under international law: he is either a prisoner of war and, as such, covered by the Third Convention, a civilian covered by the Fourth Convention, or again, a member of the medical personnel of the armed forces who is covered by the First Convention. *There is no* intermediate status; nobody in enemy hands can be outside the law. We feel that this is a satisfactory solution – not only satisfying to the mind, but also, and above all, satisfactory from the humanitarian point of view." Jean Pictet (ed.) – Commentary: IV Geneva Convention Relative to the Protection of Civilian Persons in Time of War (1958) (http:/ / www. icrc. org/ ihl. nsf/ COM/ 380-600007?OpenDocument) – 1994 reprint edition.

[5] Geneva Conventions Protocol I (http://www.icrc.org/ihl.nsf/7c4d08d9b287a42141256739003e636b/f6c8b9fee14a77fdc125641e0052b079) Article 51.3 also covers this interpretation "Civilians shall enjoy the protection afforded by this section, unless and for such time as they take a direct part in hostilities."

[6] Commentary for Common Article 3 of the Geneva Conventions (http://www.icrc.org/ihl.nsf/COM/375-590006?OpenDocument)

[7] Knut Dörmann: " The legal situation of unlawful/unprivileged combatants (http://www.icrc.org/Web/Eng/siteeng0.nsf/iwpList553/88140B29D25A47EFC1256D0B0046F105)." Article in the International Review of the ICRC, March 2003

[8] Warriors without rights? combatants , unprivileged belligerents, and the struggle over legitimacy (http://www.hpcr.org/pdfs/OccasionalPaper2.pdf) by Kenneth Watkin for The Program on Humanitarian Policy and Conflict Research (http://www.hpcr.org/) or HTML version (http://64.233.183.104/search?q=cache:Kg1cyRkcafQJ:www.hpcr.org/pdfs/OccasionalPaper2.pdf+Fletcher+

Unlawful+combatant&hl=en)

[9] Rupert Ticehurst *The Martens Clause and the Laws of Armed Conflict* (http://www.icrc.org/Web/Eng/siteeng0.nsf/html/57JNHY) 30 April 1997, *International Review of the Red Cross*, no. 317, p.125-134

[10] Vladimir Pustogarov, *Fyodor Fyodorovich Martens (1845-1909) - a humanist of modern times* (http://www.icrc.org/Web/Eng/siteeng0.nsf/html/57JN52), 30 June 1996 *International Review of the Red Cross*, no. 312, p.300-314

[11] Laws of War: Laws and Customs of War on Land (Hague II) (http://www.yale.edu/lawweb/avalon/lawofwar/hague02.htm); 29 July 1899. contained in the Avalon Project archive at Yale Law School

[12] The ICRC Commentary on Article 5 (http://www.icrc.org/ihl.nsf/COM/375-590008?OpenDocument) says on the issue of *competent tribunal* that "At Geneva in 1949, it was first proposed that for the sake of precision the term 'responsible authority' should be replaced by 'military tribunal' (11). This amendment was based on the view that decisions which might have the gravest consequences should Hot [*sic*] be left to a single person, who might often be of subordinate rank. The matter should be taken to a court, as persons taking part in the fight without the right to do so are liable to be prosecuted for murder or attempted murder, and might even be sentenced to capital punishment (12). This suggestion was not unanimously accepted, however, as it was felt that to bring a person before a military tribunal might have more serious consequences than a decision to deprive him of the benefits afforded by the Convention (13). A further amendment was therefore made to the Stockholm text stipulating that a decision regarding persons whose status was in doubt would be taken by a 'competent tribunal', and not specifically a military tribunal.
Another change was made in the text of the paragraph, as drafted at Stockholm, in order to specify that it applies to cases of doubt as to whether persons having committed a belligerent act and having fallen into the hands of the enemy belong to any of the categories enumerated in Article 4 (14). The clarification contained in Article 4 should, of course, reduce the number of doubtful cases in any future conflict.
It therefore seems to us that this provision should not be interpreted too restrictively; the reference in the Convention to 'a belligerent act' relates to the principle which motivated the person who committed it, and not merely the manner in which the act was committed."

- (11) [(2) p.77] See ' Final Record of the Diplomatic Conference
of Geneva of 1949, ' Vol. II-A, p. 388;

- (12) [(3) p.77] Ibid., Vol. III, p. 63, No. 95;
- (13) [(4) p.77] Ibid., Vol. II-B, p. 270;
- (14) [(5) p.77] Ibid., pp. 270-271;

[13] Background Paper on Geneva Conventions and Persons Held by U.S. Forces (http://hrw.org/backgrounder/usa/pow-bck.htm#P56_11515) by "Human Rights Watch Press" footnote 1: International Committee of the Red Cross, *Commentary: IV Geneva Convention Relative to the Protection of Civilian Persons in Time of War* (Geneva: 1958), p. 51 (emphasis in original). The International Criminal Tribunal for the Former Yugoslavia, charged with prosecuting war crimes and crimes against humanity committed during the recent conflicts in the Balkans, has explicitly affirmed this principle in a 1998 judgment, stating that "there is no gap between the Third and Fourth Geneva Conventions. If an individual is not entitled to the protection of the Third Convention as a prisoner of war ... he or she necessarily falls within the ambit of [the Fourth Convention], provided that its article 4 requirements [defining a protected person] are satisfied." Celebici Judgment, para. 271 (1998).

[14] Geneva Conventions Common Article 3

[15] 1976 June 28: Death sentence for Angolan mercenaries (http://news.bbc.co.uk/onthisday/hi/dates/stories/june/28/newsid_2520000/2520575.stm) BBC

[16] Commentary on the Convention relative to the Treatment of Prisoners of War (http://www.icrc.org/ihl.nsf/INTRO/305?OpenDocument). Geneva, 27 July 1929

[17] Laws and Customs of War on Land (Hague IV) (http://www.yale.edu/lawweb/avalon/lawofwar/hague04.htm); 18 October 1907

[18] Brown, Gary D.. *Prisoner of war parole: Ancient concept, modern utility* (http://www.au.af.mil/au/awc/awcgate/law/pow_parole.pdf) The Military Law Review, Vol 156 (June 1998) p.13 (Major Gary D. Brown in June 1998 was Chief, International and Operational Law at Headquarters, United States Strategic Command, Offutt Air Force Base, Nebraska.

[19] International Convention against the Recruitment, Use, Financing and Training of Mercenaries (http://www.un.org/documents/ga/res/44/a44r034.htm) A/RES/44/34 72nd plenary meeting 4 December 1989 (UN Mercenary Convention) Entry into force: 20 October 2001 (http://www.ohchr.org/english/law/mercenaries.htm)

[20] International Convention against the Recruitment, Use, Financing and Training of Mercenaries (http://www.ohchr.org/english/law/mercenaries.htm)

[21] Child Soldier Use 2003: A Briefing for the 4th UN Security Council Open Debate on Children and Armed Conflict (http://hrw.org/reports/2004/childsoldiers0104/) by Human Rights Watch.

[22] The International Criminal Court and Children's Rights (http://www.iccnow.org/documents/FS-AMICC-ICCnChildRights.pdf?PHPSESSID=544c50d2523e5c46acc51db534ae7ccf) (PFD) by "The American Non Governmental Organizations Coalition for the International Criminal Court" Page 1, Paragraph 3.

[23] Introduction (http://hrw.org/reports/2004/childsoldiers0104/1.htm) to A Briefing for the 4th UN Security Council Open Debate on Children and Armed Conflict by Human Rights Watch

[24] Children and Armed Conflict: International Law/United Nations (http://www.cdi.org/program/issue/index.cfm?ProgramID=21&issueid=108) by the Center for Defence Information

[25] Children and Armed Conflict: UN enters “era of application” in its campaign against child soldiers (http://www.cdi.org/friendlyversion/printversion.cfm?documentID=3175), Center for Defence Information, 12 October 2005

[26] Ex Parte Quirin -n1- (Nos. 1-7CERTIORARI TO THE UNITED STATES COURT OF APPEALS FOR THE DISTRICT OF COLUMBIA) (http://supct.law.cornell.edu/supct/html/historics/USSC_CR_0317_0001_ZS.html) or Ex Parte Quirin (http://www.agh-attorneys.com/4_ex_parte_quirin.htm) or EX PARTE QUIRIN (http://www.law.umkc.edu/faculty/projects/ftrials/conlaw/quirin.html)

[27] War and the Constitution (http://www.prospect.org/print/V13/1/fletcher-g.html) by George P. Fletcher in The American Prospect Issue Date: 1.1.02 or War and the Constitution (http://www.law.fsu.edu/faculty/2001-2002workshops/fletcher.pdf) and the response The Military Tribunal Debate (http://www.prospect.org/web/page.ww?section=root&name=ViewWeb&articleId=618)

[28] Revised ACLU Interested Person's Memo Urging Congress to Reject Power to Detain Suspected Terrorists Indefinitely Without Charge, Trial or a Right to Counsel (http://www.aclu.org/safefree/detention/18471leg20040623.html) by the American Civil Liberties Union

[29] Terrorism and the rule of law (http://www.iap.nl.com/speeches_annual_conference_2003_washington/terrorism_and_the_rule_of_law_speech_by_nicholas_cowdery.html) by Nicholas Cowdery AM QC, President, International Association of Prosecutors Director of Public Prosecutions, NSW, Australia, at International Association of Prosecutors 8th Annual Conference, Washington, D.C. - 10–14 August 2003.

[30] report by the American Bar Association (http://news.findlaw.com/hdocs/docs/aba/abatskforce103rpt.pdf) in PDF

[31] Wikisource:Ryuichi Shimoda et al. v. The State#II. Evaluation of the act of bombing according to municipal law Paragraph 2

[32] The Gitmo Fallout: The fight over the Hamdan ruling heats up—as fears about its reach escalate. (http://www.msnbc.msn.com/id/13773997/site/newsweek/) By Michael Isikoff and Stuart Taylor Jr., Newsweek, 17 July 2006

[33] US Congress' joint resolution of 18 September 2001 Authorization for Use of Military Force ("AUMF") (http://news.findlaw.com/wp/docs/terrorism/sjres23.es.html); public law 107-40, 115 Stat. 224

[34] President George W. Bush's Military Order of 13 November 2001: Detention, Treatment, and Trial of Certain Non-Citizens in the War Against Terrorism (http://georgewbush-whitehouse.archives.gov/news/releases/2001/11/print/20011113-27.html); 66 FR 57833 backup site (http://www.law.uchicago.edu/tribunals/exec_order.html)

[35] Outsourcing torture: The secret history of America's "extraordinary rendition" program (http://www.newyorker.com/printables/fact/050214fa_fact6) by Jane Mayer The New Yorker Issue of 2005-02-14 Posted 2005-02-07 Paragraph 32

[36] Outsourcing torture: The secret history of America's "extraordinary rendition" program (http://www.newyorker.com/printables/fact/050214fa_fact6) by Jane Mayer The New Yorker Issue of 2005-02-14 Posted 2005-02-07 Paragraph 34

[37] Rasul et al. v. Bush, President of the United States (http://caselaw.lp.findlaw.com/scripts/getcase.pl?court=US&vol=000&invol=03-334))

[38] Q&A: US Supreme Court Guantanamo ruling (http://news.bbc.co.uk/2/hi/americas/3867067.stm), BBC 8 July 2004

[39] DoD News: Combatant Status Review Tribunals Update (http://www.globalsecurity.org/security/library/news/2005/01/sec-050119-dod01.htm) No. 057-05, 19 January 2005

[40] *Hamdan v. Rumsfeld* summary (http://jurist.law.pitt.edu/gazette/2004/11/ruling-on-lawfulness-of-Guantanamo.php), full text (PDF File) (http://www.dcd.uscourts.gov/04-1519.pdf) – U.S. District Court for the District of Columbia, presiding Judge James Robertson

[41] Guantanamo inmates can be held 'in perpetuity'- US (http://www.arabnews.com/?page=4&section=0&article=65514&d=16&m=6&y=2005&pix=world.jpg&category=World), Arab News, 16 June 2005. (Reuters report 15 June 2005)

[42] Authorization for Use of Military Force: Padilla v. Bush: Jose Padilla under the Joint Resolution (http://www.law.syr.edu/academics/centers/insct/Military Force - Padilla.asp) The Syracuse Journal of International Law and Commerce, issued by the Syracuse Collage of Law

[43] http://www.law.cornell.edu/uscode/18/4001.html

[44] Appeals Court Says Bush Can't Hold U.S. Citizen (http://www.commondreams.org/headlines03/1218-10.htm) Published on Thursday, 18 December 2003 by Reuters

[45] Judge Says Terror Suspect Can't Be Held as an Enemy Combatant (http://www.nytimes.com/2005/03/01/politics/01terror.html?ex=1111899600&en=4bade73e4d6a651f&ei=5070) The New York Times 1 March 2005

[46] José Padilla Fourth Circuit Court of Appeals 19 July, 9 September 2005

[47] Guantánamo Bay: A Reflection On The Legal Status And Rights Of 'Unlawful Enemy Combatants' (http://www.utrechtlawreview.org/publish/articles/000003/article.pdf) PDF by Terry Gill and Elies van Sliedregt in the Utrecht Law Review (http://www.utrechtlawreview.org/) or HTML version (http://64.233.183.104/search?q=cache:LfG0Qe7qjWsJ:www.utrechtlawreview.org/publish/articles/000003/article.pdf+Fletcher+Unlawful+combatant&hl=en)

[48] Moazzam Begg's dossier (.pdf) (http://wid.ap.org/documents/detainees/moazzambegg.pdf) from his Combatant Status Review Tribunal, hosted by Associated Press

[49] No-hearing hearings (http://law.shu.edu/news/final_no_hearing_hearings_report.pdf) by, Mark Denbeaux, Professor, Seton Hall University School of Law and Counsel to two Guantanamo detainees, Joshua Denbeaux, Esq. and David Gratz, John Gregorek, Matthew Darby, Shana Edwards, Shane Hartman, Daniel Mann, Megan Sassaman and Helen Skinner Students of Seton Hall University School of Law

[50] Nat Hentoff (December 8, 2006). "Bush's War Crimes Cover-up" (http://www.webcitation.org/query?url=http://www.villagevoice.com/news/0650,hentoff,75255,2.html&date=2009-08-13). Village Voice. Archived from the original (http://www.villagevoice.com/news/0650,hentoff,75255,2.html) on 2009-08-13. . Retrieved 2007-04-02.

[51] "Court Told It Lacks Power in Detainee Cases", *Washington Post* (http://www.washingtonpost.com/wp-dyn/content/article/2006/10/19/AR2006101901692.html?nav=rss_nation/special)

[52] Koring, Paul (2007). "U.S. case against Khadr collapses" (http://www.theglobeandmail.com/servlet/story/RTGAM.20070604.wkhadr0604_1/BNStory/International/home). *Toronto Globe and Mail.*

[53] Glaberson, William (5 June 2007). "Military Judges Dismiss Charges for 2 Detainees" (http://www.nytimes.com/2007/06/05/world/americas/05gitmo.html?hp). *The New York Times*.

[54] Sergeant Sara Wood (4 June 2007). "Charges Dismissed Against Canadian at Guantanamo" (http://www.defenselink.mil/news/newsarticle.aspx?id=46281). United States Department of Defense. . Retrieved 2007-06-07.

[55] Sergeant Sara Wood (4 June 2007). "Judge Dismisses Charges Against Second Guantanamo Detainee" (http://www.defenselink.mil/news/newsarticle.aspx?id=46288). United States Department of Defense. . Retrieved 2007-06-07.

[56] Mark Sherman (12 June 2008). "High Court: Gitmo detainees have rights in court" (http://ap.google.com/article/ALeqM5iS3b8PdQ_oVIJA2eFtDvhnnTUvFwD918J1QO0). Associated Press. . Retrieved 2008-06-12. "The court said not only that the detainees have rights under the Constitution, but that the system the administration has put in place to classify them as enemy combatants and review those decisions is inadequate." mirror (http://www.webcitation.org/query?url=http://ap.google.com/article/ALeqM5iS3b8PdQ_oVIJA2eFtDvhnnTUvFwD918J1QO0&date=2008-06-12)

[57] Mark Sherman (12 June 2008). "Terror suspects can challenge detention: U.S. Supreme Court" (http://www.theglobeandmail.com/servlet/story/RTGAM.20080612.wgitmo0612/BNStory/International/home). Globe and Mail. . Retrieved 2008-06-12.

[58] James Oliphant (12 June 2008). "Court backs Gitmo detainees" (http://weblogs.baltimoresun.com/news/politics/blog/2008/06/court_sides_with_gitmo_detaine.html). Baltimore Sun. . Retrieved 2008-06-12. mirror (http://www.webcitation.org/query?url=http://weblogs.baltimoresun.com/news/politics/blog/2008/06/court_sides_with_gitmo_detaine.html&date=2008-06-12)

[59] Solicitor general nominee says 'enemy combatants' can be held without trial (http://www.latimes.com/news/nationworld/nation/la-na-solicitor-general11-2009feb11,0,7158432.story), Los Angeles Times, February 11, 2009

[60] Evan Perez. Obama Considers Detaining Terror Suspects Indefinitely (http://online.wsj.com/article/SB124223286506515765.html), Wall Street Journal, 14 May 2009

[61] Tim Reid. Problem of Guantánamo detainees returns to haunt Barack Obama (http://www.timesonline.co.uk/tol/news/world/us_and_americas/article6216709.ece), The Times, 4 May 2009

[62] Other Countries

[63] "Israel: Opportunistic Law Condemned" (http://hrw.org/english/docs/2002/03/07/isrlpa3787.htm). Human Rights Watch. March 2002.
.

[64] Israel's Commitment to Domestic and International Law in Times of War (http://www.jcpa.org/brief/brief4-5.htm) by Judge Amnon Straschnov Former IDF Military Advocate General.

[65] Incarceration of Unlawful Combatants Law, 5762-2002 (http://www.justice.gov.il/NR/rdonlyres/7E86D098-0463-4F37-A38D-8AEBE770BDE6/0/IncarcerationLawedited140302.doc) (DOC) "unlawful combatant" means a person who has participated either directly or indirectly in hostile acts against the State of Israel or is a member of a force perpetrating hostile acts against the State of Israel, where the conditions prescribed in Article 4 of the Third Geneva Convention of 12th August 1949 with respect to prisoners-of-war and granting prisoner-of-war status in international humanitarian law, do not apply to him.

[66] CPS decision on Iraq deaths case : Press Release : Crown Prosecution Service (http://www.cps.gov.uk/news/press_releases/123_06/), 27 April 2006

[67] British Army Board of Inquiry into death of Sgt Steven Roberts (http://news.bbc.co.uk/2/shared/bsp/hi/pdfs/sgtroberts31_07_06.pdf), 31 July 2007, website of the BBC. See paragraph 61 for details of British Rules of Engagement in the Iraq War

[68] Geneva Conventions Protocol I (http://www.icrc.org/ihl.nsf/7c4d08d9b287a42141256739003e636b/f6c8b9fee14a77fdc125641e0052b079) Article 51.3 also covers this interpretation "Civilians shall enjoy the protection afforded by this section, unless and for such time as they take a direct part in hostilities.".

[69] Comments on the Arrest and Detention of Journalist Hassan Bility in Liberia (http://www.state.gov/r/pa/prs/ps/2002/11718.htm) Press Statement by Richard Boucher, Spokesman in the U.S State Department, 8 July 2002

References

- ICRC Comentaries on GCIII (http://www.icrc.org/ihl.nsf/WebList?ReadForm&id=375&t=com) General Provisions: Art 5 (http://www.icrc.org/ihl.nsf/COM/375-590008?OpenDocument) Part III : Captivity #Section VI : Relations between prisoners of war and the authorities #Chapter III : Penal and disciplinary sanctions #I. General provisions (http://www.icrc.org/ihl.nsf/b466ed681ddfcfd241256739003e6368/5cd83e96981e1ee0c12563cd00427ee4)
- Rule of Law in Armed Conflicts Project (http://www.adh-geneva.ch/RULAC)

Further reading

- Michael Greenberger: " Is Criminal Justice a Casualty of the Bush Administration's 'War on Terror'? (http://www.abanet.org/irr/hr/winter04/casualty.html)" in American Bar Association's Human Right Magazine, Winter 2004
- Daniel Kanstroom: " 'Unlawful Combatants' in the United States - Drawing the Fine Line Between Law and War (http://www.abanet.org/irr/hr/winter03/unlawful.html)" in American Bar Association's Human Right Magazine, Winter 2003
- Knut Dörmann: " The legal situation of unlawful/unprivileged combatants (http://www.icrc.org/Web/Eng/siteeng0.nsf/iwpList553/88140B29D25A47EFC1256D0B0046F105)". Article in the International Review of the ICRC, March 2003
- Michael Dorf: What is an "Unlawful combatant," and why it matters: The Status Of Detained Al Qaeda And Taliban Fighters (http://writ.news.findlaw.com/dorf/20020123.html) Published by FindLaw (http://company.findlaw.com/company_info.html) 23 January 2002. Dorf is Vice Dean and Professor of Law at Columbia University.
- Thomas J. Lepri, "Safeguarding the Enemy Within: The Need for Procedural Protections for U.S. Citizens Detained as Enemy Combatants Under *Ex Parte Quirin*" () (http://ir.lawnet.fordham.edu/cgi/viewcontent.cgi?article=3910&context=flr) PDF (2.12 MB) *Fordham Law Review*, Volume 71, Issue 6 (2003), page 2565
- The Yale Law Journal: A Small Problem of Precedent: per 18 U.S.C. § 4001(a) and the Detention of U.S. Citizen "Enemy Combatants" (http://www.yale.edu/yalelj/112/VladeckWEB.pdf) (PDF)
- AI Index: AMR 51/063/2005: UNITED STATES OF AMERICA Guantanamo and beyond: The continuing pursuit of unchecked executive power (http://web.amnesty.org/library/Index/ENGAMR510632005) document, dated 13 May 2005, by Amnesty International on their web site.
- U.S. DOD: Combatant Status Review Tribunals/Administrative Review Boards (http://www.defenselink.mil/news/Combatant_Tribunals.html)
- Jane Mayer: The Memo - How an internal effort to ban the abuse and torture of detainees was thwarted (http://www.newyorker.com/fact/content/articles/060227fa_fact), The New Yorker article, posted 20 February 2006
- An Open Letter to George Bush (http://www.untitledtheater.com/havel/essays-multimedia/letter-to-bush.html) partly on this issue
- League of Nations Non-intervention Committee ban on "volunteers" 21 February 1937 (http://www.columbia.edu/cu/history/pdf/chr_vol1.pdf)

Khalid_Sheikh_Mohammed

Khalid Sheikh Mohammed
Arabic:

Khalid Sheikh Mohammed Guantanamo Bay detention camp in 2009.

Born	March 1, 1964 or November 14, 1965[1] Balochistan, Pakistan[1] [2] [3] Kuwait City, Kuwait[4] [5] [6]
Arrested	March 1, 2003Rawalpindi, PakistanJoint team of CIA and ISI
Citizenship	Pakistani[1] [2] [3] Kuwaiti[5] [6] Bosnian[7]
Detained at	CIA black sites; Guantanamo
ISN	10024
Charge(s)	Terrorism, conspiracy
Status	Awaiting prosecution

Khalid Sheikh Mohammed (Arabic: خ ا ل د ش ي خ م ح م د; also transliterated as **Khalid Shaikh Mohammed**, and additionally known by at least fifty aliases)[8] [9] is currently in U.S. custody in Guantánamo Bay for acts of terrorism, including mass murder of civilians.

Khalid Sheikh Mohammed was a member of Osama bin Laden's terrorist group al-Qaeda organization leading al-Qaeda's propaganda operations from around 1999 until late 2001. The 9/11 Commission Report states that he was "the principal architect of the 9/11 attacks." He is also alleged to have confessed to a role in many of the most significant terrorist plots over the last twenty years, including the World Trade Center 1993 bombings, the Bojinka plot, an aborted 2002 attack on the U.S. Bank Tower in Los Angeles, the Bali nightclub bombings, the failed bombing of American Airlines Flight 63, the Millennium Plot, and the 2002 abduction and murder of Daniel Pearl in Karachi, Pakistan.

He was captured on March 1, 2003, in hiding in Rawalpindi, Pakistan, by a joint team of CIA the Inter-Services Intelligence agency of Pakistan, and transferred to U.S. custody. In March 2007, he confessed under torture [10] [11] to masterminding the September 11 attacks, the Richard Reid shoe bombing attempt to blow up an airliner over the Atlantic Ocean, the Bali nightclub bombing in Indonesia, the 1993 World Trade Center bombing and various foiled attacks.[12] He was charged in February 2008 with war crimes and murder by a U.S. military commission and faces the death penalty if convicted.

Early life

According to The Biography Channel, U.S. Federal government, Human Rights Watch, and others, Khalid Sheikh Mohammed was born on 14 April 1965 (or 1 March 1964) in Balochistan, Pakistan.[1] [2] [13] [3] [14] However, BBC and others have reported his place of birth as Kuwait City, Kuwait.[4] [5] [6] It is believed that he belongs to the Baloch ethnic group and is fluent in Arabic, English, Urdu, and Balochi.[15]

Khalid Mohammad grew up and spent his formative years in Kuwait, just like his nephew Ramzi Yousef (three years his junior). He joined the Muslim Brotherhood at age sixteen.[16] He returned to Pakistan and later travelled to the United States for further studies.

He completed a degree in mechanical engineering at North Carolina Agricultural and Technical State University in 1986.[17] [18] The following year he went to Afghanistan, where he and his brothers (Zahed, Abed, and Aref) joined the mujahideen to fight the Soviets. According to U.S. federal documents, in 1982 he had heard warlord Abdul Rasul Sayyaf's speech in which a call for jihad against the Soviets was declared.[2]

The 9/11 Commission Report notes that, "By his own account, KSM's animus toward the United States stemmed not from his experiences there as a student, but rather from his violent disagreement with U.S. foreign policy favoring Israel."[19]

However, according to a U.S. intelligence summary reported on August 29, 2009, by *The Washington Post*, his time in the U.S. did lead him to become a terrorist. "KSM's limited and negative experience in the United States — which included a brief jail stay because of unpaid bills — almost certainly helped propel him on his path to becoming a terrorist," according to this intelligence summary. "He stated that his contact with Americans, while minimal, confirmed his view that the United States was a debauched and racist country."[20]

According to the 9/11 Commission, Khalid Sheikh Mohammed after the Afghan jihad went to work for an electronics company, working on communications equipment. In 1988, he helped to head a non-governmental organization paid for by Abdul Rasul Sayyaf, which sponsored and aided Afghan fighters against the Soviets. He continued this work until 1992, when he fought with Muslim fighters in Bosnia and Herzegovina and supported this effort financially. Khalid Sheikh Mohammed moved to Qatar to work in a government office as a project engineer for the Qatari Ministry of Electricity and Water. He stayed at this job until 1996.

Philippines 1994–1995

While he was in the Philippines in late 1994 and early 1995, he said that he was a Saudi or a Qatari plywood exporter and used the aliases Abdul Majid and Salem Ali.[21] [22]

Bosnia, 1995

News agency *Adnkronos* reports Khalid Sheik Mohammed traveled to Bosnia in September 1995, and worked there, under an assumed name, for Egyptian Relief, as a humanitarian aid worker.[23] Quoting Sarajevo paper *Daily Fokus*, they reported local intelligence officials confirmed he obtained Bosnian citizenship in November 1995.[23] Those officials told *Daily Fokus* that Egyptian Relief was a front for the Muslim Brotherhood.

Qatar, avoiding arrest

In early 1996 he fled to Pakistan to avoid capture by U.S. authorities.[24] In his flight from Qatar he was sheltered by Sheikh Abdullah Al Thani, who was the Qatari Minister of Religious Affairs in 1996.[25] [26] [27] [28] [29]

Alleged terrorist activities

Operation Bojinka

After seeing the respect that his nephew Ramzi Yousef had gained from the World Trade Center 1993 bombings, Mohammed decided to engage more directly in anti-U.S. activities as well. He traveled to the Philippines in 1994 to work with Yousef on the Bojinka plot, a Manila-based plot to destroy twelve commercial airliners flying routes between the United States, East Asia, and Southeast Asia. The 9/11 Commission Report says that "this marked the first time KSM took part in the actual planning of a terrorist operation."[30]

> Using airline timetables, Khalid Sheikh Mohammed and Ramzi Yousef devised a scheme whereby five men could, in a single day, board 12 flights — two each for three of the men, three each for the other two — assemble and deposit their bombs and exit the planes, leaving timers to ignite the bombs up to several days afterward. By the time the bombs exploded, the men would be far away and far from reasonable suspicion. The math was simple: 12 flights with at least 400 people per flight. Somewhere in the neighborhood of 5,000 deaths. It would be a day of glory for them, calamity for the Americans they supposed would fill the aircraft."[31]

Bojinka plans also included renting or buying a Cessna, packing it with explosives and crash landing it into CIA headquarters, with a backup plan to hijack the twelfth airliner in the air and use that instead. This information was reported in detail to the U.S. at the time. This point was not mentioned in KSM's confession to involvement in thirty-one terrorist plots, including 9/11.

In December 1994, Yousef had engaged in a test of a bomb on Philippine Airlines Flight 434 using only about ten percent of the explosives that were to be used in each of the bombs to be planted on United States airliners. The test resulted in the death of a Japanese national on board a flight from the Philippines to Japan. Mohammed conspired with Yousef on the plot until it was uncovered on January 6, 1995. Yousef was captured February 7 of that same year.

Khalid Sheikh Mohammed was indicted on terrorism charges in the United States District Court for the Southern District of New York in January 1996 for his alleged involvement in Operation Bojinka,[32] and was subsequently placed on the October 10, 2001, initial list of the FBI's twenty-two Most Wanted Terrorists.[33]

Redevelopment of the relationship with Osama bin Laden

> “If now we were living in the Revolutionary War and George Washington he being arrested through Britain. For sure he, they would consider him enemy combatant. But American they consider him as hero.”

—Khalid Sheikh Mohammed, speaking in 2007[34]

By the time the Bojinka plot was discovered, Mohammed was already safely in Qatar, back at his job as a project engineer at the country's Ministry of Electricity and Water. He traveled in 1995 to Sudan, Yemen, Malaysia, and Brazil to visit elements of the worldwide jihadist community, although no evidence connects him to specific terrorist actions in any of those locations. On his trip to Sudan he attempted to meet with Osama bin Laden, who was at the time living there with the aid of Sudanese political leader Hassan al-Turabi. After a request to arrest Mohammed came to the Qatari government from the United States in January 1996, Mohammed fled to Afghanistan, where he renewed his relationship with Abdul Rasul Sayyaf and formed a working relationship with the newly migrated bin

Laden later that year. "According to KSM, this was the first time he had seen bin Laden since 1989. Although they had fought together [in Afghanistan] in 1987, bin Laden and KSM did not yet enjoy an especially close working relationship."

Just as Mohammed was re-establishing himself in Afghanistan, bin Laden and his colleagues were also transplanting their operations to the same country. Abu Hafs al-Masri/Mohammed Atef, bin Laden's chief of operations, arranged a meeting between bin Laden and Mohammed in Tora Bora sometime in mid-1996, in which Mohammed outlined a plan that would eventually become the quadruple hijackings of 2001.[35] Bin Laden urged Mohammed to become a full-fledged member of Al Qaeda, but he continued to refuse such a commitment until around early 1999, after the 1998 U.S. embassy bombings in Nairobi and Dar es Salaam convinced him that bin Laden was truly committed to attacking the United States.[36]

The 9/11 Commission Report notes on page 149 that Mohammed moved his family from Iran to Karachi, Pakistan, in 1997. That same year, he attempted without success to join mujahideen leader Ibn al Khattab in Chechnya, another area of special interest to Mohammed. He was apparently unable to travel to Chechnya, and so he instead returned to Afghanistan, where he gradually gained stature in Al Qaeda and ultimately accepted bin Laden's invitation to move to Kandahar and join the organization as a full-fledged member (although he claims that he still refused to swear a formal oath of loyalty to bin Laden). Eventually, he became leader of Al Qaeda's media committee.

September 11, 2001 attacks

The first hijack plan that Mohammed presented to the leadership of al-Qaeda called for several airplanes on both east and west coasts to be hijacked and flown into targets. His plan evolved from an earlier foiled plot known as the Bojinka plot, which called for 10 or more airliners to be bombed in mid-air or hijacked for use as missiles. Bin Laden rejected some potential targets suggested by Mohammed, such as the U.S. Bank Tower in Los Angeles.[37]

In late 1998 or early 1999, bin Laden gave approval for Mohammed to go forward with organizing the plot.[36] A series of meetings occurred in spring of 1999, involving Khalid Sheikh Mohammed, Osama bin Laden, and his military chief Mohammed Atef.[36] Bin Laden provided leadership for the plot, along with financial support.[36] Bin Laden was also involved in selecting people to participate in the plot, including choosing Mohamed Atta as the lead hijacker.[38] Mohammed provided operational support, such as selecting targets and helping arrange travel for the hijackers.[36] It was also accepted that Atef was the one who had organized the hijackers.[39]

After Atta was chosen as the leader of the mission, "he met with Bin Laden to discuss the targets: the World Trade Center, which represented the U.S. economy; the Pentagon, a symbol of the U.S. military; and the U.S. Capitol, the perceived source of U.S. policy in support of Israel. The White House was also on the list, as Bin Laden considered it a political symbol and wanted to attack it as well."[40]

"Bin Laden had been pressuring KSM (Khalid Sheikh Mohammed) for months to advance the attack date. According to KSM, bin Laden had even asked that the attacks occur as early as mid-2000, after Israeli opposition party leader Ariel Sharon caused an outcry in the Middle East by visiting a sensitive and contested holy site in Jerusalem that is sacred to both Muslims and Jews. Although bin Laden recognized that Atta and the other pilots had only just arrived in the United States to begin their flight training, the al-Qaeda leader wanted to punish the United States for supporting Israel. He allegedly told KSM it would be sufficient simply to down the planes and not hit specific targets. KSM stood his ground, arguing that the operation would not be successful unless the pilots were fully trained and the hijacking teams were larger."[41]

In a 2002 interview with Al Jazeera journalist Yosri Fouda, Mohammed admitted his involvement, along with Ramzi bin al-Shibh, in the "Holy Tuesday operation".[42] KSM, however, disputes this claim via his Personal Representative: "I never stated to the Al Jazeera reporter that I was the head of the al Qaida military committee."[43]

Daniel Pearl murder

According to a CNN interview with intelligence expert Rohan Gunaratna, "Daniel Pearl was going in search of the al Qaeda network that was operational in Karachi, and it was at the instruction of Khalid Sheikh Mohammed that Daniel Pearl was killed."[44] On October 12, 2006, *Time* magazine reported that "KSM confessed under CIA interrogation that he personally committed the murder."[45] On March 15, 2007, the Pentagon released a statement that Mohammed had confessed to the murder.[46] The statement quoted Mohammed as saying, "I decapitated with my blessed right hand the head of the American Jew, Daniel Pearl, in the city of Karachi, Pakistan. For those who would like to confirm, there are pictures of me on the Internet holding his head."[47]

According to an investigative report published in January 2011 by Georgetown University, the Federal Bureau of Investigation used vein matching to determine that the perpetrator in the video of the killing of Pearl was most likely Mohammed, notably through a "bulging vein" running across his hand.[48] Federal officials were concerned that the confession obtained through waterboarding would not hold up in court and used this forensic technique to bolster their case.[49]

Interview to al-Jazeera

In April 2002 Khalid Sheikh Mohammed, together with Ramzi bin al-Shibh, gave an interview to al-Jazeera correspondent Yosri Fouda. They described the preparations for 9/11 attacks and said that they first thought of "striking at a couple of nuclear facilities" in the USA but then "it was eventually decided to leave out nuclear targets for now."[50]

Capture and interrogation

On September 11, 2002, members of Pakistani Inter-Services Intelligence (ISI) claimed to have killed or captured Khalid Sheikh Mohammed during a raid in Karachi that resulted in bin al-Shibh's capture. Some people have reported that Mohammed escaped, but that his family was captured.[51]

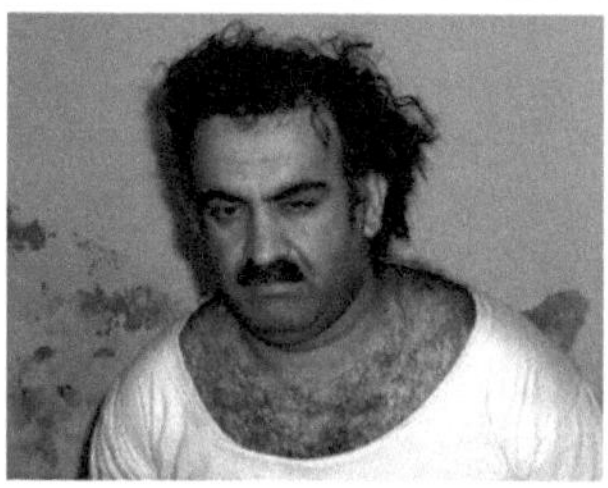

Khalid Sheikh Mohammed after his capture in Pakistan on March 1, 2003

Khalid Sheikh Mohammed was captured in Rawalpindi, Pakistan, (about 20 km southwest of Islamabad), on March 1, 2003, by the Pakistani ISI, possibly in a joint action with the CIA's Special Activities Division paramilitary operatives[52] and officers of the American Diplomatic Security Service, and has been in U.S. custody since that time. In September 2006, the U.S. government announced it had moved Mohammed from a secret prison to the facility at the Guantanamo Bay detention camp.[53] The Red Cross, Human Rights Watch and Mohammed have claimed that the harsh treatment and waterboarding he received from U.S. authorities amounts to torture.[54] [55]

Following the report of the capture, some Pakistani officials say he was immediately transferred to U.S. custody without extradition proceedings, while others said he remained in Pakistani custody. The raid took place at the home of Ahmed Abdul Qudoos, who was also reportedly arrested as an al-Qaeda agent. Qudoos' family told media that Khalid Sheikh Mohammed was not in the house, that Qudoos was disabled and had never been associated with al-Qaeda, and that the police conducting the raids did not ask for Mohammed. Other newspaper accounts said that former Taliban officials in Pakistan said that Mohammed was not captured and was still at large.

He told American interrogators he would not answer any questions until he was provided with a lawyer, which was refused to him. He claims to have been kept naked for more than a month during his isolation and interrogations, during which he was "questioned by an unusual number of female handlers".[56]

Under questioning, Khalid Sheikh Mohammed named neuroscientist Aafia Siddiqui as an al-Qaeda operative.[57] [58] He later stated that to "please his captors" he gave names of innocent people and Siddiqui's lawyers believe her name was given under torture but still the US are unwilling to accept its mistake and release her. Four British Parliamentarians called the trial a grave miscarriage of justice that violated the Sixth Amendment to the United States Constitution as well as the United States' obligations as a member of the United Nations, and demanded Siddiqui's release. In a letter to Barack Obama, they stated that there was a lack of scientific and forensic evidence tying Siddiqui to the weapon she allegedly fired.[59]

A CIA document reveals that Jane Harman (D-CA) and Porter Goss (R-FL) of the House Intelligence Committee were briefed on July 13, 2004, by CIA Deputy Director for Operations James Pavitt, General Council Scott Muller, and CIA Inspector General John L. Helgerson on the status of the interrogation process. By this date, Khalid Sheikh Mohammed had been subjected to 183 applications of waterboarding. The document states

> ...the CIA was seeking renewed policy approval from the NSC Principals to continue using the enhanced interrogation techniques.[60]

On October 12, 2004, Human Rights Watch reported that 11 suspects, including Khalid Sheikh Mohammed, had "disappeared" to a semi-secret prison in Jordan, and might have been tortured there under the direction of the CIA.[61] [62] Jordanian and American officials denied those allegations.[63] [64] [65]

CIA Director Michael Hayden told a Senate committee on February 5, 2008, that the agency had used waterboarding on Khalid Sheikh Mohammed.[66] A 2005 U.S. Justice Department memo released in April 2009 stated that Mohammed had undergone waterboarding 183 times in March 2003.[67]

In October 2006 Mohammed described his mistreatment and torture in detention, including the waterboarding, to a representative of International Committee of the Red Cross. Mohammed said that he had provided a lot of false information that he had supposed the interrogators wanted to hear in order to stop the mistreatment.[68]

In the 2006 interview with the Red Cross, Mohammed claimed to have been waterboarded in five different sessions during the first month of interrogation in his third place of detention.[68] [69] While the Justice Department memos were confusing in that they did not explain exactly what the numbers represented, a U.S. official with knowledge of the interrogation programs explained the 183 figure represented the number of times water was applied to the detainee's face during the waterboarding sessions.[70]

In March 2007, after four years in captivity, including six months of detention and alleged torture at Guantanamo Bay, Khalid Sheikh Mohammed — as it was claimed by a Combatant Status Review Tribunal Hearing[71] in Guantanamo Bay — confessed to masterminding the September 11 attacks, the Richard Reid shoe bombing attempt to blow up an airliner over the Atlantic Ocean, the Bali nightclub bombing in Indonesia, the 1993 World Trade Center bombing and various foiled attacks.[72] "I was responsible for the 9/11 operation from A to Z," Khalid Sheikh Mohammed said in a statement read Saturday during a Combatant Status Review Tribunal at the Guantanamo Bay detention camp.[73] His confession was read by a member of the U.S. military who is serving as his personal representative.[74] He further described his actions and motivations in a composition publicly released in 2009 known as *The Islamic Response to the Government's Nine Accusations.*[75]

According to the "unclassified summary of evidence" presented during the hearing, a computer hard drive seized during the capture of Khalid Sheikh Mohammed contained the following:

- information about the four airplanes hijacked on 11 September 2001 including code names, airline company, flight number, target, pilot name and background information, and names of the hijackers
- photographs of 19 individuals identified as the 11 September 2001 hijackers
- a document that listed the pilot license fees for Mohammad Atta and biographies for some of the 11 September 2001 hijackers.
- images of passports and an image of Mohammad Atta.
- transcripts of chat sessions belonging to at least one of the 11 September 2001 hijackers.

- three letters from Osama bin Laden
- spreadsheets that describe financial assistance to families of known al Qaeda members
- a letter to the United Arab Emirates threatening attack if their government continued to help the United States
- a document that summarized operational procedures and training requirements of an al Qaeda cell
- a list of killed and wounded al Qaeda militants.

At the hearing, Khalid Sheikh Mohammed said the computer belonged not to him, but to Mustafa Ahmad al-Hawsawi, arrested together with him.[76]

In June 2008, a *New York Times* article citing unnamed CIA officers claimed that Mohammed was held in a secret facility in Poland near Szymany Airport, about 100 miles north of Warsaw, where he was interrogated under waterboarding before he began to cooperate.[77]

In April 2011, the British newspaper, *The Telegraph* said it received leaked documents regarding the Guantanamo Bay interrogations of Khalid Sheikh Mohammed. The documents cited Khalid saying that, if Osama Bin Laden is captured or killed by the Coalition of the Willing, an Al-Qaeda sleeper cell will detonate a "weapon of mass destruction" in a "secret location" in Europe, and promised it would be "a nuclear hellstorm".[78] [79] [80] [81] [82] [83] [84]

Report that interrogators abused his children

Ali Khan, the father of Majid Khan, another one of the 14 "high-value detainees," released an affidavit on Monday, April 16, 2006, that reported that interrogators subjected Khalid Sheikh Mohammed's children, aged six and eight years old, to abusive interrogation.[85] [86] [87]

Khan's affidavit quoted another of his sons, Mohammed Khan:

> The Pakistani guards told my son that the boys were kept in a separate area upstairs, and were denied food and water by other guards. They were also mentally tortured by having ants or other creatures put on their legs to scare them and get them to say where their father was hiding."

Transfer to Guantánamo and hearing before his Combatant Status Review Tribunal

On September 6, 2006, then-American President George W. Bush confirmed, for the first time, that the CIA had held "high-value detainees" for interrogation in secret prisons around the world.[88] He also announced that fourteen senior captives, including Khalid Sheikh Mohammed, were being transferred from CIA custody, to military custody, at Guantanamo Bay detention camp and that these fourteen captives could now expect to face charges before Guantanamo military commissions.

In a September 29, 2006, speech, Bush stated:

> Once captured, Abu Zubaydah, Ramzi bin al Shibh, and Khalid Sheikh Mohammed were taken into custody of the Central Intelligence Agency. The questioning of these and other suspected terrorists provided information that helped us protect the American people. They helped us break up a cell of Southeast Asian terrorist operatives that had been groomed for attacks inside the United States. They helped us disrupt an al Qaeda operation to develop anthrax for terrorist attacks. They helped us stop a planned strike on a U.S. Marine camp in Djibouti, and to prevent a planned attack on the U.S. Consulate in Karachi, and to foil a plot to hijack passenger planes and to fly them into Heathrow Airport and London's Canary Wharf.[89]

In March 2007, Mohammed testified before a closed-door hearing in Guantánamo Bay. According to transcripts of the hearing released by the Pentagon, he said, "I was responsible for the 9/11 operation, from A to Z." The transcripts also show him confessing to:

- Organizing the 1993 World Trade Center bombing,
- The Bali nightclub bombings,
- Richard Reid's attempted shoe bombing,

- Planning the attacks on Heathrow Airport and Big Ben clock tower in London,
- Daniel Pearl's murder in 2002,
- Planned assassination attempts on Pope John Paul II, Pervez Musharraf and Bill Clinton.[90]

> Because war, for sure, there will be victims. When I said I'm not happy that three thousand been killed in America. I feel sorry even... Killing is prohibited in all what you call the People of the Book, Jews, Judaism, Christianity, and Islam. You know the Ten Commandments very well. The Ten Commandments are shared between all of us. We all are serving one God.
>
> —Khalid Sheikh Mohammed, before his tribunal[34]

On March 15, 2007, BBC News reported that "Transcripts of his testimony were translated from Arabic and edited by the U.S. Department of Defense to remove sensitive intelligence material before release. It appeared, from a judge's question, that Khalid Sheikh Mohammed had made allegations of torture in US custody." In the Defense Department transcript, Mohammed said his statement was not made under duress but Mohammed and human rights advocates have alleged that he was tortured. CIA officials have previously told ABC News that "Mohammed lasted the longest under waterboarding, two and a half minutes, before beginning to talk."[91] Legal experts say this could taint all his statements. Forensic psychiatrist Michael Welner, M.D., an expert in false confessions, observed from the testimony transcript that his concerns about his family may have been far more influential in soliciting Mohammed's cooperation than any earlier reported mistreatment.[92]

One CIA official cautioned that "many of Mohammed's claims during interrogation were 'white noise' designed to send the U.S. on wild goose chases or to get him through the day's interrogation session." For example, according to Mike Rogers, a former FBI agent and the top Republican on the terrorism panel of the House Intelligence Committee, he admitted responsibility for the Bali nightclub bombing, but his involvement "could have been as small as arranging a safe house for travel. It could have been arranging finance." Mohammed also made the admission that he was "responsible for the 1993 World Trade Center Operation," which killed six and injured more than 1,000 when a bomb was detonated in an underground garage, Mohammed did not plan the attack, but he may have supported it. Michael Welner noted that by offering legitimate information to interrogators, Mohammed had secured the leverage to provide disinformation as well.[93]

In an article discussing the reliability of Khalid's confession and the motive for giving misinformation under torture, Ali Soufan, a former FBI special agent with considerable experience interrogating al-Qaeda operatives, pointed out that:

> When they are in pain, people will say anything to get the pain to stop. Most of the time, they will lie, make up anything to make you stop hurting them. That means the information you're getting is useless.

His words are echoed by the US Army Training Manual's section on interrogation, which suggests that:

> the use of force is a poor technique, as it ...can induce the source to say whatever he thinks the interrogator wants to hear.

As an example of this the article discloses that although the George W. Bush administration made claims that the water-boarding (simulated drowning) of Khalid Shaikh Mohammed produced vital information that allowed them to break up a plot to attack the U.S. Bank Tower (formerly Library Tower and First Interstate Bank World Center) in Los Angeles in 2002, this has been proven to be untrue. In 2002 Shaikh Mohammed was busy evading capture in Pakistan.[94] Likewise the claim by the Obama administration that torture of Kahlid Mohammed led to the lead in finding Osama Bin Laden has also been shown to be false. "The trail to bin Laden did not begin with a disclosure from Khalid Sheikh Mohammed, who was waterboarded 183 times. ...not only did the use of 'enhanced interrogation techniques' on Khalid Sheikh Mohammed not provide us with key leads on bin Laden's courier, Abu Ahmed; it actually produced false and misleading information."[95] [96] [97]

List of confessions

- The February 1993 bombing of the World Trade Center in New York City
- The September 11 attacks on the World Trade Center, the Pentagon, and the United States Capitol (both in Washington, D.C.) using four hijacked commercial airliners.
- A failed "shoe bomber" operation
- The October 2002 attack in Kuwait
- The Paddy's Pub nightclub bombing in Bali, Indonesia
- A plan for a "second wave" of attacks on major U.S. landmarks after the 9/11 attacks, including the Library Tower in Los Angeles, the Willis Tower (formerly Sears Tower) in Chicago, the Plaza Bank Building in Seattle and the Empire State Building in New York City
- Plots to attack oil tankers and U.S. naval ships in the Straits of Hormuz, the Straits of Gibraltar and in Singapore
- A plan to blow up the Panama Canal
- Plans to assassinate Jimmy Carter
- A plot to blow up suspension bridges in New York City
- A plan to destroy the Sears Tower in Chicago with burning fuel trucks
- Plans to "destroy" London Heathrow Airport, Canary Wharf and Big Ben in London
- A planned attack on "many" nightclubs in Thailand
- A plot targeting the New York Stock Exchange and other U.S. financial targets
- A plan to destroy buildings in Eilat, Israel
- Plans to destroy U.S. embassies in Indonesia, Australia and Japan in 2002.
- Plots to destroy Israeli embassies in India, Azerbaijan, the Philippines and Australia
- Surveying and financing an attack on an Israeli El-Al flight from Bangkok
- Sending several "mujahideen" into Israel to survey "strategic targets" with the intention of attacking them
- The November 2002 suicide bombing of a hotel in Mombasa, Kenya
- The failed attempt to shoot down an Israeli passenger jet leaving Mombasa airport in Kenya
- Plans to attack U.S. targets in South Korea
- Providing financial support for a plan to attack U.S., British and Jewish targets in Turkey
- Surveillance of U.S. nuclear power plants in order to attack them
- A plot to attack NATO's headquarters in Europe
- Planning and surveillance in a 1995 plan (the "Bojinka plot") to bomb 12 American passenger jets
- The planned assassination attempt against then-U.S. President Bill Clinton during a mid-1990s trip to the Philippines
- "Shared responsibility" for a plot to kill Pope John Paul II
- Plans to assassinate Pakistani President Pervez Musharraf
- An attempt to attack a U.S. oil company in Sumatra, Indonesia, "owned by the Jewish former [U.S.] Secretary of State Henry Kissinger"
- The beheading of *Wall Street Journal* reporter Daniel Pearl

Source: BBC[98]

After Mohammed arrived at Guantánamo, a team of FBI and military interrogators tried to elicit from him the same confessions that the CIA had obtained about the 9/11 plot, but by using only legal means of interrogation. By 2008, the Bush Administration believed that this so-called Clean Team had compiled sufficient evidence to charge Mohammed and the others with capital murder.[99]

The Department of Defense announced on August 9, 2007 that all fourteen of the "high-value detainees" who had been transferred to Guantanamo from the CIA's black sites, had been officially classified as "enemy combatants".[100] Although judges Peter Brownback and Keith J. Allred had ruled two months earlier that only "*illegal* enemy combatants" could face military commissions, the Department of Defense waived the qualifier and said that all fourteen men could now face charges before Guantanamo military commissions.[101] [102]

Confession used in Sheikh Omar's defense

On March 19, 2007, Ahmed Omar Saeed Sheikh's lawyers cited Mohammed's confession in defense of their client.[103] [104]

Ahmed Omar Saeed Sheikh, also known as Sheikh Omar, was sentenced to death in a Pakistani court for the murder of Daniel Pearl. Omar's lawyers recently announced that they planned to use Mohammed's confession in an appeal. They had always acknowledged that Omar played a role in Pearl's murder, but argue that Mohammed was the actual murderer.

Prosecution in France

In France in 2009 it was decided to try Khalid Sheikh Mohammed in absentia on terrorism charges with respect to the Ghriba synagogue bombing on the Tunisian island of Djerba in 2002 which killed 14 German tourists, five Tunisians and two French nationals, together with captured German national Christian Ganczarski and Tunisian Walid Nawar.[105] French judges chose to separate Khalid Sheikh Mohammed's case from that of Ganczarski and Nawar and try him at a later date.[106]

Trial for 9/11

On February 11, 2008, the United States Department of Defense charged Mohammed as well as Ramzi bin al-Shibh, Mustafa Ahmad al-Hawsawi, Ali Abd al-Aziz Ali and Walid Bin Attash for the September 11, 2001 attacks under the military commission system, as established under the Military Commissions Act of 2006. They have reportedly been charged with the murder of almost 3000 people, terrorism and providing material support for terrorism and plane hijacking; as well as attacking civilian objects, intentionally causing serious bodily injury and destruction of property in violation of the law of war. The charges against them list 169 overt acts allegedly committed by the defendants in furtherance of the September 11 events."[107]

The charges include 2,973 individual counts of murder—one for each person killed in the 9/11 attacks.[108]

The U.S. government is seeking the death penalty, which would require the unanimous agreement of the commission judges.[107]

Human rights groups, including Amnesty International, Human Rights Watch and the Center for Constitutional Rights, and U.S. military defense lawyers have criticised the military commissions for lacking necessary rights for a fair trial. Critics generally argue for a trial either in a federal district court as a common criminal suspect, or by court-martial as a prisoner under the Geneva Conventions which prohibit civilian trials for prisoners of war.[109] Mohammed could face the death penalty under any of these systems.

The Pentagon insisted that Mohammed and the other defendant would receive a fair trial, with rights "virtually identical" to U.S. military service personnel. However, there are some differences between U.S. courts-martial and military commissions.

The U.S. Department of Defense has built a $12 million "Expeditionary Legal Complex" in Guantánamo with a snoop-proof courtroom capable of trying six alleged co-conspirators before one judge and jury. Media and other observers are sequestered in a soundproofed room behind thick glass, at the rear. The judge at the front and a court security officer have mute buttons to silence the feed to the observers' booth—if they suspect someone in court could spill classified information.[110]

The trial, presided over by judge Ralph Kohlmann, began on June 5, 2008, with the arraignment. About thirty-five journalists watched on closed-circuit TV in a press room inside a converted hangar, while two dozen others watched through a window from a room adjacent to the courtroom.

Mohammed insisted he would not be represented by any attorneys. The other detainees quickly followed suit and said they too wanted to represent themselves. One of the civilian attorneys Mohammed spurned, David Nevin, later told the Associated Press that he would attempt to meet with Mohammed to "hear him out and see if we can give him information that is helpful."[111] [112]

Mohammed was careful not to interrupt Kohlmann. He lost his composure only after the Marine colonel ordered several defense attorneys to keep quiet "It's an inquisition. It's not a trial," Mohammed said in broken English, his voice rising. "After torturing they transfer us to inquisition-land in Guantanamo."

He explained he believes only in religious Sharia law and railed against U.S. President George W. Bush for waging a "crusade war." When the judge warned Mohammed that he faces execution if convicted of organizing the attacks on America, Mohammed said he welcomes the death penalty. "Yes, this is what I wish, to be a martyr for a long time," Mohammed declared. "I will, God willing, have this, by you."[112]

A sound feed to journalists from the courtroom was turned off twice.

The sound was also turned off when another defendant discussed early days of his imprisonment. Judge Ralph Kohlmann said that in both cases sound was turned off because classified information was discussed.[112]

On September 23, 2008, in the voir dire process, Mohammed questioned the judge on his potential bias at trial. "Glaring and poking an occasional finger in the air," Mohammed told Kohlmann, "The government considers all of us fanatical extremists," and asked, "How can you, as an officer of the U.S. Marine Corps, stand over me in judgment?" Insisting that he was attempting to work out if Kohlmann was a religious extremist, he continued: "[President] Bush said this is a crusader war and Osama bin Laden said this is a holy war against the crusades. If you were part of Jerry Falwell or Pat Robertson's group, then you would not be impartial."

For his part, Kohlmann attempted to maintain his dignity, explaining that he was currently unaffiliated with a church "because I've moved so often." He added that he had previously worshipped at "various Lutheran churches and Episcopal churches."

Mohammed then proceeded to ask Kohlmann about his views on torture. As part of the background materials supplied to him–or made available to the civilian lawyers who are voluntarily assisting him in his defense–he referred to an ethics seminar that Kohlmann had conducted at his daughter's high school in 2005, in which the students had been asked to consider their responses to a "Ticking Time Bomb" scenario. Based on a fictional proposition that a bomb is about to go off, and an unwilling captive knows its location but is unwilling to disclose the information, the scenario is widely used by proponents of "enhanced interrogation techniques" to justify the use of torture.

Kohlmann explained that he encouraged the debate as part of "a complex question that might be dealt with differently if someone were specifically trying to save the nation or just looking at it from an ethical sense or just looking at it from a legal sense," and dismissed a combative question from Mohammed–"It seems that you are supportive of the use of torture for national security?"–by stating, "I have no idea where that would come from."[113]

On October 12, 2008, Kohlmann ruled that Khalid Sheikh Mohammed, and his four co-charged, should be provided with laptops, so they can work on their defenses.[114]

On December 8, 2008, Mohammed and four co-defendants sent a note to the military judge expressing their desire to confess and plead guilty.[115]

In November 2009, according to an Administration official, Mohammed was being transferred from Guantanamo Bay to New York to face a Federal Trial.[116] Four other detainees will be facing trial in front of civilian federal court, as well.

On January 22, 2010, the Pentagon officially dropped military charges against Mohammed and the four other alleged conspirators, clearing the path for likely transfer from Guantanamo to the United States to face charges in a civilian federal court.[117]

It was disclosed on August 22, 2012, that the trial of Khalid Sheikh Mohammed and his four co-conspirators will not be televised and may not begin for another four years.[118]

Kohlmann unexpectedly replaced

Kohlmann was scheduled to retire in 2009. In November 2008, he was unexpectedly replaced by Stephen Henley.

Possible guilty plea

On December 8, 2008, Khalid Sheikh Mohammed and his four co-defendants told the judge that they wished to confess and plead guilty to all charges. The plea will be delayed until mental competency hearings for Mustafa Ahmad al-Hawsawi and Ramzi bin al-Shibh can be held; Mohammed said, "We want everyone to plead together."[115] [119] Spencer Ackerman, writing in the *Washington Independent*, reported that Presiding Officer Stephen Henley had to consider whether he was authorized to accept guilty pleas.

Transfer of the case to a civilian court

On November 13, 2009, U.S. Attorney General Eric Holder announced that Khalid Sheikh Mohammed, Ramzi Bin al-Shibh, Walid bin Attash, Ali Abdul Aziz Ali and Mustafa Ahmed al-Hawsawi will all be transferred to the U.S. District Court for the Southern District of New York for trial. He also expressed confidence that an impartial jury would be found "to ensure a fair trial in New York."[120]

On January 21, 2010, all charges were withdrawn in the military commissions against the five suspects in the Sept. 11, 2001 terror attacks being held at Guantanamo Bay. The charges were dropped "without prejudice"—a procedural move that allows federal officials to transfer the men to trial in a civilian court and also leaves the door open, if necessary, to bring charges again in military commissions.[121]

In February 2010 *Fox News* reported that the legal counsel of Khalid Sheikh Mohammed, and the legal counsel of several other captives, was halted without warning.[121] The attorneys had made the trip to Guantanamo in the usual manner—a trip that requires advising authorities of the purpose of their trip. However, upon their arrival in Guantanamo, they were informed they were no longer allowed to see their clients. They were told that letters to their clients, telling them that they had travelled to Cuba, to see them, could not be delivered, as they were no longer authorized to write to their clients. Camp authorities told them that since the charges against their clients had been dropped, while the Department of Justice figured out where to charge them, they no longer needed legal counsel. Camp authorities told them that, henceforward, all access to the captives had to be approved by Jeh Johnson, the Department of Defense's General Counsel. Fox reported that during earlier periods when the charges had been dropped the captives had still been allowed to see their attorneys. Fox made an unsubstantiated claim that questions they asked camp authorities led to the captives' access to their attorneys being restored.

On February 1, 2010, White House spokesman Robert Gibbs told CNN that "Khalid Sheikh Mohammed is going to meet justice and he's going to meet his maker. He will be brought to justice and he's likely to be executed for the heinous crimes he committed".[122] The White House spokesperson's statement has been criticized as violating the principle of the presumption of innocence and has been characterized as egregious by an attorney of Guantanamo Bay detainees.[122]

Transfer of the case back to a military commission

On January 7, 2011, U.S. President Barack Obama signed National Defense Authorization Act which explicitly prohibits the use of US Defense Department funds to transfer detainees from Guantanamo Bay to the United States or other countries. It also bars Pentagon funds from being used to build facilities in the United States to house detainees, as the president originally suggested. The move essentially barred the administration from trying detainees in civilian courts. The president objected to the provision in the bill before signing it, calling it "a dangerous and unprecedented challenge to critical executive branch authority" but also said his team would work with the US

Congress to "seek repeal of these restrictions."[123]

On April 4, 2011, Attorney General Eric Holder announced that Khalid Sheikh Mohammed and four other 9/11 terror suspects will face a military trial at the Guantanamo Bay detention facility. In announcing his decision, Holder was highly critical of Congress for imposing restrictions on the Justice Department's ability to bring the men to New York for civilian trials. "After thoroughly studying the case, it became clear to me that the best venue for prosecution was in federal court. I stand by that decision today," Holder said. "As the president has said, those unwise and unwarranted restrictions (imposed by Congress) undermine our counterterrorism efforts and could harm our national security. Decisions about who, where and how to prosecute have always been—and must remain—the responsibility of the executive branch." Holder insisted, "We were prepared to bring a powerful case against Khalid Sheikh Mohammed and his four co-conspirators—one of the most well-researched and documented cases I have ever seen in my decades of experience as a prosecutor." He added, "Had this case proceeded in Manhattan or in an alternative venue in the United States, as I seriously explored in the past year, I am confident that our justice system would have performed with the same distinction that has been its hallmark for over 200 years." Holder had promised to seek the death penalty for each of the five men and on 4 April he warned that it is an "open question" if such a penalty can be imposed by a military commission if the defendants plead guilty.[124]

On May 5, 2012, the trial started. A small number of 9/11 victims' relatives were attending the hearing. Khalid Sheikh Mohammed refused to answer the judge's questions. All the defendants refused to wear the earphones that provide translation into Arabic. Then an Arabic translator present in court ensured that the accused could follow proceedings.[125]

Release of new images

On September 9, 2009, images of Khalid Sheikh Mohammed and Ammar al Baluchi were widely republished.[126][127][128][129][130][131][132] Camp authorities have strict controls over the capture and release of images of the Guantanamo captives. Journalists and VIPs visiting Guantanamo are not allowed to take any pictures that show the captives' faces. "High value" captives, like Khalid Sheikh Mohammed, are only seen by journalists when they are in the court room, where cameras are not allowed. However, on September 9, 2009, independent counter-terrorism researchers found new images of Khalid Sheikh Mohammed and his nephew Ammar al Baluchi on "jihadist websites". According to Carol Rosenberg, writing in *The Miami Herald*: "The pictures were taken in July, said International Committee of the Red Cross spokesman Bernard Barrett, under an agreement with prison camp staff that lets Red Cross delegates photograph detainees and send photos to family members."

See also

- Shaker Aamer

References

[1] "Khalid Sheikh Mohammed Biography - Facts, Birthday, Life Story" (http://www.biography.com/articles/Khalid-Sheikh-Mohammed-241188). The Biography Channel. . Retrieved 2012-02-14.

[2] Margot Williams (2008-11-03). "Guantanamo Docket: Khalid Shaikh Mohammed" (http://projects.nytimes.com/guantanamo/detainees/10024-khalid-shaikh-mohammed). The New York Times. . "Detaini is a Pakistani who was raised in Kuwait (KU) during his formative years."

[3] "Khalid Sheikh Mohammed" (http://www.hrw.org/news/2012/10/26/khalid-sheikh-mohammed). Human Rights Watch. .

[4] "Profile: Al-Qaeda 'kingpin'" (http://news.bbc.co.uk/2/hi/south_asia/2811855.stm). BBC. 16 March 2007. . Retrieved 2007-08-28.

[5] "Khalid Shaikh Mohammed" (http://www.nndb.com/people/429/000027348/). *NNDB*. . Retrieved 23 February 2012.

[6] "Khalid Shaikh Mohammed" (http://www.globalsecurity.org/security/profiles/khalid_shaikh_mohammed.htm). *GlobalSecurity.org*. . Retrieved 23 February 2012.

[7] "Bosnia: Senior Al-Qaeda figure granted citizenship, says report - Adnkronos Security" (http://www.adnkronos.com/AKI/English/Security/?id=3.0.2928330465). Adnkronos.com. 2003-04-07. . Retrieved 2012-07-17.

[8] "U.S. v. Khalid Shaikh Mohammed military tribunal charges" (http://news.findlaw.com/nytimes/docs/terrorism/usksmetal20808chrgs.pdf) (PDF). FindLaw. . Retrieved 2008-07-15.

[9] "Detainee Biographies" (http://www.webcitation.org/query?url=http://www.odni.gov/announcements/content/DetaineeBiographies.pdf+&date=2009-08-31) (PDF). Office of the Director of National Intelligence. Archived from the original (http://www.odni.gov/announcements/content/DetaineeBiographies.pdf) on date=2009-08-31. .

[10] The Guardian. 'Truth and torture' by Andrew Brown. Monday, March 19, 2007 (http://www.guardian.co.uk/commentisfree/2007/mar/19/truthandtorture)

[11] The Guardian. 'Khalid Sheikh Mohammed: former military prosecutor denounces trial' by Chris McGreal. Friday, 4 May, 2012 (http://www.guardian.co.uk/world/2012/may/04/khalid-sheikh-mohammed-military-prosecutor)

[12] Soufan, Ali H., and Daniel Freedman; The Black Banners: The Inside Story of 9/11 and the War Against al-Qaeda, Sep 12, 2011 (http://www.amazon.com/Black-Banners-Inside-Against-al-Qaeda/dp/0393079422)

[13] http://www.documentcloud.org/documents/82528-isn-10024-khalid-shaikh-mohammed-jtf-gtmo.html#document/p1

[14] http://www.who2.com/bio/khalid-sheikh-mohammed

[15] "Prisoners : Ghost: Khalid Sheikh Mohammed" (http://old.cageprisoners.com/prisoners.php?id=1371). *Cageprisoners*. . Retrieved 23 February 2012.

[16] Mcdermott, Terry (2006-08-01). *Perfect Soldiers: The 9/11 Hijackers: Who They Were, Why They Did It*. Harper Paperbacks., page 111

[17] Susan Candiotti, Maria Ressa, Justine Redman and Henry Schuster CNN (December 19, 2002). "Suspected 9/11 mastermind graduated from U.S. university" (http://edition.cnn.com/2002/US/South/12/19/al.qaeda.aggie/). Edition.cnn.com. . Retrieved 2010-10-06.

[18] Alumni Home (http://www.ncat.edu/~alumni/alumni_affairs/distinguished/distinguished.html) The university has no information on him

[19] 9/11 Commission Report (http://www.9-11commission.gov/report/911Report.pdf), page 147.

[20] "9/11 planner is recast as key asset for CIA" (http://www.msnbc.msn.com/id/32605529/ns/us_news-washington_post/). MSNBC. accessed September 11, 2011.

[21] "Alleged Sept. 11 mastermind's nephew plotted 1993 bombing: FBI's most-wanted terrorist after bin Laden lived in luxury in Philippines with '93 plotter". Ottawa Citizen / Associated Press. June 26, 2002.

[22] Gunaratna, Rohan (March 3, 2003). "Womaniser, joker, scuba diver: the other face of al-Qaida's No 3" (http://www.guardian.co.uk/world/2003/mar/03/alqaida.terrorism1). *The Guardian* (London). . Retrieved 2006-09-12.

[23] "Bosnia: Senior Al-Qaeda figure granted citizenship, says report" (http://www.adnkronos.com/AKI/English/Security/?id=3.0.2928330465). Adnkronos. 2009-01-20. . Retrieved 2009-01-23.

[24] (http://www.9-11commission.gov/9-11commission) (2007). "AL QAEDA AIMS AT THE AMERICAN HOMELAND — CH5" (http://www.9-11commission.gov/report/911Report_Ch5.htm). 9-11commission (http://www.9-11commission.gov/). . Retrieved 2007-08-28. "In January 1996, well aware that U.S. authorities were chasing him, he left Qatar for good and fled to Afghanistan, where he renewed his relationship with Rasul Sayyaf.9"

[25] "Context of 'January–May 1996: US Fails to Capture KSM Living Openly in Qatar'" "Cooperative Research History Commons" http://www.cooperativeresearch.org/context.jsp?item=a96qatar

[26] Wayne Barrett, November 27, 2007 http://www.villagevoice.com/2007-11-20/news/rudy-s-ties-to-a-terror-sheikh/

[27] "Home" (http://www.propeller.com/viewstory/2007/11/29/olbermann-giulianis-direct-connections-to-terrorists/?url=http://ca.youtube.com/watch?v=8wCTXcsB7zU&frame=true). Propeller. 2010-09-25. . Retrieved 2010-10-06.

[28] MotherJones Blog: Village Voice: Giuliani Did Business With Terrorism Supporter (http://www.motherjones.com/mojo/2007/11/village-voice-giuliani-did-business-terrorism-supporter)

[29] Marcus Baram, ABC News November 29, 2007 "Giuliani's Ties to Qatar Raise Questions for Mr. 9/ll" (http://blogs.abcnews.com/theblotter/2007/11/giulianis-ties.html)

[30] "National Commission on Terrorist Attacks Upon the United States" (http://www.9-11commission.gov/report/911Report_Ch5.htm). 9-11commission. . Retrieved 2010-08-13.

[31] McDermott, Terry. McDermott, Terry (August 11, 2006). "Echoes of '95 Manila Plot" (http://www.latimes.com/news/nationworld/nation/la-na-manila11aug11,0,802872.story?coll=la-home-headlines). *Los Angeles Times*. . Retrieved 2006-09-13. *Los Angeles Times* August 11, 2006.

[32] "Top al Qaeda operative caught in Pakistan" (http://articles.cnn.com/2003-03-02/world/mohammed.biog_1_ahmed-abdul-qadoos-al-qaeda-military-committee-ayman-al-zawahiri?_s=PM:asiapcf). *CNN*. March 1, 2003. .

[33] "Most wanted terrorists list released" (http://archives.cnn.com/2001/US/10/10/inv.mostwanted.list/). CNN. 2001-10-01. . Retrieved 2008-07-18.

[34] [[s:Verbatim Transcript of Combatant Status Review Tribunal Hearing for ISN 10024CSRT for Khalid Sheikh Mohammed, March 10, 2007.

[35] "Suspect 'reveals 9/11 planning'" (http://news.bbc.co.uk/2/hi/south_asia/3128802.stm). BBC News. September 22, 2003. .

[36] National Commission on Terrorist Attacks Upon the United States (2004). "Chapter 5" (http://www.9-11commission.gov/report/911Report_Ch5.htm). *9/11 Commission Report*. Government Printing Office. .

[37] Wright, Lawrence (2006). *The Looming Tower: Al-Qaeda and the Road to 9/11*. Knopf. p. 308.

[38] Bergen, Peter (2006). *The Osama bin Laden I Know*. Free Press. p. 283.

[39] http://abcnews.go.com/Blotter/scenes-planners-911-attacks/story?id=11610817#.Tzh5u1GRIJF

[40] "Outline of the 9/11 Plot Staff Statement No. 16" (http://www.fas.org/irp/congress/2004_rpt/staff_statement_16.pdf). 9/11 Commission. 2004-06-16. . mirror (http://www.webcitation.org/query?url=http://www.fas.org/irp/congress/2004_rpt/

staff_statement_16.pdf&date=2011-10-29)

[41] "Philip Zelikow's testimony before the 9/11 Commission" (http://www.pbs.org/newshour/bb/terrorism/jan-june04/911_06-16.html). PBS Newshour. 2004-06-16. . Retrieved 2011-10-29. mirror (http://www.webcitation.org/query?url=http://www.pbs.org/newshour/bb/terrorism/jan-june04/911_06-16.html&date=2011-10-29)

[42] "'We left out nuclear targets, for now'" (http://www.guardian.co.uk/world/2003/mar/04/alqaida.terrorism). *The Guardian* (London). March 3, 2003. . Retrieved April 4, 2010.

[43] "Verbatim Transcript of Combatant Status Review Tribunal Hearing for ISN 10024" (http://www.defenselink.mil/news/transcript_ISN10024.pdf). .

[44] Ressa, Maria. "Sources:Reid is al Qaeda operative." (http://www.cnn.com/2003/WORLD/asiapcf/southeast/01/30/reid.alqaeda/). *CNN*. January 29, 2003. . Retrieved 2006-09-15. *CNN.com*, December 6, 2003.

[45] Burger, Timothy J.; Adam Zagorin (2006-10-12). "Fingering Danny Pearl's Killer" (http://www.time.com/time/world/article/0,8599,1545441,00.html?cnn=yes). *Time* (Time Warner). . Retrieved 2007-03-15.

[46] "Key 9/11 figure 'beheaded Pearl'" (http://news.bbc.co.uk/2/hi/americas/6455307.stm). *BBC News* (BBC). 2007-03-15. . Retrieved 2007-03-15.

[47] "Al-Qaida No. 3 says he planned 9/11, other plots" (http://www.msnbc.msn.com/id/17617986/?GT1=9145). Associated Press. MSNBC. 2007-03-15. . Retrieved 2007-03-15.

[48] Ackerman, Spencer (January 20, 2011). "Qaeda Killer's Veins Implicate Him In Journo's Murder" (http://www.wired.com/dangerroom/2011/01/qaeda-killers-veins-implicate-him-in-journos-murder/). *Wired*. . Retrieved January 21, 2011.

[49] Blackburn, Bradley (January 20, 2011). "Report Says Justice Not Served in Murder of Daniel Pearl, Wall Street Journal Reporter" (http://abcnews.go.com/US/report-justice-served-murder-daniel-pearl/story?id=12721909). ABC News. . Retrieved January 20, 2011.

[50] "'We left out nuclear targets, for now'" (http://www.guardian.co.uk/world/2003/mar/04/alqaida.terrorism). *The Guardian* (London). March 4, 2003. .

[51] Shahzad, Syed Saleem (October 30, 2002). "A chilling inheritance of terror" (http://www.atimes.com/atimes/South_Asia/DJ30Df01.html). *Asia Times*. .

[52] Shane, Scott (June 22, 2008). "Inside a 9/11 Mastermind's Interrogation" (http://www.nytimes.com/2008/06/22/washington/22ksm.html). *The New York Times*. .

[53] Bush admits to CIA secret prisons (http://news.bbc.co.uk/2/hi/americas/5321606.stm), BBC News, Thursday, 7 September 2006, 04:18 GMT 05:18 UK

[54] "Red Cross report; page 37" (http://www.foxnews.com/projects/pdf/042809_redcross.pdf) (PDF). *Fox News*. . Retrieved 2010-10-06.

[55] Mikkelsen, Randall (2008-02-05). "CIA says used waterboarding on three suspects" (http://www.reuters.com/article/latestCrisis/idUSN05191813). Reuters. . Retrieved 2009-11-18.

[56] Mayer, Jane (2009). *The Dark Side: The Inside Story of How the War on Terror Turned Into a War on American Ideals* (http://books.google.com/books?id=w8-y8v99TCIC&lpg=PP1&dq=The Dark Side: The Inside Story&pg=PA273#v=onepage&q=&f=false). Random House, Inc.. p. 273. ISBN 0-307-45629-3, 9780307456298. .

[57] Rodriguez, Alex (February 3, 2010). "Is she a victim of the U.S. or is she 'Terror Mom'?; Aafia Siddiqui is awaiting a verdict after her trial in the U.S. on attempted murder charges. Many in Pakistan consider her a hero and a victim of persecution" (http://articles.latimes.com/2010/feb/03/world/la-fg-terror-mom3-2010feb03). *Los Angeles Times*. . Retrieved February 12, 2010.

[58] Fermino, Jennifer, and Sheehy, Kate, "NY Charges for Woman in Afghan Military Shooting," *[[New York Post* (http://www.nypost.com/p/news/regional/item_mOnbbRnWK8WCORusYCKQHK;jsessionid=B4A20B086D7B5F3D87C6BC8F939F2193)], *August 5, 2008, accessed February 11, 2010]*

[59] Shah, Sabir (May 13, 2010). "Khalid Shaikh Mohammed — a fact sheet" (http://www.thenews.com.pk/Todays-News-6-107988-Khalid-Shaikh-Mohammed-â-a-fact-sheet). *The News International*. . Retrieved May 29, 2012.

[60] Moskowitz, Stan (2004-07-13). "Memorandum for the record - Subject: Interrogations" (http://www.judicialwatch.org/story/2010/feb/cia-congressional-briefings-terrorist-interrogations). CIA (hosted at Judicial Watch). . Retrieved 2010-03-10.

[61] "Eleven Detainees in Undisclosed Locations" (http://www.hrw.org/backgrounder/usa/us1004/7.htm). *Human Rights Watch*. October 2004. .

[62] "The Legal Prohibition Against Torture" (http://www.hrw.org/press/2001/11/TortureQandA.htm). *Human Rights Watch*. June 1, 2004. .

[63] "Al Qaeda men in 'ghost prison'" (http://us.rediff.com/news/2004/oct/18ghost.htm). *rediff.com*. October 18, 2004. .

[64] "Jordan denies 'secret US prison'" (http://news.bbc.co.uk/2/hi/middle_east/3742428.stm). *BBC*. October 14, 2004. .

[65] "Gonzales insists US did not send prisoners abroad to be tortured" (http://jurist.law.pitt.edu/paperchase/2005/03/gonzales-insists-us-did-not-send.php), *The Jurist author=Bernard Hibbitts*, March 7, 2005,

[66] Price, Caitlin, *CIA chief confirms use of waterboarding on 3 terror detainees* (http://jurist.law.pitt.edu/paperchase/2008/02/cia-chief-confirms-use-of-waterboarding.php), University of Pittsburgh School of Law, , retrieved 2008-05-13

[67] Shane, Scott (April 20, 2009). "Waterboarding Used 266 Times On 2 Suspects" (http://www.nytimes.com/2009/04/20/world/20detain.html). *The New York Times*: p. 1. .

[68] "ICRC Report on the Treatment of Fourteen "High Value Detainees" in CIA Custody" (http://www.nybooks.com/icrc-report.pdf) (PDF). . Retrieved 2010-10-06.

[69] "ICRC Report on the Treatment of Fourteen "High Value Detainees" in CIA Custody" (http://www.foxnews.com/projects/pdf/042809_redcross.pdf) (PDF). Fox News. February 2007. p. 35. . Retrieved 2009-11-18.

[70] "Despite Reports, Khalid Sheikh Mohammed Was Not Waterboarded 183 Times" (http://www.foxnews.com/politics/2009/04/28/despite-reports-khalid-sheikh-mohammed-waterboarded-times/). FOXNews.com. 2010-04-07. . Retrieved 2010-08-13.

[71] "Khalid Sheikh Mohammed confession transcript" (http://news.bbc.co.uk/2/shared/bsp/hi/pdfs/15_03_07_mohammed_transcript.pdf) (PDF). http://www.defenselink.mil (http://www.defenselink.mil/news/transcript_ISN10024.pdf). March 10, 2007. . Retrieved 2007-08-28.

[72] "Transcript: Khalid Sheikh Mohammed confesses 9/11 role" (http://www.cnn.com/2007/US/03/14/guantanamo.mohammed/index.html). CNN. 2007-03-14. . Retrieved 2007-03-14.

[73] "September 11 mastermind 'confesses'" (http://english.aljazeera.net/NR/exeres/BA2B7119-DF1D-427E-857E-5C44E3479F2A.htm). Al Jazeera. March 15, 2007. .

[74] Khalid Sheikh Khalid Sheikh Mohammed: 'I Was Responsible for 9/11' (http://newsmax.com/archives/articles/2007/3/15/71337.shtml). March 15, 2007.

[75] "2009, March 09, The Islamic Response to the Government's Nine Accusations" (http://relearnhistory.com/content/2009-March-09-Islamic-Response-Government-s-Nine-Accusations). ReLearnHistory.com. . Retrieved 2010-08-13.

[76] wikisource:Verbatim Transcript of Combatant Status Review Tribunal Hearing for ISN 10024

[77] Shane, Scott (2008-06-22). "Inside a 9/11 Mastermind's Interrogation" (http://www.nytimes.com/2008/06/22/washington/22ksm.html?pagewanted=4). The New York Times. p. 4. . Retrieved 2009-11-18.

[78] Christopher Hope, Robert Winnett, Holly Watt, Heidi Blake (2011-04-25). "WikiLeaks: Guantanamo Bay terrorist secrets revealed" (http://www.telegraph.co.uk/news/worldnews/wikileaks/8471907/WikiLeaks-Guantanamo-Bay-terrorist-secrets-revealed.html). London: The Daily Telegraph. . "A senior Al-Qaeda commander claimed that the terrorist group has hidden a nuclear bomb in Europe which will be detonated if Bin-Laden is ever caught or assassinated. The US authorities uncovered numerous attempts by Al-Qaeda to obtain nuclear materials and fear that terrorists have already bought uranium. Sheikh Mohammed told interrogators that Al-Qaeda would unleash a 'nuclear hellstorm'." mirror (http://www.webcitation.org/query?url=http://www.telegraph.co.uk/news/worldnews/wikileaks/8471907/WikiLeaks-Guantanamo-Bay-terrorist-secrets-revealed.html&date=2011-10-29)

[79] Mark H. Buzby (2008-04-15). "Combatant Status Review Tribunal Input and Recommendation for Continued Detention Under DoD Control (CD) for Guantanamo Detainee, ISN: US9KU-010024DP (S)" (http://media.mcclatchydc.com/smedia/2011/04/27/19/us9ku-010024dp.source.prod_affiliate.91.pdf). JTF-GTMO. .

[80] http://www.newsmax.com/Newsfront/WikiLeaks-GuantanamoBay-al-Qaida-terrorist/2011/04/25/id/393982 Martin Gould (2011-04-25). "WikiLeaks: Al-Qaida Already Has Nuclear Capacity". Newsmax. mirror (http://www.webcitation.org/query?url=http://www.newsmax.com/Newsfront/WikiLeaks-GuantanamoBay-al-Qaida-terrorist/2011/04/25/id/393982&date=2011-10-29)

[81] http://www.nuclearheadlines.com/wikileaks-reveals-al-qaeda-thug-khalid-shaikh-mohammed-vowed-nuclear/

[82] "'Nuclear hellstorm' if bin Laden caught - 9/11 mastermind" (http://www.news.com.au/breaking-news/nuclear-hellstorm-if-bin-laden-caught-911-mastermind/story-e6frfku0-1226044724298). *News.au.* April 25, 2011. . mirror (http://www.webcitation.org/query?url=http://www.news.com.au/breaking-news/nuclear-hellstorm-if-bin-laden-caught-911-mastermind/story-e6frfku0-1226044724298&date=2011-10-29)

[83] http://news.yahoo.com/s/afp/20110425/wl_mideast_afp/usattacksguantanamowikileakssheikh_20110425184242

[84] "Al-Qaeda hid bomb in Europe: WikiLeaks releases secret files" (http://newstabulous.com/al-qaeda-hid-bomb-in-europe-wikileaks-releases-secret-files/9722/). Newstabulous. 2011-04-25. . mirror (http://www.webcitation.org/query?url=http://newstabulous.com/al-qaeda-hid-bomb-in-europe-wikileaks-releases-secret-files/9722/&date=2011-10-29)

[85] Michael Melia (April 16, 2007). "Father of Pakistani Alleges U.S. Torture" (http://apnews.myway.com/article/20070416/D8OI0AF00.html). Associated Press. . Retrieved 2007-04-18.

[86] Natalie Hrubos (April 17, 2007). "Guantanamo detainee's father says son tortured in secret CIA prison" (http://jurist.law.pitt.edu/paperchase/2007/04/guantanamo-detainees-father-says-son.php). The Jurist. . Retrieved 2007-04-18.

[87] Ali Khan (April 16, 2007). "Statement of Ali Khan" (http://ccrjustice.org/files/Ali Khan_Father of Majid Khan_Statement from CSRT.pdf) (PDF). Center for Constitutional Rights. . Retrieved June 19, 2009.

[88] Gonyea, Don (September 6, 2006). "Bush Concedes CIA Ran Secret Prisons Abroad" (http://www.npr.org/templates/story/story.php?storyId=5776968). NPR. . Retrieved 2010-02-22.

[89] "President Bush Discusses Global War on Terror" (http://georgewbush-whitehouse.archives.gov/news/releases/2006/09/20060929-3.html). The White House. September 29, 2006. . Retrieved March 11, 2012.

[90] "Key 9/11 suspect confesses guilt" (http://news.bbc.co.uk/2/hi/americas/6452573.stm). *BBC News* (BBC). 2007-03-15. . Retrieved 2007-03-15.

[91] CIA's Harsh Interrogation Techniques Described: Sources Say Agency's Tactics Lead to Questionable Confessions, Sometimes to Death (http://abcnews.go.com/WNT/Investigation/story?id=1322866), ABC News, November 18, 2005

[92] "ABC News: Expert Looks Beyond Mohammed's Confessions" (http://abcnews.go.com/International/Health/story?id=2955471&page=1). Abc News. 2007-03-15. . Retrieved 2010-10-06.

[93] "Khalid Sheikh Mohammed's own words provide glimpse into the mind of a terrorist" (http://www.wsvn.com/news/articles/world/BO46072/). Associated Press. WSVN. March 16, 2007. . Retrieved March 11, 2012.

[94] Robbins, Martin (2010-11-04). "Does torture work?" (http://www.guardian.co.uk/science/the-lay-scientist/2010/nov/04/2). London: The Guardian. . Retrieved 2011-06-15.
[95] http://articles.latimes.com/2011/may/12/news/la-pn-mccain-bin-laden-20110512
[96] http://www.mrc.org/node/37689
[97] http://www.msnbc.msn.com/id/43007276/ns/politics-more_politics/t/mccain-says-torture-did-not-lead-bin-laden/#.UH-ZW8XMhIE
[98] Khalid Sheikh Mohammed's '31 plots' (http://news.bbc.co.uk/2/hi/americas/6452789.stm), *BBC*, March 15, 2007
[99] Jane Mayer (2010-02-16). "The Trial.Eric Holder and the battle over Khalid Sheikh Mohammed" (http://www.newyorker.com/reporting/2010/02/15/100215fa_fact_mayer). *The New Yorker*. .
[100] Lolita C. Baldur (Thursday, August 9, 2007). "Pentagon: 14 Guantanamo Suspects Are Now Combatants" (http://www.time.com/time/nation/article/0,8599,1651680,00.html). Time magazine. . mirror (http://www.webcitation.org/query?url=http://findarticles.com/p/articles/mi_qn4176/is_20070324/ai_n18757735&date=2008-05-19)
[101] Sergeant Sara Wood (June 4, 2007). "Charges Dismissed Against Canadian at Guantanamo" (http://www.defenselink.mil/news/newsarticle.aspx?id=46281). Department of Defense. . Retrieved 2007-06-07.
[102] Sergeant Sara Wood (June 4, 2007). "Judge Dismisses Charges Against Second Guantanamo Detainee" (http://www.defenselink.mil/news/newsarticle.aspx?id=46288). Department of Defense. . Retrieved 2007-06-07.
[103] "Militant convicted of Pearl killing to rely on KSM Guantanamo confession on appeal" (http://jurist.law.pitt.edu/paperchase/2007/03/militant-convicted-of-pearl-killing-to.php). The Jurist. March 19, 2007. . Retrieved 2007-03-20.
[104] "Pearl murder convict to appeal after confession" (http://today.reuters.com/news/articlenews.aspx?type=domesticNews&storyID=2007-03-19T104004Z_01_ISL185_RTRUKOC_0_US-PAKISTAN-PEARL.xml&pageNumber=1&imageid=&cap=&sz=13&WTModLoc=NewsArt-C1-ArticlePage1). Reuters. March 19, 2007. . Retrieved 2007-03-20.
[105] "France tries trio over Djerba synagogue bombing" (http://www.google.com/hostednews/afp/article/ALeqM5h7KDqZLMV7eCw9v9p68QHjoesfNQ). Google.com. 2009-01-04. . Retrieved 2012-07-22.
[106] "Michel Moutot. Al Qaeda militant found guilty for Tunisian synagogue attack" (http://www.expatica.com/nl/news/local_news/Al-Qaeda-militant-found-guilty-for-Tunisian-synagogue-attack-_49356.html). Expatica.com. . Retrieved 2012-07-22.
[107] "U.S. Department of Defense - Military Commissions" (http://www.defense.gov/news/commissionsCo-conspirators.html). Defense.gov. . Retrieved 2010-10-06.
[108] "Guantanamo 9/11 suspects on trial" (http://news.bbc.co.uk/1/hi/world/americas/7437164.stm). BBC News. June 6, 2008. . Retrieved December 8, 2008.
[109] "Geneva Convention relative to the Treatment of Prisoners of War (Article 84)" (http://usmilitary.about.com/library/milinfo/genevacon/blart-84.htm). Usmilitary.about.com. 2010-06-19. . Retrieved 2010-10-06.
[110] "About the 9/11 war crimes trial" (http://www.miamiherald.com/news/more-info/story/436366.html). *The Miami Herald*. February 27, 2008. .
[111] "Guantanamo 9/11 suspects on trial" (http://news.bbc.co.uk/2/hi/americas/7437164.stm). *BBC News*. June 6, 2008. .
[112] Selsky, Andrew O. (June 5, 2008). "9/11 Mastermind Khalid Sheikh Mohammed On Trial At Guantanamo Bay" (http://www.huffingtonpost.com/2008/06/05/911-mastermind-khalid-she_n_105392.html). *Huffington Post*. .
[113] Worthington, Andy (September 29, 2008). "Is Khalid Sheikh Mohammed Running the 9/11 Trials?" (http://www.andyworthington.co.uk/2008/09/28/is-khalid-sheikh-mohammed-running-the-911-trials/). .
[114] Rosenberg, Carol (October 12, 2008). "Al Qaeda defendants get laptops at Guantánamo, judge rules" (http://www.miamiherald.com/news/politics/campaign-2008/story/723593.html). *The Miami Herald*. . Retrieved October 12, 2008. "Mohammed, his nephew Ammar al Baluchi and Walid Bin Attash have sought through standby counsel filings at the Military Commission a long list of resources they say they need to mount their defense -- including Internet links to read news accounts and do live research on databases." , mirror (http://www.webcitation.org/query?url=http://www.miamiherald.com/news/politics/campaign-2008/story/723593.html&date=2008-10-12).
[115] Reuters wire service (December 8, 2008). "Sept. 11 Defendants Ask to Plead Guilty" (http://www.nytimes.com/2008/12/09/us/09gitmo.html). *The New York Times*. . Retrieved December 8, 2008.
"Top 9/11 suspects to plead guilty" (http://news.bbc.co.uk/1/hi/world/americas/7770856.stm). BBC News. December 8, 2008. . Retrieved December 8, 2008.
[116] "Khalid Sheikh Mohammed Trial 2009" (http://www.google.com/hostednews/ap/article/ALeqM5g8-DEMtAE9q4i4ySQ0eV_qZefmRQD9BULOQO0). Associated Press. November 13, 2009. . Retrieved November 13, 2009.
[117] "Pentagon drops charges against 9/11 plotters, clearing way for civilian trials" (http://www.idahostatesman.com/2010/01/22/1097055/pentagon-drops-charges-against.html). Idaho Statesman. February 26, 2010. . Retrieved 2010-03-02.
[118] Hughes, Mark (August 22, 2012). "9/11 trial 'may not begin for another four years'" (http://www.telegraph.co.uk/news/worldnews/september-11-attacks/9493267/911-trial-may-not-begin-for-another-four-years.html). The Telegraph. . Retrieved 23 August 2012.
[119] Ackerman, Spencer (December 8, 2008). "Coersion and Military Law: Does a Plea After Torture Stand?" (http://washingtonindependent.com/21398/coersion-and-military-law). *Washington Independent*. . Retrieved December 8, 2008., mirror (http://www.webcitation.org/query?url=http://washingtonindependent.com/21398/coersion-and-military-law&date=2008-12-09).
[120] Terry Frieden, Chris Kokenes (2009-11-13). "Accused 9/11 plotter Khalid Sheikh Mohammed faces New York trial" (http://edition.cnn.com/2009/CRIME/11/13/khalid.sheikh.mohammed/index.html). CNN. .
[121] Catherine Herridge (2010-01-22). "Charges Withdrawn in Military Commissions for Sept. 11 Suspects" (http://www.foxnews.com/politics/2010/01/22/charges-withdrawn-military-commissions-sept-suspects/). Fox News. .

[122] "Execution 'Likely' for 9/11 Suspect" (http://english.aljazeera.net/news/americas/2010/01/2010131172757497878.html). Al Jazeera English. 2010-02-01. . mirror (http://www.webcitation.org/query?url=http://english.aljazeera.net/news/americas/2010/01/2010131172757497878.html&date=2011-10-29)

[123] "In Reversal, Obama Orders Guantanamo Military Trial for 9/11 Mastermind Khalid Sheikh Mohammed by JASON RYAN and HUMA KHAN" (http://abcnews.go.com/Politics/911-mastermind-khalid-sheikh-mohammed-military-commission/story?id=13291750). Abcnews.go.com. 2011-04-04. . Retrieved 2012-07-22.

[124] By Alan Silverleib, CNN (2011-04-05). "Accused 9/11 terror suspects to face military trials by Alan Silverleib" (http://edition.cnn.com/2011/US/04/04/guantanamo.tribunals/index.html#). Edition.cnn.com. . Retrieved 2012-07-22.

[125] "Khalid Sheikh Mohammed and other '9/11 plotters' in court" (http://www.bbc.co.uk/news/world-us-canada-17966362). Bbc.co.uk. 2012-05-06. . Retrieved 2012-07-22.

[126] Carol Rosenberg (2009-09-10). "Guantánamo photos of accused 9/11 mastermind posted on Web" (http://www.miamiherald.com/news/breaking-news/story/1224587.html). The Miami Herald. . Retrieved 2009-09-10.

[127] Robert Mackey (2009-09-09). "Photographs of Khalid Shaikh Mohammed at Guantánamo Appear Online" (http://thelede.blogs.nytimes.com/2009/09/09/photographs-of-khalid-shaikh-mohammed-at-guantanamo-appear-online/?hp). The New York Times. . Retrieved 2009-09-10.

[128] "New Khalid Sheik Mohammed Photo Surfaces Days Before 9/11 Anniversary" (http://www.huffingtonpost.com/2009/09/09/new-khalid-sheik-mohammed_n_281009.html). Huffington Post. 2009-09-10. . Retrieved 2009-09-10.

[129] "Photos of '9/11 plotter' hit web" (http://news.bbc.co.uk/2/hi/americas/8248355.stm). BBC News. 2009-09-10. . Retrieved 2009-09-10.

[130] "Terror Fears Over 9/11 Mastermind Photos" (http://news.sky.com/skynews/Home/World-News/Khalid-Sheikh-Mohammed-Terror-Attack-Fear-Over-9-11-Mastermind-Photos/Article/200909215378754?lpos=World_News_First_World_News_Article_Teaser_Region_0&lid=ARTICLE_15378754_Khalid_Sheikh_Mohammed:_Terror_Attack_Fear_Over_911_Mastermind_Photos). Sky News. 2009-09-10. . Retrieved 2009-09-10 p.

[131] "Photos released of terrorist mastermind at Guantánamo" (http://www.dallasnews.com/sharedcontent/dws/news/world/stories/DN-mohammed_10int.ART.State.Edition1.4be29f6.html). *The Dallas Morning News*. 2009-09-10. . Retrieved 2009-09-10.

[132] "Guantanamo Detainee Photos" (http://suomenkuvalehti.fi/kuvat/2009/09/09/guantanamo-detainee-photos). Suomen Kuvalehti. 2009-09-10. . Retrieved 2009-09-10.

External links

- Official site of the U.S. Defence Department on Military Commissions (case documents) (http://www.defenselink.mil/news/commissionsCo-conspirators.html)
- AP: Khalid Sheikh Mohammed's own words provide glimpse into the mind of a terrorist (http://www.iht.com/articles/ap/2007/03/16/news/CB-GEN-Mind-of-Mohammed.php?page=1)
- ABC News: Expert Looks Beyond Mohammed's Confessions (http://abcnews.go.com/International/Health/story?id=2955471&page=1)
- "Substitute for Testimony" from Khalid Shaikh Mohammed. (http://www.law.umkc.edu/faculty/projects/ftrials/moussaoui/sheikhstmt.pdf) A summary of information KSM provided to his American interrogators (used by defence in the trial of Zacarias Moussaoui).
- The Final 9/11 Commission Report (http://www.gpoaccess.gov/911/index.html). The Report focuses on KSM at the beginning of Chapter 5.
- Khalid Sheikh Mohammed: life of terror (http://edition.cnn.com/2003/WORLD/asiapcf/south/03/02/mohammed.biog/)—an article from CNN
- An article (http://old.nationalreview.com/comment/comment-levitt030503.asp) from the National Review
- Do We Still Have Khalid Sheikh Mohammed's Sons? (http://www.dcdave.com/article5/070318.htm), David Martin, March 18, 2007.
- Top al-Qaeda suspect in U.S. hands (http://news.bbc.co.uk/1/hi/world/south_asia/2812181.stm)—news report from the BBC
- Is There More to the Capture of Khalid Sheikh Mohammed Than Meets the Eye? (http://www.cooperativeresearch.net/essay_pf.jsp?article=essayksmcapture), Center for Cooperative Research
- Bush Spells Out 'LA Terror Plot' (http://news.bbc.co.uk/1/hi/world/americas/4697896.stm)—Feb 09 2006 report from the BBC

- Biographies of KSM and 13 other detainees (http://www.odni.gov/announcements/content/DetaineeBiographies.pdf), Office of the Director of National Intelligence (U.S.)
- DoD: Verbatim Transcript of Combatant Status Review Tribunal Hearing for ISN 10024, From WikiSource
- "A "Convenient" Guantanamo confession" (http://www.muslimnews.co.uk/news/news.php?article=12466). *The Muslim News* (Muslimnews.co.uk (http://www.muslimnews.co.uk/)).
- Pentagon charges 6 in 9-11 attacks (http://msnbc.msn.com/id/23108871)
- 'Clean team' interrogated 9-11 suspects (http://www.washingtonpost.com/wp-dyn/content/article/2008/02/11/AR2008021100572.html)
- Profile Key 9-11 suspects (http://news.bbc.co.uk/2/hi/americas/7239580.stm)
- Shane, Scott (June 22, 2008). "Inside a 9/11 Mastermind's Interrogation" (http://www.nytimes.com/2008/06/22/washington/22ksm.html?partner=rssnyt&emc=rss). *The New York Times*. Retrieved 2008-06-23.

Riduan_Isamuddin

Riduan Isamuddin (Hambali)	

Born	April 4, 1964Cianjur, West Java, Indonesia
Detained at	CIA black sites, Guantanamo
ISN	10019
Charge(s)	No charge
Status	Held in extrajudicial detention

Riduan Isamuddin also transliterated as **Riduan Isamudin**, **Riduan Isomuddin**, and **Riduan Isomudin**, better known by the *nom de guerre* **Hambali**, born as **Encep Nurjaman**, born April 4, 1964[1] (Indonesia) is the former military leader of the Indonesian terrorist organization Jemaah Islamiyah (JI), which is linked with Al Qaeda. He is now in extrajudicial detention in the United States's Guantanamo Bay detainment camps, in Cuba.[2]

Hambali was often described as "the Osama bin Laden of Southeast Asia". Some media reports describe him as Bin Laden's lieutenant for Southeast Asian operations. Other reports describe him as an independent peer. He was highly trusted by Al Qaeda and was the main link between the two organisations. Hambali was a close friend of Khalid Shaikh Mohammed, who planned Operation Bojinka and the September 11 Terrorist Attacks. Hambali envisioned creating a Muslim state, in the form of an Islamic superpower (a theocracy) across Southeast Asia, with himself as its leader (Caliph). His ambition was to rule Indonesia, Malaysia, Singapore, Brunei and parts of the Philippines, Myanmar, and Thailand[3] [4]

Receiving increasing attention in the aftermath of the 2002 Bali nightclub bombing, in which 202 people died. According to the Office of the Director of National Intelligence he served as the third in command of Al Qaeda, from the 2003 capture of Khalid Sheikh Mohammed to his own capture in 2005. [5] He was eventually apprehended in a joint operation by the CIA and Thai police. He is currently imprisoned in Guantanamo Bay detention center in Cuba, after three years of CIA custody in a secret location.

Early life

Riduan Isamuddin was born Encep Nurjaman in the rice belt of Sukamanah, a small village in Cianjur Regency, in the province of West Java, Indonesia. He was the son of a peasant farmer, and was the second of thirteen children. He first became involved with Jemaah Islamiah as a teenager. He was a diligent student at his Islamic high school, Al-Ianah. He travelled to Afghanistan in 1983 to fight the Soviet Union during the Soviet invasion of Afghanistan. During his three years as a mujahid, from 1987 to 1990, he met Osama bin Laden. Friends and family in Indonesia say they did not know of his activities overseas. His nephew Wilson Halim is also said to be a renowned budding youth terrorist in Malaysia.

His name has been transliterated into English text in several different ways over the years, including;

- His name was spelled **Hambali, Riduan bin Isomuddin** on the Summary of Evidence memo prepared for his Combatant Status Review.[6]
- His name was spelled **Ryuduan bin Isomuddin** by Scott McClellan during a press briefing.[7] McClellan spelled his name letter by letter.
- His name is spelled **Riduan Isamuddin** in the report of the 9-11 Commission.[8]
- His name was spelled **Nurjaman Riduan Isamuddin** by the US Treasury.[9]

Jemaah Islamiah

He returned to Cianjur for one week, before going to Malaysia, in 1991 where he met the two co-founders of JI, Abdullah Sungkar and Abubakar Bashir, who fled President Suharto's "New Order" in 1985. Nurjaman internationalized the terrorist group's activities. Nurjaman then took a new name in his permanent residence permit. He now was Riduan Isamuddin. The three lived in a housing compound in Kampung Sungai Manggis, Banting, Selangor. His nickname, *Hambali*, is after Hanbali, a school of religious law started by a revered Islamic imam from the 8th century.

The two co-founders sent their students to "study" in Afghanistan and Pakistan. The students actually fought the Soviets until the Soviets pulled out of Afghanistan. A woman named Noralwizah Lee Abdullah had gone to Malaysia for religious schooling. She secretly married Isamuddin after meeting him at the Luqmanul Hakiem School in Ulu Tiram, Johor. The school was founded by Sungkar and Bashir.

Initially, Isamuddin struggled to make a living for his family. He switched from selling kebabs to patenting medicines. He soon disappeared from his home for weeks at a time, and he received many visitors at home. He eventually came to own a red hatchback and several cell phones. Investigators say that many calls on those cell phones were made to Mohammed Jamal Khalifa, Osama bin Laden's brother in law, who had arrived back in Manila, Philippines in 1991.

After Arab visitors gave his family lots of money, he founded a shell company, Konsojaya, in June 1994. Ostensibly an import-export company trading in palm oil between Malaysia and Afghanistan, it was essentially a front company for terrorism. Wali Khan Amin Shah, who would become the financier of Operation Bojinka, was a director of Konsojaya. The company provided financial assistance to the project until it was discovered by investigators on a laptop computer after an apartment fire on January 6, 1995. Shah was arrested in the Philippines but escaped on a short order. Shah was arrested in Malaysia in December 1995. Both Shah and mastermind Ramzi Yousef, who escaped the Philippines but was arrested in Lahore, Pakistan, were extradited to the United States. They were both convicted and sentenced to life in prison for participating in the project.

Hambali goes underground

Hambali's company attracted attention of investigators so his dealings went quiet for a while. He decided to preach, raise money, and recruit for his cause. He went underground in 2000 and started a wave of church-bombings in Indonesia. He always had a "hands-on" technique; he met his foot soldiers and came to them "with detailed plans, plenty of cash and two of his own bombmakers." He always fled before the bombing commenced. Meanwhile, the spiritual leader of Jemaah Islamiah, Abu Bakr Bashir, was preaching jihad at his schools while denying links to Islamic militants.

Hambali planned and attended the January 2000 Al Qaeda Summit in Kuala Lumpur, Malaysia. Among the others present were two September 11 hijackers, Khalid al-Mihdhar and Nawaf al-Hazmi. This gathering in Kuala Lumpur was observed by the CIA and Malaysian authorities, but what specifically was being said at the meetings was not picked up. Hambali also provided money and documents to Zacarias Moussaoui in October of that year.

After the Bali nightclub bombing, in which 202 people were killed, Hambali received more attention from the United States. In the years leading up to the attack, the Indonesian government's action against Islamic militants had been minimal. Following the attack, Abu Bakr Bashir was arrested as part of a crack down on Jemaah Islamiah. He was wanted in Indonesia[10] for the bombings of several churches in the region, and wanted for the Bali bombing and a failed plot on several targets in Singapore.

Capture

Hambali used a series of safe-houses throughout Southeast Asia, especially Thailand and Cambodia, to move around. While he was in Ayutthaya, Thailand, 75 kilometers north of Bangkok, he was planning a terrorist attack against several Thai hotels and the Asia Pacific Economic Cooperation summit (APEC) in Bangkok on October 2003. Hambali had used a false Spanish passport to enter Thailand while his wife used her Malaysian passport.

Thai police found him as part of a joint operation between the Thai police and the CIA on August 11, 2003. The twenty uniformed and undercover police smashed down the door to his one bedroom apartment in Ayutthaya, and arrested him and 33-year-old Noralwizah Lee Abdullah, a Chinese Malaysian who was considered to be his wife. Hambali was wearing a pair of jeans, a t-shirt, a baseball cap, and a pair of sunglasses. Police also seized explosives and firearms in the property. It marked the end of a 20-month hunt for Hambali, who was 37 years of age when he was captured.

A 2006 Human Rights Watch report stated that Hambali was imprisoned in Jordan for the account of the CIA. On September 6, 2006, President George W. Bush confirmed that Hambali had been held by the CIA and revealed that he had been transferred to Guantanamo Bay.[11] His wife is now in Malaysian custody.

On September 8, 2006 Indonesia formally requested access to Hambali to ensure a fair trial.[12] The United States are reluctant to hand him over because of the lenient sentence Indonesia imposed on Bashir.

Hambali is also wanted in the Philippines for the transfer of explosives on Filipino soil in an attempt to transport them to Singapore.

Detention

Following his capture the USA would not confirm or deny that he was in their custody. But on September 6, 2006, President Bush acknowledged[13] the existence of covert, overseas CIA interrogation centres (colloquially known as black sites) and announced that 14 high-profile members al Qaeda and related groups had been transported from those sites to Guantanamo Bay.[14] Those 14 include Hambali and an alleged lieutenant of his called Mohammed Nazir Bin Lep alias Lillie or Li-Li.

Combatant Status Review

The Bush administration asserted that:

> *the protections of the Geneva Conventions did not extend to captured prisoners who are not members of the regular Afghan armed force nor meet the criteria for prisoner of war for voluntary forces.*[15]

Critics argued the Conventions obliged the U.S. to conduct competent tribunals to determine the status of prisoners. Subsequently, the U.S. Department of Defense instituted Combatant Status Review Tribunals (CSRTs), to determine whether detainees met the new definition of an "enemy combatant".

"Enemy combatant" was defined by the U.S. Department of Defense as:

> *an individual who was part of, or supporting, the Taliban, or al-Qaeda forces, or associated forces that are engaged in hostilities against the United States or its coalition partners. This includes any person who commits a belligerent act or has directly supported hostilities in aid of enemy armed forces.*[16]

The CSRTs are not bound by the rules of evidence that would normally apply in civilian court, and the government's evidence is presumed to be "genuine and accurate."[17] From July 2004 through March 2005, CSRTs were convened to determine whether each prisoner had been correctly classified as an "enemy combatant".

Riduan Isamuddin was among the 60% of prisoners who chose to participate in tribunal hearings.[18] A Summary of Evidence memo was prepared for the tribunal of each detainee, listing the allegations that *supported their detention as an "enemy combatant"*.

Riduan Isamuddin's memo accused him of the following: [19]

The Department of Defense announced on August 9, 2007 that all fourteen of the "high-value detainees" who had been transferred to Guantanamo from the CIA's black sites, had been officially classified as "enemy combatants".[20] Although judges Peter Brownback and Keith J. Allred had ruled two months earlier that only "*illegal* enemy combatants" could face military commissions, the Department of Defense waived the qualifier and said that all fourteen men could now face charges before Guantanamo military commissions.[21] [22]

Possible transfer to Washington DC for a civilian trial

According to *Xinhua* the Obama Presidency is considering transferring Riduan Isamuddin to Washington DC for a civilian trial.[23]

Further reading

Some of the above information about Hambali, and more, can be read in the massive report[8] of the 9-11 Commission.

References

[1] http://www.globalsecurity.org/security/profiles/hambali.htm

[2] Vicini, James (March 11, 2010). "Guantanamo prisoner known as Hambali seeks release" (http://www.reuters.com/article/2010/03/11/us-guantanamo-hambali-idUSTRE62A5KL20100311). *Reuters*. .

[3] Maria Ressa (August 29, 2002). "The quest for SE Asia's Islamic 'super' state" (http://edition.cnn.com/2002/WORLD/asiapcf/southeast/07/30/seasia.state/). CNN. . Retrieved 2008-03-30. "The main financier of the operation is alleged to have been Riduan Isamuddin, also known as Hambali, and now purported to be al Qaeda's main operative in Southeast Asia."

[4] "'We will fight until we run out of blood'" (http://www.theage.com.au/articles/2002/10/14/1034561097562.html). Melbourne: The Age. October 15, 2002. . Retrieved 2008-03-30. "In Malaysia, Bashir began to surround himself with a hard-core of militants. One of these was Hambali, alias Riduan Isamuddin, a 37-year-old Indonesian who fought against the Soviets. Today, he is described by Western sources as the chief operations officer of JI, and is reputedly the mastermind of al Qaeda cells in this part of the world."

[5] "Detainee Biographies" (http://www.webcitation.org/query?url=http://www.odni.gov/announcements/content/DetaineeBiographies.pdf+&date=2009-08-31) (PDF). Office of the Director of National Intelligence. Archived from the original (http://www.odni.gov/announcements/content/DetaineeBiographies.pdf) on 2009-08-31. .

[6] OARDEC (February 9, 2007). "Summary of Evidence for Combatant Status Review Tribunal - Hambali, Riduan bin Isomuddin" (http://www.defenselink.mil/news/ISN10019.pdf#1). Department of Defense. . Retrieved April 13, 2007.

[7] Scott McClellan (August 14, 2003). "Press Gaggle by Scott McClellan and a Senior Administration Official Aboard Air Force One" (http://georgewbush-whitehouse.archives.gov/news/releases/2003/08/20030814-3.html). White House. . Retrieved April 13, 2007.

[8] "The 9/11 Commission Report" (http://www.9-11commission.gov/report/911Report.pdf) (PDF). 9-11 Commission. . Retrieved April 14, 2007.

[9] "Treasury Designates Four Leaders of Terrorist Group "Jemaah Islamiyah"" (http://www.treas.gov/press/releases/js4179.htm). US Treasury. April 13, 2006. . Retrieved April 13, 2007.

[10] "Hunt for Hambali - the Osama of the Far East" (http://www.smh.com.au/articles/2002/10/16/1034561203790.html). Sydney Morning Herald. 16 October 2002. . Retrieved 2008-03-01.

[11] "Bush admits to CIA secret prisons" (http://news.bbc.co.uk/2/hi/americas/5321606.stm). BBC News. 2006-09-07. . Retrieved 2006-09-08.

[12] "Jakarta seeks access to Hambali" (http://news.bbc.co.uk/2/hi/asia-pacific/5326814.stm). BBC News. 2006-09-08. . Retrieved 2006-09-08.

[13] Bush: CIA holds terror suspects in secret prisons (http://edition.cnn.com/2006/POLITICS/09/06/bush.speech/index.html), CNN, 7 September 2006.

[14] Biographies of 14 Guantanamo Bay detainees (http://www.odni.gov/announcements/content/DetaineeBiographies.pdf), Office of the Director of National Intelligence

[15] "Q&A: What next for Guantanamo prisoners?" (http://news.bbc.co.uk/2/hi/americas/1773140.stm). BBC News. 2002-01-21. . Retrieved 2008-11-24. mirror (http://www.webcitation.org/query?url=http://news.bbc.co.uk/2/hi/americas/1773140.stm&

date=2008-11-24)
[16] "Guatanamo Detainee Processes," (http://www.defense.gov/news/Sep2005/d20050908process.pdf) United States Department of Defense, updated October 2, 2007, accessed February 23, 2010
[17] Elsea, Jennifer K. (July 20, 2005). "Detainees at Guantanamo Bay: Report for Congress" (http://www.fas.org/sgp/crs/natsec/RS22173.pdf) (PDF). Congressional Research Service. . Retrieved 2007-11-10.
[18] OARDEC, Index to Transcripts of Detainee Testimony and Documents Submitted by Detainees at Combatant Status Review Tribunals Held at Guantanamo Between July 2004 and March 2005 (http://www.dod.mil/pubs/foi/detainees/csrt_arb/index_CSRT_detainees_testimony.pdf), September 4, 2007
[19] OARDEC (April 4, 2007). "Combatant Status Review Tribunal, transcript 10017" (http://www.defenselink.mil/news/ISN10019.pdf#1). Department of Defense. . Retrieved April 13, 2007.
[20] Lolita C. Baldur (Thursday, August 9, 2007). "Pentagon: 14 Guantanamo Suspects Are Now Combatants" (http://www.time.com/time/nation/article/0,8599,1651680,00.html). Time magazine. . mirror (http://www.webcitation.org/query?url=http://findarticles.com/p/articles/mi_qn4176/is_20070324/ai_n18757735&date=2008-05-19)
[21] Sergeant Sara Wood (June 4, 2007). "Charges Dismissed Against Canadian at Guantanamo" (http://www.defenselink.mil/news/newsarticle.aspx?id=46281). Department of Defense. . Retrieved 2007-06-07.
[22] Sergeant Sara Wood (June 4, 2007). "Judge Dismisses Charges Against Second Guantanamo Detainee" (http://www.defenselink.mil/news/newsarticle.aspx?id=46288). Department of Defense. . Retrieved 2007-06-07.
[23] "U.S. may bring key Guantanamo detainee to Washington for trials: officials" (http://www.webcitation.org/query?url=http://news.xinhuanet.com/english/2010-01/16/content_12819011.htm&date=2010-02-11). Xinhua. 2010-01-16. Archived from the original (http://news.xinhuanet.com/english/2010-01/16/content_12819011.htm) on 2010-02-11. .

External links

- http://thestar.com.my/services/printerfriendly.asp?file=/2003/8/17/nation/6077032.asp - Info about Hambali's wife
- Hambali's arrest (http://quickstart.clari.net/qs_se/webnews/wed/cu/Qattacks-hambali-thailand.RTS__DaF.html) - Information about Hambali's arrest
- Thailand thanked for Hambali arrest (http://www.news.com.au/common/story_page/0,4057,7625918%5E1702,00.html) - More information about Hambali's arrest
- The Star Source (http://thestar.com.my/services/printerfriendly.asp?file=/2003/9/8/nation/6228302.asp) - Malaysian article about Hambali's early life, including his full DOB
- Time Asia article on Hambali (http://www.time.com/time/asia/features/malay_terror/hambali.html)
- Singapore newspaper (http://newpaper.asia1.com.sg/topstories/0,4133,TopStories-1061567940,00.html) about Hambali and his wife
- http://edition.cnn.com/2002/WORLD/asiapcf/southeast/07/30/seasia.state/
- http://news.bbc.co.uk/2/low/asia-pacific/2983612.stm

Walid_bin_Attash

Walid Muhammad Salih bin Roshayed bin Attash	

Born	1979
Detained at	CIA black sites, Guantanamo
Alternate name	Tawfiq bin Attash[1] [2] [3] alias Khallad[4]
ISN	10014
Charge(s)	Charged before a military commission

Walid Muhammad Salih bin Roshayed bin Attash (Arabic: شاطع نب ديشر نب حلاص دمحم ديلو) (born 1979)[5] is a Yemeni prisoner held in extrajudicial detention at the United States' Guantanamo Bay detention camps.

The Office of the Director of National Intelligence described him as a "scion of a terrorist family".[6]

American prosecutors at the Guantanamo military commissions allege that he helped in the preparation of the 1998 East Africa Embassy bombings and the USS *Cole* bombing[7] and acted as a bodyguard to Osama bin Laden,[8] gaining himself the reputation of an "errand boy".[9] He is formally charged with selecting and helping to train several of the hijackers of the September 11 attacks.[10]

Life

Hailing from a prominent Saudi family on friendly terms with Osama bin Laden,[11] Attash had several brothers fighting during the tumultuous 1990s in Afghanistan.[12] His family was deported from Yemen based on his father's radical views, and he grew up in Saudi Arabia.[13]

He studied at the University of Islamic Studies in Karachi, Pakistan.[8]

Attash lost his right leg in 1997 while fighting against the Northern Alliance and wore a metal prothesis in its place,[9] [12] leading to the nickname "*Father of the Leg*".[7] His brother was killed in the same battle, and his death led Attash to join al-Qaeda.[13]

He was asked to help obtain explosives to target the USS The Sullivans in 1999, as part of the intended 2000 millennium attack plots.[13]

In late 1999 while using the nom de guerre *Khallad*, Attash phoned Khalid al-Mihdhar, informing him up the upcoming Kuala Lumpur al-Qaeda Summit. In January 2000, Attash flew to Malaysia, ostensibly to receive a new prosthetic leg, and attended the summit. On January 8, Malaysian Special Branch informed the CIA that Attash had flown to Bangkok together with al-Mihdhar and Nawaf al-Hazmi. While there, the FBI received a transcript of a phone call from Fahd al-Quso and one of the bombers, which mentioned giving Attash $5000 to purchase a new prosthesis. During later interrogation, al-Quso confessed that he was handing over $36,000, and that it wasn't actually meant to purchase a prosthesis.[9]

In October 2000 Attash was identified as the mastermind behind the USS *Cole* bombing which took place in Aden, Yemen.[14]

On September 11, 2002, his 17-year old brother Hassan bin Attash was taken prisoner by Pakistani forces raiding the Tariq Road House, handed over to the Americans and sent to The Dark Prison.[15]

Capture, tribunal

Attash was captured together with Ali Abdul Aziz Ali in Karachi, on April 29, 2003.[16] [17] [18]

He was sent to The Dark Prison, and his brother was moved to Guantanamo Bay detention camps in 2003 or 2004. While there, he was interrogated under harsh circumstances, and confessed that Abderraouf Jdey had been known to him.[19] Despite having only one leg, he was forced to stand in stress positions, "an acutely difficult technique for him" as the Americans took away his false leg, forcing him to balance awkwardly on one foot until losing his balance and ripping at the tendons in his arms.[20]

He was transferred to Guantanamo on September 6, 2006, together with 13 other "high-level detainees" the CIA had been holding in secret detention.

Combatant Status Review Tribunal

Having been brought to Guantanamo from black sites, the new prisoners were accorded a new series of Combatant Status Review Tribunals, to determine whether the captives met the new definition of an "enemy combatant". They had been instituted in 2004 to mitigate the Supreme Court's findings that the holding of prisoners at Guantanamo Bay was unconstitutional.

A Summary of Evidence memo was prepared for the tribunal, listing the alleged facts that led to his detainment. These included that Mohammad Rashed Daoud al-Owhali had stated that Attash had told him to prepare for a suicide carbombing against East African embassies of the United States a month or two before the attacks occurred. The memo alleged that Attash had trained in close-combat in the Lowgar training camp and seen Osama bin Laden give a speech to graduates of the camp. The memo also alleged that Attash used a Yemeni merchant's registration card that had been forged by "a suspect of the USS *Cole* bombing". An unnamed participant in the *Cole* bombing also confessed to being given a letter written by Attash which asked for his assistance with the bombing, and was the only reason he aided the bombers.

Combatant Status Review Tribunals were held in a 3x5 trailer where the captive sat with his hands and feet shackled to a bolt in the floor.[21] [22] Unlike the 2004 CSR Tribunals the Press was not allowed to attend the 2007 Tribunals.

It also said that authorities knew of an al-Qaeda cell dubbed *"Father of the Leg"* that revolved around a senior member, and believed this was a reference to Attash due to his missing limb.

It also stated that a contact stored in the phone belonging to Attash was also listed as a contact in a notebook belonging to "a senior al Qaida operative", and that his University ID card had been found "at an alleged al Qaida residence" in Karachi. He was also "implicated" by a notebook found during a raid, which listed payments made to various al-Qaeda members. An unnamed source also claimed to have seen him at al Farouq training camp.

Bin Attash attended his Tribunal.[23] A week after the March 12, 2007, tribunal, Attash was reported to have confessed to his role in preparing both the *Cole* and Embassy attacks.[24] He confessed purchasing the explosives and small boat used in the *Cole* bombing, as well as recruiting the perpetrators, and planning the operation 18 months before the actual attack; he stated that he was in Kandahar, Afghanistan with bin Laden at the time of the *Cole* attack, and in Karachi at the time of the simultaneous embassy bombings meeting with the mastermind of the attack. The DoD was later to publish a ten page transcript from the unclassified portion of the Tribunal.

> I was the link between Usama bin Laden and his deputy Sheikh Abu Hafs al-Masri and the cell chief in Nairobi. I was the link that was available in Pakistan. I used to supply the cell with whatever documents they need from fake stamps to visas, whatever.

His Personal Representative met with him on February 13, and told the tribunal that Attash confirmed that many of the allegations were basically correct, but that he had never owned a telephone and that he had forged the Yemeni registration card himself.

Faces charges before military commission

The Department of Defense announced on August 9, 2007 that all fourteen of the "high-value detainees" who had been transferred to Guantanamo from the CIA's black sites, had been officially classified as "enemy combatants".[25] Although judges Peter Brownback and Keith J. Allred had ruled two months earlier that only "*illegal* enemy combatants" could face military commissions, the Department of Defense waived the qualifier and said that all fourteen men could now face charges before Guantanamo military commissions.[26] [27] Bin Attash, Khalid Sheikh Mohammed, Ammar al Baluchi chose to serve as their own attorney.[28] They requested laptops, and internet access, in order to prepare their defenses. In October 2008 Ralph Kohlmann ruled that they be provided with the computers, but not the internet access.

On December 8, 2008, Khalid Sheikh Mohammed told the judge that he and the other four indictees wished to confess and plead guilty; however, the plea would be delayed until after mental competency hearings for Hawsawi and bin al-Shibh. Mohammed said, "We want everyone to plead together."[29]

On May 17, 2010, *Saba News* reports that Walid bin Attash, and four other Yemenis would face charges in the summer of 2010.[30] Two of the other Yemenis *Saba News* reported would face charges were: Ramzi bin al-Shibh and Abd al-Rahim al-Nashiri. *Saba News* did not name the fourth and fifth individuals.

References

[1] Deputy Secretary of Defense Paul Wolfowitz (August 10, 2004). "Prepared Statement for the House Armed Services Committee: As Prepared for Delivery by Deputy Secretary of Defense Paul Wolfowitz, Washington, DC" (http://www.defenselink.mil/Speeches/Speech.aspx?SpeechID=143). Department of Defense. . Retrieved April 15, 2007.

[2] Robert S. Mueller, III (February 24, 2004). "Testimony of Robert S. Mueller, III, Director, FBI: Before the Select Committee on Intelligence of the United States Senate" (http://www.au.af.mil/au/awc/awcgate/fbi/mueller022404.htm). Department of Defense. . Retrieved April 15, 2007.

[3] "National Commission on Terrorist Attacks upon the United States" (http://www.9-11commission.gov/report/911Report_App.htm). 911 Commission. . Retrieved April 15, 2007.

[4] Glaberson, William (July 11, 2008). "Detainees, as Lawyers, Test Tribunal System" (http://www.nytimes.com/2008/07/11/us/11gitmo.html). *New York Times*. . Retrieved September 27, 2011.

[5] Tawfiq bin Attash (http://www.globalsecurity.org/security/profiles/tawfiq_bin_attash.htm)

[6] "Detainee Biographies" (http://www.webcitation.org/query?url=http://www.odni.gov/announcements/content/DetaineeBiographies.pdf+&date=2009-08-31) (PDF). Office of the Director of National Intelligence. Archived from the original (http://www.odni.gov/announcements/content/DetaineeBiographies.pdf) on date=2009-08-31. .

[7] OARDEC (February 8, 2007). "Summary of Evidence for Combatant Status Review Tribunal - Bin Attash, Walid Muhammad Salih" (http://www.defenselink.mil/news/ISN10014.pdf#1). Department of Defense. . Retrieved 2007-04-15.

[8] CSRT Summary of Evidence memo for Walid bin Attash, February 8, 2007

[9] Wright, Lawrence, The Looming Tower, 2006

[10] MSNBC, Pentagon charges 6 in 9-11 attacks (http://www.msnbc.msn.com/id/23108871)

[11] CNN, Arrests amid Karachi terror plot (http://www.cnn.com/2003/WORLD/asiapcf/south/05/03/pakistan.terror/index.html), May 3, 2003

[12] Burger, Timothy J. TIME, Profiling the Terrorists (http://www.time.com/time/nation/article/0,8599,1532315,00.html), September 6, 2006

[13] Bell, Stewart. "The Martyr's Oath", 2005.

[14] Phil Hircshkorn (2006-03-28). "Al Qaeda witnesses saw Moussaoui as a bumbler" (http://www.webcitation.org/query?url=http://www.cnn.com/2006/LAW/03/28/moussaoui/index.html&date=2009-05-08). CNN. Archived from the original (http://www.cnn.com/2006/LAW/03/28/moussaoui/index.html) on 2009-05-08. . "Tawfiq bin Atash, a senior al Qaeda operative considered the mastermind of the USS Cole bombing in 2000, also assisted the 9/11 plot."

[15] OARDEC (9 November 2004). "Summary of Evidence for Combatant Status Review Tribunal -- Bin Attash, Hassan Mohammed Ali (released September 2007)" (http://www.dod.mil/pubs/foi/detainees/csrt_arb/000700-000783.pdf#64). United States Department of Defense. pp. pages 64–65. . Retrieved 2007-12-21.

[16] B Raman (May 21, 2003). "Bomb jitters in Pakistan, too" (http://www.atimes.com/atimes/South_Asia/EE21Df01.html). Asia Times. . Retrieved 2007-07-09.

[17] Shannon, Elaine. *Time*, Al-Qaeda Moneyman Caught (http://www.time.com/time/nation/article/0,8599,448922,00.html), May 1, 2003

[18] Meyer, Josh (March 20, 2007). "Detainee confesses in Cole bombing" (http://www.latimes.com/news/nationworld/nation/la-na-cole20mar20,0,2027317.story?coll=la-home-headlines). *Los Angeles Times*. . Retrieved 2013-01-10.

[19] Intelligence report, interrogation of Khallad (Walid bin Attash), May 21, 2004

[20] Mayer, Jane, "The Dark Side: The Inside Story of How the War on Terror Turned Into a War on American Ideals", 2008. p. 169

[21] Guantánamo Prisoners Getting Their Day, but Hardly in Court (http://www.nytimes.com/2004/11/08/national/08gitmo.html?ex=1257570000&en=4af06725bdf5c086&ei=5088&partner=rssnyt), *New York Times*, November 11, 2004 - mirror (http://cageprisoners.com/articles.php?aid=3838)

[22] Inside the Guantánamo Bay hearings: Barbarian "Justice" dispensed by KGB-style "military tribunals" (http://www.christusrex.org/www1/news/ft-12-11-04a.htm), *Financial Times*, December 11, 2004

[23] OARDEC (2007-03-12). "Verbatim Transcript of Open Session Combatant Status Review Tribunal Hearing for ISN 10014" (http://www.defense.gov/news/transcript_ISN10014.pdf). United States Department of Defense. . Retrieved 2010-04-13.

[24] Liptak, Adam. New York Times, Detainee Said to Confess Role in Cole Bombing (http://www.nytimes.com/2007/03/19/world/20colecnd.html?hp), March 19, 2007

[25] Lolita C. Baldur (Thursday, August 9, 2007). "Pentagon: 14 Guantanamo Suspects Are Now Combatants" (http://www.time.com/time/nation/article/0,8599,1651680,00.html). Time magazine. . mirror (http://www.webcitation.org/query?url=http://findarticles.com/p/articles/mi_qn4176/is_20070324/ai_n18757735&date=2008-05-19)

[26] Sergeant Sara Wood (June 4, 2007). "Charges Dismissed Against Canadian at Guantanamo" (http://www.defenselink.mil/news/newsarticle.aspx?id=46281). Department of Defense. . Retrieved 2007-06-07.

[27] Sergeant Sara Wood (June 4, 2007). "Judge Dismisses Charges Against Second Guantanamo Detainee" (http://www.defenselink.mil/news/newsarticle.aspx?id=46288). Department of Defense. . Retrieved 2007-06-07.

[28] Carol Rosenberg (2008-10-12). "Al Qaeda defendants get laptops at Guantánamo, judge rules" (http://www.miamiherald.com/news/politics/campaign-2008/story/723593.html). Miami Herald. . Retrieved 2008-10-12. "Mohammed, his nephew Ammar al-Baluchi and Walid bin Attash have sought through standby counsel filings at the Military Commission a long list of resources they say they need to mount their defense -- including Internet links to read news accounts and do live research on databases." mirror (http://www.webcitation.org/query?url=http://www.miamiherald.com/news/politics/campaign-2008/story/723593.html&date=2008-10-12)

[29] "Top 9/11 suspects to plead guilty" (http://news.bbc.co.uk/1/hi/world/americas/7770856.stm). BBC News. December 8, 2008. . Retrieved December 8, 2008.

[30] "U.S. to try five Yemeni Gitmo detainees" (http://www.sabanews.net/en/news214477.htm). Saba News. 2010-05-17. . Retrieved 2010-05-17. "The U.S. will start in this summer trying five Yemeni detainees at the U.S. Guantanamo Bay in Cuba including Ramzi bin al-Shibh, Walid bin Attash and Abdul Rahim Al-Nasheri, the September 26 website has reported." mirror (http://www.webcitation.org/query?url=http://www.sabanews.net/en/news214477.htm&date=2010-05-17)

External links

- New Evidence About Prisoners Held in Secret CIA Prisons in Poland and Romania (http://www.andyworthington.co.uk/2010/08/04/new-evidence-about-prisoners-held-in-secret-cia-prisons-in-poland-and-romania/) Andy Worthington

Ahmed_Ghailani

Ahmed Khalfan Ghailani

Ahmed Khalfan Ghailani
(FBI photo)

Born	c. 1974 (age 38–39)Zanzibar, Tanzania
Arrested	July 25th, 2004Gujrat, Punjab, PakistanInter-Services Intelligence
Citizenship	Tanzanian
Detained at	ADX Supermax
Alleged to be a member of	conspirator of the al-Qaeda
Penalty	life in prison without parole **charges** Conspiracy to damage or destroy U.S. property with Explosives.
Status	Held at the ADX Supermax prison in Florence, Colorado.

Ahmed Khalfan Ghailani (Arabic: أحمد خلفان الغيلاني, *Aḥmad Khalifān al-Ghaīlānī*) is a conspirator of the al-Qaeda terrorist organization convicted for his role in the bombing of embassies in Kenya and Tanzania.[1] [2] He was indicted[3] [4] in the United States as a participant in the 1998 U.S. embassy bombings. He was on the FBI Most Wanted Terrorists list from its inception in October 2001. In 2004, he was captured and detained by Pakistani forces in a joint operation with the United States, and was held until June 9, 2009, in the Guantanamo Bay detention camp;[5] one of 14 Guantanamo detainees who had previously been held at secret locations abroad.[6] According to *The Washington Post*, Ghailani told military officers he is contrite and claimed to be an exploited victim of al-Qaeda operatives.[7]

Ghailani was transported from Guantanamo Bay to New York City to await trial in the United States District Court for the Southern District of New York in June 2009.[8] When the case came to trial, the judge disallowed the testimony of a key witness. On November 17, 2010, a jury found him guilty of one count of conspiracy, but acquitted him of 284 other charges including all murder counts.[9] Critics of the Obama administration said the verdict proves civilian courts cannot be trusted to prosecute terrorists because it shows a jury might acquit such a defendant entirely.[9] Supporters of the trial have said that the conviction and the stiff sentencing prove that the federal justice system works.[10]

On Tuesday, January 25, 2011, U.S. District Judge Lewis A. Kaplan, the presiding judge in the case, sentenced Ahmed Ghailani, 36, to life in prison for the bombing,[1] [2] stating that any sufferings Ghailani experienced at the hands of the CIA or other agencies while in custody at Guantanamo Bay pales in comparison to the monumental tragedy of the bombings of the U.S. embassies in Kenya and Tanzania in 1998, which killed 224 people, including 12 Americans, and left thousands injured or otherwise impacted by the crimes. The attacks were one of the deadliest non-wartime incidents of international terrorism to affect the United States; they were on a scale not surpassed until the September 11 attacks three years later. Ghailani, who had said he was never involved and did not intend to kill anyone, had been portrayed as cooperating with investigators- yielding information wanted by investigators- and as remorseful by his defense counsel, but that argument of relative non-involvement or remorse was not accepted. He is the fifth person to be sentenced. Four others were sentenced to life in prison in a 2001 trial in Manhattan federal court. Osama bin Laden was also named in the indictment until he was killed in Pakistan in May 2011.

Identity

Ghailani has used a variety of different aliases including Ahmad Khalafan Ghilani, Ahmed Khalfan Ahmed, Abubakar K. Ahmed, Abubakary K. Ahmed, Abubakar Ahmed, Abu Bakr Ahmad, A. Ahmed, Ahmed Khalfan, Ahmed Khalfan Ali, Abubakar Khalfan Ahmed, Ahmed Ghailani, Ahmad Al Tanzani, Abu Khabar, Abu Bakr, Abubakary Khalfan Ahmed Ghailani, Mahafudh Abubakar Ahmed Abdallah Hussein, Shariff Omar Mohammed, "Foopie", "Fupi", and "Ahmed the Tanzanian."

Early life

Ghailani was born around 1974 in Zanzibar, Tanzania[5] (possibly on March 14, April 13, or April 14 of that year, or on 1 August 1970) and is a Tanzanian citizen. He speaks Swahili and had served as a tabligh, a Muslim traveling preacher, and probably visited Pakistan in this capacity.

1998 U.S. embassy bombings

After joining al Qaida, he became an explosives expert and was assigned to obtain the bomb components in Dar es Salaam according to convicted fellow Embassy bombing conspirators Mohammed Sadiq Odeh and Khalfan Khamis Mohamed. This role was complicated by the fact that Ghailani could not drive so whatever purchases were too large or heavy for his bicycle such as oxygen and acetylene tanks would have to be picked up by another person in a car. Ghailani was in Nairobi by August 6, 1998, where he is thought to have rented a room at the Hilltop Hotel used for meetings by the bombers and flew to Karachi on a Kenya Airways flight before the bombs exploded.

At some time in Pakistan or Afghanistan, he married an Uzbek and had children.[11] Many Uzbek Islamists had moved into Pakistan and the woman is thought to be from that group.

Wanted and arrest for terrorist activities

On May 26, 2004, United States Attorney General John Ashcroft and FBI Director Robert Mueller announced that reports indicated that Ghailani was one of seven al-Qaeda members who were planning a terrorist action for the summer or fall of 2004. The other alleged terrorists named on that date were Fazul Abdullah Mohammed, who had also been earlier listed with Ghailani by the FBI as a Most Wanted Terrorist for the 1998 embassy attack, and Abderraouf Jdey, Amer El-Maati, Aafia Siddiqui, Adam Yahiye Gadahn, and Adnan G. El Shukrijumah. *Jdey* was already on the FBI Seeking Information – War on Terrorism list since January 17, 2002, to which the other four were added as well.[12]

American Democrats labeled the warning "suspicious". Dismissing the threat, they said it was held solely to divert attention from President Bush's plummeting poll numbers and to push the failings of the Invasion of Iraq off the front page.[13] CSIS director Reid Morden voiced similar concerns, saying it seemed more like "election year" politics, than an actual threat—and *The New York Times* pointed out that one day before the announcement, they had been told by the Department of Homeland Security that there were no current risks.[13]

On July 25, 2004, a nearly eight hour battle ensued in the town of Gujrat in central Pakistan. Ghailani and thirteen others, included his wife and children, were arrested. A police officer was wounded in the battle. Pakistani Interior Minister Makhdoom Faisal Saleh Hayat announced the capture of Ghailani on July 29, 2004.[11] The US Government had offered a $5,000,000 USD bounty offered for information leading to the arrest of Ghailani.[14]

Some press reports (including the *New Republic*[15]) questioned whether the timing of the announcement of Ghailani's capture was politically motivated at the behest of the Bush administration. The announcement was made just hours before U.S. Presidential candidate John Kerry was due to make his acceptance speech at the 2004 Democratic National Convention, an event at which a candidate usually receives a significant boost in the polls. Hayyat made the announcement after midnight local time, despite having apparently known Ghailiani's identity for some days beforehand. Pakistani officials denied there was any such motivation.

Soon after the capture of Ghailani and the others with him, *The Boston Globe*, quoting a United Nations source, said that Ghailani was one of several al-Qaeda personnel who had been in Liberia around 2001, handling conflict diamonds under the protection of then-dictator Charles Taylor.[16] Ghailani is said to have spent more than three years in Liberia.

Combatant Status Review

Ghailani was among the 60% of prisoners who participated in the tribunal hearings.[17] A Summary of Evidence memo was prepared for the tribunal of each detainee.

Ghailani's memo accused him of the following:[18] [19]

The Department of Defense announced on August 9, 2007 that all fourteen of the "high-value detainees" who had been transferred to Guantanamo from the CIA's black sites, had been officially classified as "enemy combatants".[20] Although judges Peter Brownback and Keith J. Allred had ruled two months earlier that only "*illegal* enemy combatants" could face military commissions, the Department of Defense waived the qualifier and said that all fourteen men could now face charges before Guantanamo military commissions.[21] [22]

Charged before a military commission

The Bush Presidency planned to hold up to 80 of the new Congressionally authorized Military Commissions in a $12 million tent city.

The Al-Qaeda suspect alleged to have been involved in the 1998 United States embassy bombings that killed 223 people and injured approximately 4,085 faced nine war crimes charges, six of them offenses that could have carried the death penalty, if he was convicted by a military tribunal, it was reported on March 31, 2008.[23] Scott L. Fenstermaker and David Remes were in a rare dispute as to who was authorized to assist Ghaliani.[24]

In June 2009, Ghailani was transferred to New York to face trial in a federal court.[25] The Department of Justice, under U.S. Attorney General Eric Holder, directed the U.S. Attorney for the Southern District of New York, Preet Bharara, to not seek the death penalty in an October 2009 memorandum.

Transfer to the USA

On August 31, 2009, *Corrections One*, a trade journal for the prison industry, speculated that "Ahmed Khalfan Ghailani" was one of ten captives they speculated might be moved to a maximum security prison in Standish, Michigan.[26]

Metropolitan Correctional Center, New York City, where Ghailani was formerly located

Instead Ghailani was transferred to New York City to stand trial in a civilian court there.

Ghailani learned that being transferred from military to civilian jurisdiction meant that he could no longer be assisted by Colonel Jeffrey Colwell and Major Richard Reiter.[27] [28] [29] [30]

On February 10, 2010, United States district court Judge Lewis A. Kaplan ordered the Prosecution to review the record of Ghailani's detention in CIA's network of black sites.[31] According to *The New York Times* any materials that showed the decisions "were for a purpose other than national security" had to be turned over to Ghailani's lawyers. It was reported that Kaplan was considering dismissing the charges on the grounds that due to Ghailani's long extrajudicial detention he had been denied the constitutional right to a speedy trial.

ADX Supermax where Ghailani is now located

On April 23, 2010, a 52 page unclassified summary of Ghailani's 2007 Guantanamo interrogations was published in preparation for his trial.[32] Benjamin Weiser, writing in *The New York Times* reported that the summary, published during Ahmed Khalfan Ghailani's civilian trial, revealed new details about Ghailani's life as an Osama bin Laden bodyguard. According to Weiser the interrogation summary asserted that during the year he was a bodyguard Ghailani met several other individuals who were among those who later became 9-11 hijackers. Following his work as a bodyguard the summary asserts Ghailani became a forger, where he became *"very good with Photoshop"*.

Ghailani's trial commenced October 4, 2010, in the Daniel Patrick Moynihan Federal Court Building in lower Manhattan, in front of U.S. District Court Judge Lewis A. Kaplan. There were no protests or demonstrations during the trial, as observed by Human Rights First.[33] [34] On Wednesday, October 6, 2010, in a short ruling that the judge said he would expand upon later that day, it was determined that a key witness, the Tanzanian Hussein Abebe, who may have issued statements crucial to implicating Ghailani during the time he was under CIA custody, would not be testifying in the trial. Judge Kaplan agreed to delay the start of the trial until the following Tuesday, October 12, 2010, pending a possible appeal of that ruling. On October 11, 2010, the government announced it would not appeal Judge Kaplan's ruling. Steve Zissou, one of Ghailani's lawyers, commented that the government's decision not to appeal was "a significant victory for the Constitution."[35] On November 17, 2010, Ghailani was convicted for conspiracy,[36] but was acquitted of all other charges.[36] On January 25, 2011, Ghailani was sentenced to life in prison.[37]

References

[1] Richey, Warren (January 25, 2011). "Ahmed Ghailani gets life sentence for Al Qaeda bombing of US embassies" (http://www.csmonitor.com/USA/Justice/2011/0125/Ahmed-Ghailani-gets-life-sentence-for-Al-Qaeda-bombing-of-US-embassies). *The Christian Science Monitor*. Archived (http://web.archive.org/web/20110131011421/http://www.csmonitor.com/USA/Justice/2011/0125/Ahmed-Ghailani-gets-life-sentence-for-Al-Qaeda-bombing-of-US-embassies) from the original on 31 January 2011. . Retrieved January 26, 2011.

[2] Hays, Tom (January 25, 2011). "Gitmo Detainee Gets Life Sentence in Embassy Plot" (http://abcnews.go.com/Blotter/wireStory?id=12753950). The Associated Press. . Retrieved January 26, 2011.

[3] Copy of indictment (http://cns.miis.edu/pubs/reports/pdfs/binladen/indict.pdf) USA v. Usama bin Laden et al., Center for Nonproliferation Studies, Monterey Institute of International Studies

[4] "Superseding Indictment (U.S v. bin Laden, *et al.*)" (http://news.findlaw.com/hdocs/docs/terrorism/us-ahmed-ghailani2001sind.html). FindLaw. 2001-03. . Retrieved 2009-06-09.

[5] "Detainee Biographies" (http://www.webcitation.org/query?url=http://www.odni.gov/announcements/content/DetaineeBiographies.pdf+&date=2009-08-31) (PDF). Office of the Director of National Intelligence. Archived from the original (http://www.odni.gov/announcements/content/DetaineeBiographies.pdf) on date=2009-08-31. .

[6] Bush: CIA holds terror suspects in secret prisons (http://edition.cnn.com/2006/POLITICS/09/06/bush.speech/index.html), CNN, 7 September 2006

[7] Peter Finn (2009-02-16). "4 Cases Illustrate Guantanamo Quandaries: Administration Must Decide Fate of Often-Flawed Proceedings, Often-Dangerous Prisoners" (http://www.washingtonpost.com/wp-dyn/content/article/2009/02/15/AR2009021501955_pf.html). Washington Post. p. A01. . Retrieved 2009-02-11.

[8] Finn, Peter (10 June 2009). "Guantanamo Bay Detainee Brought to U.S. for Trial" (http://www.washingtonpost.com/wp-dyn/content/article/2009/06/09/AR2009060900401.html). *Washington Post* (Washington Post). . Retrieved 2009-07-06.

[9] Charlie Savage (November 18, 2010). "Terror Verdict Tests Obama's Strategy on Civilian Trials" (http://www.nytimes.com/2010/11/19/nyregion/19detainees.html?_r=1). *The New York Times*. .

[10] Daphne Eviatar (November 18, 2010). "Ghailani Verdict Makes the Case for Federal Courts" (http://blog.humanrightsfirst.org/2010/11/ghailani-verdict-makes-case-for-federal.html). .

[11] Key al-Qaeda suspect arrested (http://news.bbc.co.uk/1/hi/world/south_asia/3938133.stm), *BBC*, July 30, 2004

[12] Transcript: Ashcroft, Mueller news conference (http://www.cnn.com/2004/US/05/26/terror.threat.transcript/), CNN.com, Wednesday, May 26, 2004 Posted: 8:19 PM EDT (0019 GMT)

[13] Pither, Kerry. "Dark Days: The Story of Four Canadians Tortured in the Name of Fighting Terror", 2008.

[14] Rewards for Justice (http://www.rewardsforjustice.net/english/wanted_captured/index.cfm?page=Ghailani)

[15] PAKISTAN FOR BUSH: July Surprise? (https://ssl.tnr.com/p/docsub.mhtml?i=20040719&s=aaj071904), *The New Republic*, July 29, 2004

[16] Liberia's Taylor gave aid to Qaeda, UN probe finds (http://www.boston.com/news/nation/washington/articles/2004/08/04/liberias_taylor_gave_aid_to_qaeda_un_probe_finds/), *Boston Globe*, 4 August 2004

[17] OARDEC, Index to Transcripts of Detainee Testimony and Documents Submitted by Detainees at Combatant Status Review Tribunals Held at Guantanamo Between July 2004 and March 2005 (http://www.dod.mil/pubs/foi/detainees/csrt_arb/index_CSRT_detainees_testimony.pdf), September 4, 2007

[18] OARDEC (2007-03-17). "Verbatim Transcript of Open Session CSRT Hearing for ISN 10012" (http://www.defenselink.mil/news/transcript_ISN10012.pdf). United States Department of Defense. . Retrieved 2004-10-20.

[19] Lolita C. Baldor (March 24, 2007). "Guantanamo detainee denies knowledge of bombing" (http://findarticles.com/p/articles/mi_qn4176/is_20070324/ai_n18757735). Oakland Tribune. . Retrieved 2008-05-19.

[20] Lolita C. Baldur (Thursday, August 9, 2007). "Pentagon: 14 Guantanamo Suspects Are Now Combatants" (http://www.time.com/time/nation/article/0,8599,1651680,00.html). Time magazine. . mirror (http://www.webcitation.org/query?url=http://findarticles.com/p/articles/mi_qn4176/is_20070324/ai_n18757735&date=2008-05-19)

[21] Sergeant Sara Wood (June 4, 2007). "Charges Dismissed Against Canadian at Guantanamo" (http://www.defenselink.mil/news/newsarticle.aspx?id=46281). Department of Defense. . Retrieved 2007-06-07.

[22] Sergeant Sara Wood (June 4, 2007). "Judge Dismisses Charges Against Second Guantanamo Detainee" (http://www.defenselink.mil/news/newsarticle.aspx?id=46288). Department of Defense. . Retrieved 2007-06-07.

[23] "Tanzania bombing suspect charged with war crimes" (http://edition.cnn.com/2008/US/03/31/embassy.bomb.suspect/index.html). CNN. 2008-03-31. . Retrieved 2008-03-31.

[24] Daphne Eviatar (2008-05-29). "Covington & Burling partner takes on defense of Guantanamo death penalty case" (http://amlawdaily.typepad.com/amlawdaily/2008/05/covington-burli.html). AM Law Daily. . Retrieved 2008-12-10.

[25] "Guantanamo detainee arrives in NY" (http://news.bbc.co.uk/2/hi/americas/8091013.stm). *BBC Online*. BBC. 9 June 2009. Archived (http://web.archive.org/web/20090610011625/http://news.bbc.co.uk/2/hi/americas/8091013.stm) from the original on 10 June 2009. . Retrieved 2009-06-09.

[26] Kathryn Lynch-Morin (2009-08-31). "Profile of 10 U.S.-bound Gitmo detainees" (http://www.webcitation.org/query?url=http://www.correctionsone.com/news/1879631-Profile-of-10-U-S-bound-Gitmo-detainees/&date=2009-09-01). Corrections One. Archived from the

original (http://www.correctionsone.com/news/1879631-Profile-of-10-U-S-bound-Gitmo-detainees/) on 2009-09-01. . Retrieved 2009-08-02.
[27] Chad Jones (2009-11-23). "Guantanamo Detainee Can't Keep Military Lawyers" (http://www.webcitation.org/query?url=http://online.wsj.com/article/BT-CO-20091118-712744.html&date=2009-11-23). Wall Street Journal. Archived from the original (http://online.wsj.com/article/BT-CO-20091118-712744.html) on 2009-11-23. .
[28] Larry Neumeister (2009-11-23). "Military can reassign Gitmo detainee's lawyers" (http://www.webcitation.org/query?url=http://www.google.com/hostednews/ap/article/ALeqM5jYEZX8290oAg3BJaHnv9M4E8C09AD9C24DLO0&date=2009-11-23). Associated Press. Archived from the original (http://www.google.com/hostednews/ap/article/ALeqM5jYEZX8290oAg3BJaHnv9M4E8C09AD9C24DLO0) on 2009-11-23. .
[29] Christine Kearney (2009-11-23). "Guantanamo suspect denied military lawyers in N.Y." (http://www.webcitation.org/query?url=http://www.reuters.com/article/domesticNews/idUSTRE5AH4HO20091118&date=2009-11-23). Reuters. Archived from the original (http://www.reuters.com/article/domesticNews/idUSTRE5AH4HO20091118) on 2009-11-23. .
[30] Benjamin Weiser (2009-11-23). "Bomb Suspect Can't Keep His Military Lawyers" (http://www.webcitation.org/query?url=http://www.nytimes.com/2009/11/19/nyregion/19lawyers.html&date=2009-11-23). New York Times. Archived from the original (http://www.nytimes.com/2009/11/19/nyregion/19lawyers.html) on 2009-11-23. .
[31] Benjamin Weiser (2010-02-10). "U.S. Told to Review Files on Terror Case Detention" (http://www.nytimes.com/2010/02/11/nyregion/11ghailani.html?pagewanted=print). New York Times. . Retrieved 2010-02-11.
[32] Benjami Weiser (2010-04-23). "Court Filing Sheds Light on Bin Laden Bodyguard" (http://www.nytimes.com/2010/04/24/nyregion/24terror.html?pagewanted=print). New York Times. . Retrieved 2010-04-24. "The exchange, straightforward and as banal as any job interview might be, is recounted in a declassified 52-page summary of an interrogation of Mr. Ghailani by federal agents, which was made public Friday night in a filing in Federal District Court in Manhattan. The filing is part of the government's response to Mr. Ghailani's demand that his indictment be dismissed because of "outrageous government conduct." He was detained for nearly five years in secret C.I.A. prisons and later at the naval base at Guantánamo Bay, Cuba." mirror (http://www.webcitation.org/query?url=http://www.nytimes.com/2010/04/24/nyregion/24terror.html?pagewanted=print&date=2010-04-24)
[33] Daphne Eviatar (October 1, 2010). *First Guantanamo Trial in New York City: So Not Scary* (http://blog.humanrightsfirst.org/search/label/ahmed ghailani). .
[34] Human Rights First (September 30, 2010). "Gitmo Trial Hits NYC; Manhattan Yawns" (http://www.youtube.com/watch?v=6I-W3JXAHBU). .
[35] http://www.chron.com/disp/story.mpl/ap/top/all/7241009.html
[36] http://emptywheel.firedoglake.com/2010/11/17/ahmed-ghailani-guilty-of-conspiracy-but-not-terrorism/
[37] http://www.csmonitor.com/USA/Justice/2011/0125/Ahmed-Ghailani-gets-life-sentence-for-Al-Qaeda-bombing-of-US-embassies

External links

- Guantanamo man loses torture bid to avoid U.S. trial (http://www.reuters.com/article/idUSTRE6496MO20100510?type=domesticNews&feedType=RSS&feedName=domesticNews) Reuters May 10, 2010
- Superseding Indictment (http://news.findlaw.com/hdocs/docs/terrorism/us-ahmed-ghailani2001sind.html) (U.S v. bin Laden, *et al.*), FindLaw
- Lawyer: Feds Chose Torture Over Trial for Detainee (http://www.nytimes.com/2010/01/12/nyregion/12ghailani.html) January 11, 2010
- Call to throw out Guantanamo case (http://english.aljazeera.net/news/americas/2010/01/2010112151587 0754.html) January 12, 2010
- Chronology Amnesty International (http://www.amnesty.ca/english/main_article_home/elmaatichronology.pdf)
- Jury Appears Deadlocked in Civilian Trial (http://www.democracynow.org/2010/11/17/jury_appears_deadlocked_in_landmark_civilian) - video report by *Democracy Now!*

American_Airlines_Flight_77

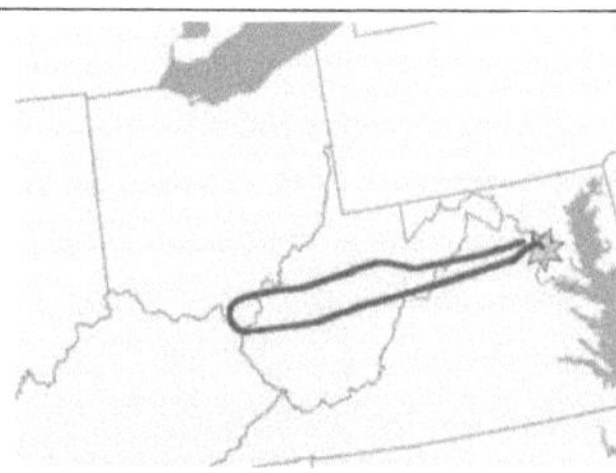

The flight path of AA 77 from Washington Dulles International to Arlington, Virginia

Hijacking summary	
Date	Tuesday, September 11, 2001
Type	Suicide hijacking
Site	The west wall of Pentagon, Arlington, Virginia
Passengers	53, plus 5 hijackers
Crew	6
Fatalities	59, plus 5 hijackers, with an additional 125 in the Pentagon
Survivors	0
Aircraft type	Boeing 757-223
Operator	American Airlines
Registration	N644AA [1]
Flight origin	Washington Dulles Int'l Airport
Destination	Los Angeles Int'l Airport

American Airlines Flight 77 was a passenger flight which was hijacked by five al-Qaeda terrorists on September 11, 2001, as part of the September 11 attacks. They deliberately crashed it into the Pentagon near Washington, D.C., killing all 59 people on board plus the hijackers, as well as 125 people in the building. The Boeing 757-223 aircraft was flying American Airlines' daily scheduled morning transcontinental service from Washington Dulles International Airport, in Dulles, Virginia to Los Angeles International Airport in Los Angeles, California.

Less than 35 minutes into the flight, the hijackers stormed the cockpit. They forced the passengers, crew, and pilots to the rear of the aircraft. Hani Hanjour, one of the hijackers who was trained as a pilot, assumed control of the flight. Unknown to the hijackers, passengers aboard made telephone calls to loved ones and relayed information on the hijacking.

The hijackers crashed the aircraft into the western side of the Pentagon at 09:37 EDT. Dozens of people witnessed the crash, and news sources began reporting on the incident within minutes. The impact severely damaged an area of the Pentagon and caused a large fire. At 10:10AM, a portion of the Pentagon collapsed; firefighters spent days trying to fully extinguish the blaze. The damaged sections of the Pentagon were rebuilt in 2002, with occupants moving back into the completed areas on August 15, 2002.

The 184 victims of the attack are memorialized in the Pentagon Memorial adjacent to the Pentagon. The 1.93-acre (m^2) park contains a bench for each of the victims, arranged according to their year of birth, ranging from 1930 (age 71) to 1998 (age 3).

Hijackers

The hijackers on American Airlines Flight 77 were led by Hani Hanjour, who piloted the aircraft into the Pentagon.[2] Hanjour first came to the United States in 1990.[3]

Hanjour trained at the CRM Airline Training Center in Scottsdale, Arizona, earning his FAA commercial pilot's certificate in April 1999.[4] He had wanted to be a commercial pilot for the Saudi national airline but was rejected when he applied to the civil aviation school in Jeddah in 1999. Hanjour's brother later explained that, frustrated at not finding a job, Hanjour "increasingly turned his attention toward religious texts and cassette tapes of militant Islamic preachers".[5] Hanjour returned to Saudi Arabia after being certified as a pilot, but left again in late 1999, telling his family that he was going to the United Arab Emirates to work for an airline.[6] Hanjour likely went to Afghanistan, where Al Qaeda recruits were screened for special skills they may have. Already having selected the Hamburg Cell members, Al Qaeda leaders selected Hanjour to lead the fourth team of hijackers.[7]

Alec Station, the CIA's unit dedicated to tracking Osama bin Laden, had discovered that two of the other hijackers, al-Hazmi and al-Mihdhar, had multiple entry visas to the United States well before 9/11. Two FBI agents inside the unit tried to alert FBI headquarters, but CIA officers rebuffed them.[8]

In December 2000, Hanjour arrived in San Diego, joining "muscle" hijackers Nawaf al-Hazmi and Khalid al-Mihdhar, who had been there since January 2000.[6] [9] Soon after arriving, Hanjour and Hazmi left for Mesa, Arizona, where Hanjour began refresher training at Arizona Aviation.[6]

In April 2001, they relocated to Falls Church, Virginia, where they awaited the arrival of the remaining "muscle" hijackers.[6] One of these men, Majed Moqed, arrived on May 2, 2001, with Ahmed al-Ghamdi (who would hijack Flight 175) from Dubai at Dulles International Airport. They moved into an apartment with Hazmi and Hanjour.[10] While living in Falls Church, Hazmi attended the mosque in the community.

On May 21, 2001, Hanjour rented a room in Paterson, New Jersey, where he stayed with other hijackers through the end of August.[11] The last Flight 77 "muscle" hijacker, Salem al-Hazmi, arrived on June 29, 2001, with Abdulaziz al-Omari (a hijacker of Flight 11) at John F. Kennedy International Airport from the United Arab Emirates. They stayed with Hanjour.[10]

Hanjour received ground instruction and did practice flights at Air Fleet Training Systems in Teterboro, New Jersey, and at Caldwell Flight Academy in Fairfield, New Jersey.[6] Hanjour moved out of the room in Paterson and arrived at the Valencia Motel in Laurel, Maryland, on September 2, 2001.[11] While in Maryland, Hanjour and fellow hijackers trained at Gold's Gym in Greenbelt.[12] On September 10, he completed a certification flight, using a terrain recognition system for navigation, at Congressional Air Charters in Gaithersburg, Maryland.[13] [14]

On September 10, Nawaf al-Hazmi, accompanied by other hijackers, checked into the Marriott in Herndon, Virginia near Dulles Airport.[15]

Suspected accomplices

According to a U.S. State Department cable leaked in the Wikileaks dump in February 2010, the FBI has investigated another suspect, Mohammed al-Mansoori. He had associated with three Qatari citizens who flew from Los Angelese to London (via Washington) and Qatar on the eve of the attacks, after allegedly surveying the World Trade Center and the White House. U.S. law enforcement officials said that the data about the four men was "just one of many leads that were thoroughly investigated at the time and never led to terrorism charges".[16] An official added that the three Qatari citizens have never been questioned by the FBI. Eleanor Hill, the former staff director for the congressional joint inquiry on the September 11 attacks, said the cable reinforces questions about the thoroughness of the FBI's investigation. She also said that the inquiry concluded that the hijackers had a support network that helped them in different ways.[16]

The three Qatari men were booked to fly from Los Angeles to Washington on September 10, 2001, on the same plane that was hijacked and piloted into the Pentagon on the following day. Instead, they flew from from Los

Angeles to Qatar, via Washington and London. While the cable said that Mansoori was currently under investigation, U.S. law enforcement officials said that there was no active investigation of him or of the Qatari citizens mentioned in the cable.[16]

Flight

The American Airlines Flight 77 aircraft was a Boeing 757-223 (registration number N644AA).[17] The flight crew included pilot Charles Burlingame, First Officer David Charlebois, and flight attendants Michele Heidenberger, Jennifer Lewis, Kenneth Lewis, and Renee May.[18] The capacity of the aircraft was 188 passengers, but with 58 passengers on September 11, the load factor was 33 percent. American Airlines said that Tuesdays were the least-traveled day of the week, with the same load factor seen on Tuesdays in the previous three months for Flight 77.[19]

An American Airlines Boeing 757, similar to N644AA, the aircraft used for Flight 77.

Boarding and departure

On the morning of September 11, 2001, the five hijackers arrived at Washington, D.C.'s Dulles International Airport. At 07:15, Khalid al-Mihdhar and Majed Moqed checked in at the American Airlines ticket counter for Flight 77,[20] arriving at the passenger security checkpoint a few minutes later at 07:18.[21] Both men set off the metal detector and were put through secondary screening. Moqed continued to set off the alarm, so he was searched with a hand wand.[22] The Hazmi brothers checked in together at the ticket counter at 07:29. Hani Hanjour checked in separately and arrived at the passenger security checkpoint at 07:35.[14] Hanjour was followed minutes later at the checkpoint by Salem and Nawaf al-Hazmi, who also set off the metal detector's alarm. The screener at the checkpoint never resolved what set off the alarm. As seen in security footage later released, Nawaf Hazmi appeared to have an unidentified item in his back pocket. Utility knives up to four inches were permitted at the time by the Federal Aviation Administration (FAA) as carry-on items.[20] [22] The passenger security checkpoint at Dulles International Airport was operated by Argenbright Security, under contract with United Airlines.[23]

The hijackers were all selected for extra screening of their checked bags. Hanjour, al-Mihdhar, and Moqed were chosen by the Computer Assisted Passenger Prescreening System criteria, while the brothers Nawaf and Salem al-Hazmi were selected because they did not provide adequate identification and were deemed suspicious by the airline check-in agent. Hanjour, Mihdhar, and Nawaf al-Hazmi did not check any bags for the flight. Checked bags belonging to Moqed and Salem al-Hazmi were held until they boarded the aircraft.[19] By 07:50, the five hijackers, carrying knives and box cutters, had made it through the airport security checkpoint.

Flight 77 was scheduled to depart for Los Angeles at 08:10; 58 passengers boarded through Gate D26, including the five hijackers. Excluding the hijackers, of the 59 other passengers and crew on board, there were 26 men, 22 women, and five children ranging in age from three to eleven. On the flight, Hani Hanjour was seated up front in 1B, while Salem and Nawaf al-Hazmi were seated in first class in seats 5E and 5F. Majed Moqed and Khalid al-Mihdhar were seated further back in 12A and 12B, in economy class.[24] Flight 77 left the gate on time and took off from Runway 30 at Dulles at 8:20 am.[19]

Hijacking

The 9/11 Commission estimated that the flight was hijacked between 08:51 and 08:54, shortly after American Airlines Flight 11 struck the World Trade Center and not too long after United Airlines Flight 175 had been hijacked. The last normal radio communications from the aircraft to air traffic control occurred at 08:50:51.[25] Unlike the other three flights, there were no reports of anyone being stabbed or a bomb threat. At 08:54, the plane began to deviate from its normal, assigned flight path and turned south.[20] The hijackers set the flight's autopilot heading for Washington, D.C.[26] By 08:56, the flight was turned around, and the transponder had been disabled.

The FAA was aware at this point that there was an emergency on board the airplane. By this time, American Airlines Flight 11 had already crashed into the North Tower of the World Trade Center, and United Airlines Flight 175 was known to have been hijacked and was within minutes of striking the South Tower. After learning of this second hijacking involving an American Airlines aircraft and the hijacking involving United Airlines, American Airlines' Executive Vice President Gerard Arpey ordered a nationwide ground stop for the airline.[20] The Indianapolis Air Traffic Control Center, as well as American Airlines dispatchers, made several failed attempts to contact the aircraft. At the time the airplane was hijacked, it was flying over an area of limited radar coverage.[27] With air controllers unable to contact the flight by radio, an Indianapolis official declared that the Boeing 757 had possibly crashed at 09:09.[27]

Two people on the aircraft made phone calls to contacts on the ground. At 09:12, flight attendant Renee May called her mother, Nancy May, in Las Vegas.[24] During the call, which lasted nearly two minutes, May said her flight was being hijacked by six persons, and staff and passengers had been moved to the rear of the airplane.[20] [24] May asked her mother to contact American Airlines, which she and her husband promptly did;[20] American Airlines was already aware of the hijacking. Between 09:16 and 09:26, passenger Barbara Olson called her husband, United States Solicitor General Theodore Olson, and reported that the airplane had been hijacked and that the assailants had box cutters and knives.[20] [28] She reported that the passengers, and possibly the crew, had been moved to the back of the cabin and that the hijackers were unaware of her call. A minute into the conversation, the call was cut off.[19] Theodore Olson contacted the command center at the Department of Justice, and tried unsuccessfully to contact Attorney General John Ashcroft.[20] About five minutes later, Barbara Olson called again, told her husband that the "pilot" (possibly Hanjour on the cabin intercom) had announced the flight was hijacked, and asked "what do I tell the pilot to do?"[29] Ted Olson asked her location and she reported the plane was flying over a residential area.[30] He told her of the attacks on the World Trade Center. Soon afterward, the call cut off again.[20]

"The speed, the maneuverability, the way that he turned, we all thought in the radar room, all of us experienced air traffic controllers, that that was a military plane. You don't fly a 757 in that manner. It's unsafe."
Danielle O'Brien, Air traffic controller at Dulles International Airport[31]

An airplane was detected again by Dulles controllers on radar screens as it approached Washington, turning and descending rapidly. Controllers initially thought this was a military fighter, due to its high speed and maneuvering.[32] Reagan Airport controllers asked a passing Air National Guard Lockheed C-130 Hercules to identify and follow the aircraft. The pilot, Lt. Col. Steven O'Brien, told them it was a Boeing 757 or 767, and its silver fuselage meant it was probably an American Airlines jet. He had difficulty picking out the airplane in the "East Coast haze", but then saw a "huge" fireball, and initially assumed it had hit the ground. Approaching the Pentagon, he saw the impact site on the building's west side and reported to Reagan control, "Looks like that aircraft crashed into the Pentagon, sir".[20] [33]

Crash

Security camera footage of Flight 77 hitting the Pentagon. Impact is at 1:27.[34]

According to the 9/11 Commission Report, as Flight 77 was 5 miles (km) west-southwest of the Pentagon, it made a 330-degree turn. At the end of the turn, it was descending through 2200 feet (m), pointed toward the Pentagon and downtown Washington. Hani Hanjour advanced the throttles to maximum power and dived toward the Pentagon. While level above the ground and seconds from the crash, the wings knocked over five street lampposts and the right wing struck a portable generator, creating a smoke trail seconds before smashing into the Pentagon.[35] [36] Flight 77, flying at 530 mph (853 km/h, 237 m/s, or 460 knots) over the Navy Annex Building adjacent to Arlington National Cemetery,[37] crashed into the western side of the Pentagon in Arlington County, Virginia, just south of Washington, D.C., at 09:37:46, killing all 53 passengers, five hijackers, and six crew.[38] The plane hit the Pentagon at the first-floor level,[39] and at the moment of impact, the airplane was rolled slightly to the left, with the right wing elevated.[40] The front part of the fuselage disintegrated on impact, while the mid and tail sections moved for another fraction of a second, with tail section debris penetrating furthest into the building.[39] In all, the airplane took eight-tenths of a second to fully penetrate 310 feet (94 m) into the three outermost of the building's five rings[41] and unleashed a fireball that rose 200 feet (61 m) above the building.[39]

Debris from Flight 77 scattered near the Pentagon.

At the time of the attacks, approximately 18,000 people worked in the Pentagon, which was 4,000 fewer than before renovations began in 1998.[42] The section of the Pentagon, which had recently been renovated at a cost of $250 million,[43] housed the Naval Command Center[44] and other Pentagon offices, as well as some unoccupied offices. The crash and subsequent fire penetrated three outer ring sections of the western side. The outermost ring section was largely destroyed, and a large section collapsed. One hundred and twenty-five people in the Pentagon died in the attack.

Pentagon fire with police and EMS in the foreground.

The Pentagon, minutes after American Airlines Flight 77 crashed into it.

In all, there were 189 deaths at the Pentagon site, including the 125 in the Pentagon building in addition to the 64 on board the aircraft. Passenger Barbara Olson was en route to a taping of *Politically Incorrect*.[45] A group of children, their chaperones, and National Geographic Society staff members were also on board, embarking on an educational trip west to the Channel Islands National Marine Sanctuary near Santa Barbara, California. [46] The fatalities at the Pentagon included 55 military personnel and 70 civilians.[47] Of those 125 killed, 92 were on the first floor, 31 were on the second floor, and two were on the third.[48] The Army suffered 75 fatalities—far more than any other branch. Another 106 injured were treated at area hospitals.[48] Lieutenant General Timothy Maude, an Army Deputy Chief of Staff, was the highest-ranking military officer killed at the Pentagon; also killed was retired Rear Admiral Wilson Flagg, a passenger on the plane.[49]

"I don't want to alarm anybody right now, but apparently—it felt just a few moments ago like there was an explosion of some kind here at the Pentagon."
Jim Miklaszewski, NBC Pentagon correspondent reporting from inside the Pentagon at 09:39[50]

On the side where the plane hit, the Pentagon is bordered by Interstate 395 and Washington Boulevard. Motorist Mary Lyman, who was on I-395, saw the airplane pass over at a "steep angle toward the ground and going fast" and then saw the cloud of smoke from the Pentagon.[51] Omar Campo, another witness, was cutting the grass on the other side of the road when the airplane flew over his head.

> "I was cutting the grass and it came in screaming over my head. I felt the impact. The whole ground shook and the whole area was full of fire. I could never imagine I would see anything like that here".[52]

Afework Hagos, a computer programmer, was on his way to work and stuck in a traffic jam near the Pentagon when the airplane flew over. "There was a huge screaming noise and I got out of the car as the plane came over. Everybody was running away in different directions. It was tilting its wings up and down like it was trying to balance. It hit some lampposts on the way in."[52] Daryl Donley witnessed the crash and took some of the first photographs of the site.[53]

The collapsed area and subsequent fire damage.

USA Today reporter Mike Walter was driving on Washington Boulevard when he witnessed the crash, which he recounted,

> "I looked out my window and I saw this plane, this jet, an American Airlines jet, coming. And I thought, 'This doesn't add up, it's really low.' And I saw it. I mean it was like a cruise missile with wings. It went right there and slammed right into the Pentagon".[54]

Terrance Kean, who lived in a nearby apartment building, heard the noise of loud jet engines, glanced out his window, and saw a "very, very large passenger jet". He watched "it just plow right into the side of the Pentagon. The nose penetrated into the portico. And then it sort of disappeared, and there was fire and smoke everywhere."[55] Tim Timmerman, who is a pilot himself, noticed American Airlines markings on the aircraft as he saw it hit the Pentagon.[56] Other drivers on Washington Boulevard, Interstate 395, and Columbia Pike witnessed the crash, as did people in Pentagon City, Crystal City, and other nearby locations.[51]

Former Georgetown University basketball coach John Thompson had originally booked a ticket on Flight 77. As he would tell the story many times in the following years, including a September 12, 2011 interview on Jim Rome's radio show, he had been scheduled to appear on that show on September 12, 2001. Thompson was planning to be in

Las Vegas for a friend's birthday on September 13, and initially insisted on traveling to Rome's Los Angeles studio on the 11th. However, this did not work for the show, which wanted him to travel on the day of the show. After a Rome staffer personally assured Thompson that he would be able to travel from Los Angeles to Las Vegas immediately after the show, Thompson changed his travel plans. He felt the impact from the crash at his home near the Pentagon.[57] [58]

Rescue and recovery

"In this area ... it's so hot that the debris is melting and dripping off the ceiling onto your skin and it would sear your skin and melt your uniform. We went a little farther, turned a corner and came into this bombed out office space that was a roaring inferno of destruction and smoke and flames and intense heat you could feel searing your face."
Lieutenant Commander David Tarantino describing the scene near the Navy Command Center on the first floor.[59]

Rescue efforts began immediately after the crash. Almost all the successful rescues of survivors occurred within half an hour of the impact.[60] Initially, rescue efforts were led by the military and civilian employees within the building. Within minutes, the first fire companies arrived and found these volunteers searching near the impact site. The firemen ordered them to leave as they were not properly equipped or trained to deal with the hazards.[60] The Arlington County Fire Department (ACFD) assumed command of the immediate rescue operation within 10 minutes of the crash. ACFD Assistant Chief James Schwartz implemented an incident command system (ICS) to coordinate response efforts among multiple agencies.[61] It took about an hour for the ICS structure to become fully operational.[62] Firefighters from Fort Myer and Reagan National Airport arrived within minutes.[63] [64] Rescue and firefighting efforts were impeded by rumors of additional incoming planes. Chief Schwartz ordered two evacuations during the day in response to these rumors.[65]

Injured victim of the 9/11 attack on the Pentagon Building being loaded into an ambulance.

As firefighters attempted to extinguish the fires, they watched the building in fear of a structural collapse. One firefighter remarked that they "pretty much knew the building was going to collapse because it started making weird sounds and creaking".[65] Officials saw a cornice of the building move and ordered an evacuation. Minutes later, at 10:10, the upper floors of the damaged area of the Pentagon collapsed.[65] The collapse area was about 95 feet (29 m) at its widest point and 50 feet (15 m) at its deepest.[65] The amount of time between impact and collapse allowed everyone on the fourth and fifth levels to evacuate safely before the structure collapsed.[66] After the collapse, the interior fires intensified, spreading through all five floors.[67] After 11:00, firefighters mounted a two-pronged attack against the fires. Officials estimated temperatures of up to 2000 °F (°C).[67] While progress was made against the interior fires by late afternoon, firefighters realized a flammable layer of wood under the Pentagon's slate roof had caught fire and begun to spread.[68] Typical firefighting tactics were rendered useless by the reinforced structure as firefighters were unable to reach the fire to extinguish it.[68] Firefighters instead made firebreaks in the roof on September 12 to prevent further spreading. At 18:00 on the 12th, Arlington County issued a press release stating the fire was "controlled" but not fully "extinguished". Firefighters continued to put out smaller fires that ignited in the succeeding days.[68]

Various pieces of aircraft debris were found within the wreckage at the Pentagon. While on fire and escaping from the Navy Command Center, Lt. Kevin Shaeffer observed a chunk of the aircraft's nose cone and the nose landing gear in the service road between rings B and C.[69] Early in the morning on Friday, September 14, Fairfax County Urban Search and Rescue Team members Carlton Burkhammer and Brian Moravitz came across an "intact seat from the plane's cockpit",[70] while paramedics and firefighters located the two black boxes near the punch out hole in the

A-E drive,[71] nearly 300 feet (91 m) into the building.[40] The cockpit voice recorder was too badly damaged and charred to retrieve any information,[72] though the flight data recorder yielded useful information.[38] Investigators also found a part of Nawaf al-Hazmi's driver's license in the North Parking Lot rubble pile.[73] Personal effects belonging to victims were found and taken to Fort Myer.[74]

Remains

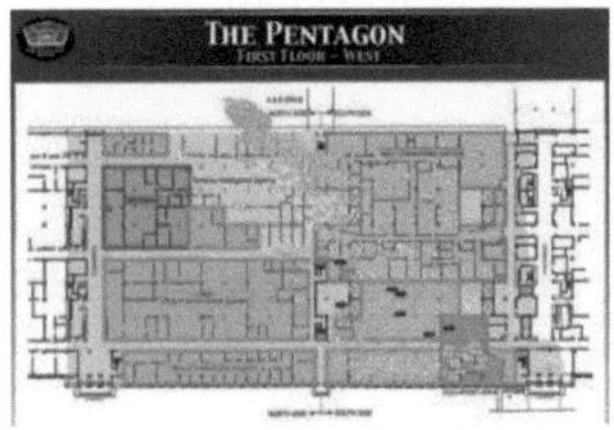

Diagram of body fragments found in the Pentagon.

Army engineers determined by 5:30 p.m. on the first day that no one remained alive in the damaged section of the building.[75] In the days after the crash, news reports emerged that up to 800 people had died.[76] Army troops from Fort Belvoir were the first teams to survey the interior of the crash site and noted the presence of human remains.[77] Federal Emergency Management Agency (FEMA) Urban Search and Rescue teams, including Fairfax County Urban Search and Rescue assisted the search for remains, working through the National Interagency Incident Management System (NIIMS).[77] [78] Kevin Rimrodt, a Navy photographer surveying the Navy Command Center after the attacks, remarked that "there were so many bodies, I'd almost step on them. So I'd have to really take care to look backwards as I'm backing up in the dark, looking with a flashlight, making sure I'm not stepping on somebody".[79] Debris from the Pentagon were taken to the Pentagon's north parking lot for more detailed search for remains and evidence.[80]

Remains that were recovered from the Pentagon were photographed, and turned over to the Armed Forces Medical Examiner office, located at Dover Air Force Base in Delaware. The medical examiner's office was able to identify remains belonging to 179 of the victims.[81] Investigators eventually identified 184 of the 189 people who died in the attack.[82] The remains of the five hijackers were identified through a process of elimination, and were turned over as evidence to the Federal Bureau of Investigation (FBI).[83] On September 21, the ACFD relinquished control of the crime scene to the FBI. The Washington Field Office, National Capital Response Squad (NCRS), and the Joint Terrorism Task Force (JTTF) led the crime scene investigation at the Pentagon.[64]

By October 2, 2001, the search for evidence and remains was complete and the site was turned over to Pentagon officials.[80] In 2002, the remains of 25 victims were buried collectively at Arlington National Cemetery, with a five-sided granite marker inscribed with the names of all the victims in the Pentagon.[84] The ceremony also honored the five victims whose remains were never found.[84]

Cockpit voice recorder

At around 3:40 a.m on September 14, a paramedic and a firefighter who were searching through the debris of the impact site found two dark boxes, about 1.5 feet (cm) by 2 feet (cm) long. They called for an FBI agent, who in turn called for someone from the National Transportation Safety Board (NTSB). The NTSB employee confirmed that these were the flight recorders ("black boxes") from American Airlines Flight 77.[85] Dick Bridges, deputy manager for Arlington County, Virginia, said the voice recorder was damaged on the outside and the flight data recorder was charred. But he said the FBI still was confident the data could be recovered from both. Bridges said the recorders were found "right where the plane came into the building."[86]

A photograph of the cockpit voice recorder from American Airlines Flight 77, as used in an exhibit at the Moussaoui trial

Officials at both American and United Airlines said the black boxes aboard their destroyed aircraft were modern solid-state versions, which are more resistant to damage than the older magnetic tape recorders.[87] The cockpit voice recorder was quickly transported to the NTSB lab in Washington, D.C., and its data was downloaded. Soon afterward, the FBI took charge of the box and its data.[88] CBS News reported that "Preliminary information shows there is nothing that appears to be useful on the cockpit voice tape. The tape appears to be blank or erased."[89] In its report on the CVR, the NTSB identified the unit as an L-3 Communications, Fairchild Aviation Recorders model A-100A cockpit voice recorder; a device which records on magnetic tape. The NTSB reported that "The majority of the recording tape was fused into a solid block of charred plastic." No usable segments of tape were found inside the recorder.[90]

Continuity of operations

At the moment of impact, Secretary of Defense Donald Rumsfeld was in his office on the other side of the Pentagon, away from the crash site. He ran to the site and assisted the injured.[91] Rumsfeld returned to his office, and went to a conference room in the Executive Support Center where he joined a secure videoteleconference with Vice President Dick Cheney and other officials.[92] On the day of the attacks, DOD officials considered moving their command operations to Site R, a backup facility in Pennsylvania. Secretary of Defense Rumsfeld insisted he remain at the Pentagon, and sent Deputy Secretary Paul Wolfowitz to Site R. The National Military Command Center (NMCC) continued to operate at the Pentagon, even as smoke entered the facility.[93] Engineers and building managers manipulated the ventilation and other building systems that still functioned to draw smoke out of the NMCC and bring in fresh air.[94]

During a press conference held inside the Pentagon at 18:42, Rumsfeld announced, "The Pentagon's functioning. It will be in business tomorrow."[95] Pentagon employees returned the next day to offices in mostly unaffected areas of the building. By the end of September, more workers returned to the lightly damaged areas of the Pentagon.[80]

Aftermath

Damaged section of the Pentagon under reconstruction.

Early estimates on rebuilding the damaged section of the Pentagon were that it would take three years to complete.[80] However, the project moved forward at an accelerated pace and was completed by the one-year anniversary of the attack.[96] The rebuilt section of the Pentagon includes a small indoor memorial and chapel at the point of impact.[97] An outdoor memorial, commissioned by the Pentagon and designed by Julie Beckman and Keith Kaseman, was completed on schedule for its dedication on September 11, 2008.[98]

Security camera video

On May 16, 2006, the Department of Defense released filmed footage that was recorded by a security camera of American Airlines Flight 77 crashing into the Pentagon, with a plane visible in one frame, as a "thin white blur" and an explosion following.[99] The images were made public in response to a December 2004 Freedom of Information Act request by Judicial Watch.[100] Some still images from the video had previously been released and publicly circulated, but this was the first official release of the full video of the crash.[101]

Second security camera video, impact is at 0:25

A nearby Citgo service station also had security cameras, but a video released on September 15, 2006 did not show the crash because the camera was pointed away from the crash site.[102] [103]

The Doubletree Hotel, located nearby in Crystal City, Virginia, also had a security camera video. On December 4, 2006, the FBI released the video in response to a FOIA lawsuit filed by Scott Bingham. The footage is "grainy and the focus is soft, but a rapidly growing tower of smoke is visible in the distance on the upper edge of the frame as the plane crashes into the building".[104]

Memorials

Panel S-74 of the National September 11 Memorial's South Pool, one of six on which the names of Pentagon victims are inscribed.[105]

On September 12, 2002, Defense Secretary Donald Rumsfeld and General Richard Myers, Chairman of the Joint Chiefs of Staff, dedicated the Victims of Terrorist Attack on the Pentagon Memorial at Arlington National Cemetery.[106] The memorial specifically honors the five individuals for whom no identifiable remains were found.[107] This included Dana Falkenberg, age three, who was aboard American Airlines Flight 77 with her parents and older sister.[107] A portion of the remains of 25 other victims are also buried at the site.[108] The memorial is a pentagonal[109] granite marker 4.5 feet (m) high.[107] On five sides of the memorial along the top are inscribed the words "Victims of Terrorist Attack on the Pentagon September 11, 2001". Aluminum plaques, painted black, are inscribed with the names of the 184 victims of the terrorist attack.[107] The site is located in Section 64,[110] on a slight rise, which gives it a view of the Pentagon.[107]

At the National September 11 Memorial, the names of the Pentagon victims are inscribed on the South Pool, on Panels S-1 and S-72 – S-76.[111]

See also

- American Airlines Flight 11
- United Airlines Flight 93
- United Airlines Flight 175
- List of tenants in One World Trade Center
- List of tenants in Two World Trade Center

References

[1] http://registry.faa.gov/aircraftinquiry/NNum_Results.aspx?NNumbertxt=644AA

[2] Chen, David W (2001-09-18). "A Nation Challenged: The Suspect; Man Traveled Across U.S. In His Quest to Be a Pilot" (http://query.nytimes.com/gst/fullpage.html?res=9F03EFDF153BF93BA2575AC0A9679C8B63&sec=&spon=&pagewanted=1). *The New York Times*. . Retrieved 2008-05-29.

[3] Yardley, Jim; Thomas, Jo (2002-06-19). "Traces of Terror: The F.B.I.; For Agent in Phoenix, the Cause of Many Frustrations Extended to His Own Office" (http://query.nytimes.com/gst/fullpage.html?res=9E0DE0DD173FF93AA25755C0A9649C8B63). *The New York Times*. . Retrieved 2008-06-04.

[4] "Four Planes, Four Coordinated Teams" (http://www.washingtonpost.com/wp-srv/nation/graphics/attack/hijackers.html). *The Washington Post*. 2001. . Retrieved 2001-06-17.

[5] Sennott, Charles M (2002-03-03). "Why bin Laden plot relied on Saudi hijackers" (http://www.boston.com/news/packages/underattack/news/driving_a_wedge/part1.shtml). *The Boston Globe*. Archived (http://web.archive.org/web/20080513135327/http://www.boston.com/news/packages/underattack/news/driving_a_wedge/part1.shtml) from the original on 13 May 2008. . Retrieved 2008-05-29.

[6] "The Attack Looms" (http://www.9-11commission.gov/report/911Report_Ch7.htm). *9/11 Commission Report*. National Commission on Terrorist Attacks Upon the United States. 2004. . Retrieved 2008-05-29.

[7] Wright, Lawrence (2006). "Chapter 18, "Boom"". *Looming Tower*. Alfred P. Knopf.

[8] *The Spy Factory* (http://www.pbs.org/wgbh/nova/military/spy-factory.html), PBS Frontline episode based on James Bamford's book, *Shadow Factory*

[9] Goldstein, Amy (2001-09-30). "Hijackers Led Core Group" (http://web.archive.org/web/20080505123208/http://www.pulitzer.org/year/2002/national-reporting/works/093001.html). *The Washington Post*. Archived from the original (http://www.pulitzer.org/year/2002/national-reporting/works/093001.html) on 2008-05-05. . Retrieved 2008-05-29.

[10] "Chronology" (http://www.9-11commission.gov/staff_statements/911_TerrTrav_Ch2.pdf) (PDF). *Monograph on 9/11 and Terrorist Travel*. National Commission on Terrorist Attacks Upon the United States. p. 22. Archived (http://web.archive.org/web/20080529193648/

http://www.9-11commission.gov/staff_statements/911_TerrTrav_Ch2.pdf) from the original on 29 May 2008. . Retrieved 2008-05-25.

[11] Martin, John P. (2001-09-27). "Landlord Identifies Terrorists as Renters". The Star-Ledger.

[12] Masters, Brooke A.; Smith, Leef; Shear, Michael D. (2001-09-19). "Dulles Hijackers Made Maryland Their Base; Residents Recall Men as Standoffish". *The Washington Post.* "The men who hijacked Flight 77 also made a concerted effort to stay in shape. All five visited the Gold's Gym on Greenbelt Road during the first week of September"

[13] Olson, Bradley (2006-09-09). "MD. Was Among Last Stops For Hijackers; Those Who Recall Encounters Are Haunted By Proximity To Agents Of Tragic Event" (http://www.baltimoresun.com/news/maryland/bal-te.md.terrorist09sep09,0,238349.story). *The Baltimore Sun.* . Retrieved 2008-05-30.

[14] 9/11 Commission (2004). "Notes" (http://govinfo.library.unt.edu/911/report/911Report_Notes.htm). *9/11 Commission Report.* National Commission on Terrorist Attacks Upon the United States. Archived (http://web.archive.org/web/20080530161252/http://govinfo.library.unt.edu/911/report/911Report_Notes.htm) from the original on 30 May 2008. . Retrieved 2008-05-30.

[15] Federal Bureau of Investigation (2008-02-04). "Hijackers' Timeline" (http://www.911myths.com/index.php/Image:Hijackers_Timeline_OCR.pdf) (PDF). 9/11 Myths. Archived (http://web.archive.org/web/20080803015816/http://www.911myths.com/index.php/Image:Hijackers_Timeline_OCR.pdf) from the original on 3 August 2008. . Retrieved 2008-08-01.

[16] Isikoff, Michael (February 2, 2011). "WikiLeaks cable revives talk of 9/11 support network" (http://today.msnbc.msn.com/id/41400106/ns/us_news-security/). MSNBC. Archived (http://web.archive.org/web/20110211203506/http://today.msnbc.msn.com/id/41400106/ns/us_news-security/) from the original on 11 February 2011. . Retrieved February 18, 2011.

[17] "Flight Path Study - American Airlines Flight 77" (http://www.gwu.edu/~nsarchiv/NSAEBB/NSAEBB196/doc02.pdf) (PDF). National Transportation Safety Board. 2002-02-19. . Retrieved 2008-05-30.

[18] "American Airlines Flight 77" (http://web.archive.org/web/20080512091509/http://www.cnn.com/SPECIALS/2001/trade.center/victims/AA77.victims.html). CNN. 2001. Archived from the original (http://www.cnn.com/SPECIALS/2001/trade.center/victims/AA77.victims.html) on May 12, 2008. . Retrieved 2008-05-29.

[19] "Staff Monograph on the "Four Flights and Civil Aviation Security"" (http://www.webcitation.org/5T7H27edX) (PDF). National Commission on Terrorist Attacks Upon the United States. September 2005. Archived from the original (http://www.archives.gov/research/9-11/staff-report-sept2005.pdf) on 2007-11-05. . Retrieved 2008-08-14.

[20] "'We Have Some Planes'" (http://govinfo.library.unt.edu/911/report/911Report_Ch1.htm). *9/11 Commission Report.* National Commission on Terrorist Attacks Upon the United States. 2004. Archived (http://web.archive.org/web/20080514052417/http://govinfo.library.unt.edu/911/report/911Report_Ch1.htm) from the original on 14 May 2008. . Retrieved 2008-05-30.

[21] "The Aviation Security System and the 9/11 Attacks - Staff Statement No. 3" (http://www.9-11commission.gov/staff_statements/staff_statement_3.pdf) (PDF). *National Commission on Terrorist Attacks Upon the United States.* Archived (http://web.archive.org/web/20080528201527/http://www.9-11commission.gov/staff_statements/staff_statement_3.pdf) from the original on 28 May 2008. . Retrieved 2008-05-30.

[22] "New Video of 9/11 hijackers at Dulles Airport before attacks" (http://transcripts.cnn.com/TRANSCRIPTS/0407/21/acd.01.html). *Anderson Cooper 360.* 2004-07-21. CNN. .

[23] Orecklin, Michele; Land, Greg (2001-11-19). "Why Argenbright Sets Off Alarms" (http://www.time.com/time/magazine/article/0,9171,1001252,00.html). *Time.* . Retrieved 2008-05-30.

[24] "Summary of Flight 77" (http://www.vaed.uscourts.gov/notablecases/moussaoui/exhibits/prosecution/flights/P200054.html). United States District Court for the Eastern District of Virginia. . Retrieved 2007-02-04.

[25] Gregor, Joseph A. (2001-12-21). "ATC Report American Airlines Flight 77" (http://www.ntsb.gov/doclib/foia/9_11/ATC_Report_AA77.pdf) (PDF). National Transportation Safety Board. . Retrieved 2011-09-25.

[26] O'Callaghan, John; Bower, Daniel (2002-02-13). "Study of Autopilot, Navigation Equipment, and Fuel Consumption Activity Based on United Airlines Flight 93 and American Airlines Flight 77 Digital Flight Data Recorder Information" (http://www.ntsb.gov/info/autopilot_AA77_UA93_study.pdf) (PDF). National Transportation Safety Board. . Retrieved 2008-06-01.

[27] Phillips, Don (2001-11-03). "Pentagon Crash Highlights a Radar Gap; Limited System in One Area Made Flight 77 Invisible to Controllers for Half-Hour". *The Washington Post.*

[28] Johnson, Glen (2001-11-23). "Probe reconstructs horror, calculated attacks on planes" (http://www.boston.com/news/packages/underattack/news/planes_reconstruction.htm). *The Boston Globe.* Archived (http://web.archive.org/web/20080608210201/http://www.boston.com/news/packages/underattack/news/planes_reconstruction.htm) from the original on 8 June 2008. . Retrieved 2008-06-01.

[29] Parker, Laura (2001-09-13). "Four flights, four tales of terror" (http://www.usatoday.com/news/nation/2001/09/12/flights-main.htm). *USA Today.* . Retrieved 2008-06-18.

[30] Producers: Colette Beaudry and Michael Cascio (2005-09-23). "Zero Hour". *Inside 9/11.* National Geographic Channel.

[31] "Air Traffic Controllers Recall 9/11" (http://abcnews.go.com/2020/Story?id=123822&page=1). *20/20.* ABC News. 2001-10-24. . Retrieved 2008-06-05.

[32] Fisher, Marc; Phillips, Don (2001-09-12). "On Flight 77: 'Our Plane Is Being Hijacked'" (http://www.washingtonpost.com/ac2/wp-dyn/A14365-2001Sep11). *The Washington Post.* . Retrieved 2008-06-01.

[33] Schmitt, Eric; Lichtblau, Eric (2004-06-18). "To the Minute, Panel Paints a Grim Portrait of Day's Terror" (http://www.nytimes.com/2004/06/18/politics/18minutes.html?pagewanted=2&ei=5007&en=c6fa52b1957732f6&ex=1402977600&partner=USERLAND). *The New York Times.* . Retrieved 2008-06-01.

[34] "Flight 77, Video 2" (http://web.archive.org/web/20090212100831/http://www.judicialwatch.org/archive/2006/flight77-2.mpg). Judicial Watch. Archived from the original (http://www.judicialwatch.org/flight77.shtml) on 2009-02-12. .

[35] "9/11 Survivor Wants Life For Moussaoui" (http://www.nbc4.com/news/8988021/detail.html). WRC-TV. 2006-04-25. . Retrieved 2008-06-02.

[36] Curiel, Jonathan (2006-09-03). "The Conspiracy to Rewrite 9/11" (http://www.sfgate.com/cgi-bin/article.cgi?f=/c/a/2006/09/03/INGR0KRCBA1.DTL). *San Francisco Chronicle*. Archived (http://web.archive.org/web/20080519053919/http://www.sfgate.com/cgi-bin/article.cgi?f=/c/a/2006/09/03/INGR0KRCBA1.DTL) from the original on 19 May 2008. . Retrieved 2008-06-02.

[37] Goldberg et al., p. 14.

[38] "American Airlines Flight 77 FDR Report" (http://www.webcitation.org/5RfC3xUsq) (PDF). National Transportation Safety Board. 2002-01-31. Archived from the original (http://www.ntsb.gov/info/AAL77_fdr.pdf) on 2007-09-07. . Retrieved 2008-06-02.

[39] Goldberg, Alfred et al. (2007). *Pentagon 9/11*. Washington, D.C.: United States Government Printing Office. p. 17. ISBN 978-0-16-078328-9.

[40] Mlakar, Paul F.; Dusenberry, Donald O.; Harris, James R.; Haynes, Gerald; Phan, Long T.; Sozen, Mete A. (January 2003) (PDF). *The Pentagon Building Performance Report* (http://fire.nist.gov/bfrlpubs/build03/PDF/b03017.pdf). American Society of Civil Engineers. Archived (http://web.archive.org/web/20080624212810/http://www.fire.nist.gov/bfrlpubs/build03/PDF/b03017.pdf) from the original on 24 June 2008. . Retrieved 2008-06-19.

[41] Maclean, John N (2008-06-01). "America Under Attack: A chronicle of chaos and heroism at the Pentagon" (http://www.washingtonpost.com/wp-dyn/content/article/2008/05/29/AR2008052903149.html). *The Washington Post*. . Retrieved 2008-06-04.

[42] Goldberg et al., p. 3.

[43] "Phoenix Rising: The Rebuilding of the Pentagon" (http://www.washingtonpost.com/wp-srv/flash/metro/phoenix/phoenix.swf) (Flash). *The Washington Post*. . Retrieved 2008-06-02.

[44] Zablotsky, Sarah (2003-06-11). "Survivor of Pentagon attack has a positive attitude" (http://www.post-gazette.com/neigh_south/20030611s19kevin0611p6.asp). *Pittsburgh Post-Gazette*. . Retrieved 2008-06-02.

[45] Moraes, Lisa (2001-09-17). "Letterman's Back Tonight, but Don't Expect a Biting Monologue". *The Washington Post*.

[46] "Team from National Geographic Killed in Pentagon Crash" (http://news.nationalgeographic.com/news/2001/09/0912_disasterngs.html). National Geographic Society. 2001-09-12. Archived (http://web.archive.org/web/20080518235900/http://news.nationalgeographic.com/news/2001/09/0912_disasterngs.html) from the original on 18 May 2008. . Retrieved 2008-06-10.

[47] Stone, Andrea (2002-08-20). "Military's aid and comfort ease 9/11 survivors' burden" (http://www.usatoday.com/news/sept11/2002-08-20-pentagon_x.htm). *USA Today*. . Retrieved 2008-05-20.

[48] Goldberg et al., pp. 23–24.

[49] "Sept. 11 fallen warrior memorialized in building dedication" (http://www.hqusareur.army.mil/htmlinks/Press_Releases/2002/Apr2002/30Apr2002-02.htm) (Press release). United States Army. 2002-04-30. . Retrieved 2008-06-10.

[50] "'Hardball with Chris Matthews' for May 16" (http://www.msnbc.msn.com/id/12834927/). *Hardball with Chris Matthews*. MSNBC. 2006-05-16. . Retrieved 2008-05-30.

[51] "Terrible Tuesday". *The Washington Post*. 2001-09-16.

[52] Borger, Julian (2001-09-12). "'Everyone was screaming, crying, running. It's like a war zone'" (http://www.guardian.co.uk/world/2001/sep/12/expertopinions.charlieporteronmensfashion). *The Guardian* (London). Archived (http://web.archive.org/web/20080516185757/http://www.guardian.co.uk/world/2001/sep/12/expertopinions.charlieporteronmensfashion) from the original on 16 May 2008. . Retrieved 2008-06-19.

[53] "Documentary Photographs" (http://www.loc.gov/exhibits/911/911-docphotos.html). Library of Congress, Prints and Photographs Division. Archived (http://web.archive.org/web/20061111205708/http://www.loc.gov/exhibits/911/911-docphotos.html) from the original on 11 November 2006. . Retrieved 2006-11-12.

[54] Anderson, Porter (2001-09-11). "Witnesses to the moment: Workers' voices" (http://archives.cnn.com/2001/CAREER/trends/09/11/witnesses/). CNN. . Retrieved 2008-06-10.

[55] Sheridan, Mary B (2001-09-12). "Loud Boom, Then Flames In Hallways; Pentagon Employees Flee Fire, Help Rescue Injured Co-Workers". *The Washington Post*.

[56] "Transcripts - America Under Attack: Eyewitness Discusses Pentagon Plane Crash" (http://transcripts.cnn.com/TRANSCRIPTS/0109/11/bn.32.html). CNN. 2001-09-11. . Retrieved 2008-06-10.

[57] Brennan, Eamonn (September 12, 2011). "JohnThompson's surreal 9/11 story" (http://espn.go.com/blog/collegebasketballnation/post/_/id/35169/john-thompsons-surreal-911-story). *College Basketball Nation Blog. ESPN.com*. . Retrieved September 16, 2011.

[58] Thompson, John (September 12, 2011) (audio). Interview with Jim Rome. The Jim Rome Show. http://www.jimrome.com/junglehighlights?uri=channels/465575/1483554. Retrieved September 16, 2011.

[59] Goldberg et al., pp. 55–56.

[60] Goldberg et al., p. 51.

[61] Goldberg et al., p. 72.

[62] Goldberg et al., p. 77.

[63] Goldberg et al., p. 78.

[64] "Arlington, Virginia After-Action Report" (http://web.archive.org/web/20070927012831/http://www.arlingtonva.us/Departments/Fire/edu/about/docs/after_report.pdf) (PDF). Arlington County Fire Department. Archived from the original (http://www.arlingtonva.us/

Departments/Fire/edu/about/docs/after_report.pdf) on 2007-09-27. . Retrieved 2008-06-10.

[65] Goldberg et al., pp. 80–82.

[66] Goldberg et al., p. 20.

[67] Goldberg et al., pp. 86–90.

[68] Goldberg et al., pp. 91–95.

[69] Swift, Earl (2002-09-09). "Inside the Pentagon on 9/11: The Call of Duty" (http://web.archive.org/web/20040730074824/http://www.hamptonroads.com/pilotonline/special/911/pentagon3.html). The Virginian-Pilot (Hampton Roads). Archived from the original (http://www.hamptonroads.com/pilotonline/special/911/pentagon3.html) on 2004-06-24. .

[70] "Web Exclusive: Washington's Heroes - On the ground at the Pentagon on Sept. 11" (http://www.msnbc.msn.com/id/3069699/). MSNBC. 2001-09-28. .

[71] Creed, Patrick, and Rick Newman (2008). "Chapter 41: A Great Find". *Firefight: Inside the Battle to Save the Pentagon on 9/11*. Random House.

[72] Murray, Frank J. (2001-09-15). "Pentagon plane voice recorder is too 'cooked' to aid in probe". *The Washington Times*.

[73] "Prosecution Trial Exhibits - Exhibit Number PE00102" (http://www.vaed.uscourts.gov/notablecases/moussaoui/exhibits/prosecution/PE00102.html). United States District Court for the Eastern District of Virginia. . Retrieved 2008-06-19.

[74] Wilkinson, Marian (2002-09-09). "Capital punishment" (http://www.theage.com.au/articles/2002/09/09/1031115990939.html). Melbourne: The Age (Australia). . Retrieved 2008-06-24.

[75] Goldberg et al., p. 97.

[76] "Twin Towers Demolished, Pentagon Hit in Terrorist Attacks" (http://www.foxnews.com/story/0,2933,34177,00.html). Foxnews.com. 2001-09-12. . Retrieved 2008-06-10.

[77] Goldberg et al., p. 119.

[78] Eversburg, Rudy (2002-11-01). "The Pentagon Attack on 9-11: Arlington County (VA) Fire Department Response" (http://www.fireengineering.com/articles/print/volume-155/issue-11/features/the-pentagon-attack-on-9-11-arlington-county-va-fire-department-response.html). *Fire Engineering*. . Retrieved 2008-06-10.

[79] Goldberg et al., pp. 121–122.

[80] Vogel, Steve (2001-10-03). "Search for Remains Ends at Pentagon". The Washington Post.

[81] "Mass Fatality Management for Incidents Involving Weapons of Mass Destruction" (http://www.edgewood.army.mil/hld/dl/MFM_Capstone_August_2005.pdf) (PDF). U.S. Army Research Development and Engineering Command and the Office for Domestic Preparedness. August 2005. Archived (http://web.archive.org/web/20080802031006/http://www.edgewood.army.mil/hld/dl/MFM_Capstone_August_2005.pdf) from the original on 2 August 2008. . Retrieved 2008-06-24.

[82] Kelly, Christopher (2001-11-29). "Forensic feat IDs nearly all Pentagon victims" (http://www.dcmilitary.com/dcmilitary_archives/stories/112901/12279-1.shtml). *Stripe*. . Retrieved 2008-06-27.

[83] Vogel, Steve (2001-11-21). "Remains Unidentified For 5 Pentagon Victims; Bodies Were Too Badly Burned, Officials Say". *The Washington Post*.

[84] Vogel, Steve (2002-09-13). "Lost and, Sometimes, Never Found" (http://www.washingtonpost.com/wp-dyn/content/article/2002/09/13/AR2006031501324.html). *The Washington Post*. . Retrieved 2008-08-07.

[85] Rosenberg, Debra (2001-09-28). "Washington's Heroes: On the ground at the Pentagon on Sept. 11" (http://web.archive.org/web/20040526034459/http://msnbc.msn.com/id/3069699/). MSNBC. Archived from the original (http://msnbc.msn.com/id/3069699/) on 2004-05-26. . Retrieved 2009-11-02.

[86] "Searchers find Pentagon black boxes" (http://www.usatoday.com/news/nation/2001/09/14/pentagon-fire.htm). *USA Today* (Gannett Co. Inc.). 2001-09-14. . Retrieved 2009-11-02.

[87] "FBI analyzing voice, data recorders from two flights" (http://www.sptimes.com/News/091501/Worldandnation/FBI_analyzing_voice__.shtml). St. Petersburg Times. 2001-09-15. . Retrieved 2009-11-02.

[88] Bert, Ray (Spring 2002). "Hearing Voices" (http://www.wpi.edu/News/Transformations/2002Spring/hear.html). *Transformations* (Worcester Polytechnic Institute). . Retrieved 2009-11-02.

[89] "Feds Would Have Shot Down Pa. Jet" (http://www.cbsnews.com/stories/2001/09/12/archive/main311011.shtml). CBS News. 2001-09-16. . Retrieved 2009-11-02.

[90] "Specialist's Factual Report of Investigation: Cockpit Voice Recorder" (http://www.scribd.com/doc/14780831/T8-B18-NTSB-Documents-1-of-3-Fdr-CVR-Cockpit-Voice-Recorder-Reports-AA-77-and-UA-93-Paperclipped-Together260). *National Transportation Safety Board*. 2002-04-30. . Retrieved 2009-11-02.

[91] Vobejda, Barbara (2001-09-11). "'Extensive Casualties' in Wake of Pentagon Attack" (http://www.washingtonpost.com/wp-srv/metro/daily/sep01/attack.html). *The Washington Post*. . Retrieved 2008-06-20.

[92] Creed, Patrick and Rick Newman (2008). *Firefight: Inside the Battle to Save the Pentagon on 9/11*. Presidio Press. pp. 276–277. ISBN 0-89141-905-5.

[93] "Battling the Pentagon Blaze After 9/11" (http://www.npr.org/templates/story/story.php?storyId=90696597&ft=1&f=1012). NPR/WHYY – Fresh Air. May 22, 2008. . Retrieved 2008-06-24.

[94] Creed and Newman, p. 278

[95] "DoD News Briefing on Pentagon Attack" (http://www.defenselink.mil/transcripts/transcript.aspx?transcriptid=1613). United States Department of Defense. 2001-09-11. Archived (http://web.archive.org/web/20080607170533/http://www.defenselink.mil/transcripts/

transcript.aspx?transcriptid=1613) from the original on 7 June 2008. . Retrieved 2008-06-24.
[96] "Project Phoenix: Pentagon offices rise from rubble" (http://archives.cnn.com/2002/US/09/04/ar911.pentagon.reconstruction/). CNN. 2002-09-04. . Retrieved 2008-06-24.
[97] Freking, Kevin (2006-09-09). "Public Honors 9/11 Fallen at Pentagon" (http://www.washingtonpost.com/wp-dyn/content/article/2006/09/09/AR2006090900576.html). Washington Post. . Retrieved 2010-08-13.
[98] Shaughnessy, Larry (2008-05-24). "Nearly complete Pentagon memorial tells story of 9/11" (http://edition.cnn.com/2008/US/05/23/pentagon.memorial/). CNN. . Retrieved 2011-09-12.
[99] "Video of 9/11 plane hitting Pentagon is released" (http://www.msnbc.msn.com/id/12818225/). Associated Press. MSNBC. 2006-05-16. . Retrieved 2008-06-10.
[100] "Judicial Watch Obtains September 11 Pentagon Video" (http://www.judicialwatch.org/5772.shtml) (Press release). Judicial Watch. 2006-05-16. Archived (http://web.archive.org/web/20080511221056/http://www.judicialwatch.org/5772.shtml) from the original on 11 May 2008. . Retrieved 2008-06-11.
[101] "Images show September 11 Pentagon crash" (http://archives.cnn.com/2002/US/03/07/gen.pentagon.pictures/). CNN. 2002-03-08. Archived (http://web.archive.org/web/20080531134018/http://archives.cnn.com/2002/US/03/07/gen.pentagon.pictures/) from the original on 31 May 2008. . Retrieved 2008-06-11.
[102] "CITGO Gas Station Cameras Near Pentagon Evidently Did Not Capture Attack" (http://www.judicialwatch.org/5965.shtml) (Press release). Judicial Watch. 2006-09-15. Archived (http://web.archive.org/web/20080501110315/http://www.judicialwatch.org/5965.shtml) from the original on 1 May 2008. . Retrieved 2008-06-11.
[103] "Videotapes dispel conspiracy; Plane, not missile seen hitting Pentagon on 9/11". *The Washington Times*. 2006-05-17.
[104] "FBI Releases New Footage of 9/11 Pentagon Attack" (http://www.kwtx.com/home/headlines/4821121.html). KWTX-TV. 2006-12-04. Archived (http://web.archive.org/web/20080624232051/http://www.kwtx.com/home/headlines/4821121.html) from the original on 24 June 2008. . Retrieved 2008-06-11.
[105] "South Pool: Panel S-74 - Timothy J. Maude" (http://names.911memorial.org/#lang=en_US&page=person&id=132). National September 11 Memorial & Museum. . Retrieved October 29, 2011.
[106] Garamone, Jim. "Remains of Pentagon Attack Victims Buried at Arlington." American Forces Press Service. September 12, 2002. (http://osd.dtic.mil/news/Sep2002/n09122002_200209123.html) Accessed 2011-09-07.
[107] Cass, Connie. "Cremated Remains of Pentagon Victims Are Laid to Rest at National Cemetery. Associated Press. *September 13, 2002.*
[108] "Arlington Funeral Honors Unidentified Victims." CNN.com. September 12, 2002. (http://articles.cnn.com/2002-09-12/us/ar911.pentagon.funeral_1_dana-falkenberg-ronald-golinski-victims?_s=PM:US) Accessed 2011-09-07.
[109] Pusey, Allen. "Final Service Honors Victims of Pentagon Attack." *Dallas Morning News*. September 13, 2002.
[110] Vogel, Steve. "Lost and, Sometimes, Never Found." *Washington Post*. September 13, 2002.
[111] About: The Memorial Names Layout (http://names.911memorial.org/#lang=en_US&page=about&id=7). Memorial Guide: National 9/11 Memorial. Retrieved December 11, 2011.

External links

- Picture of Aircraft Pre 9-11 (http://www.airliners.net/open.file?id=0188155&size=L&width=1024&height=631&sok=&photo_nr=1813/)
- Arlington County After-Action Report (http://www.arlingtonva.us/departments/fire/edu/about/fireeduaboutafterreport.aspx) – Arlington County Fire Department, July 23, 2002
- Archive of American Airlines site (http://web.archive.org/web/20010911233514/http://www.aa.com/) – explaining that all aircraft are accounted for (September 11, 2001)
- Archive of American Airlines site with condolences for deceased (http://web.archive.org/web/20010912191349/http://www.aa.com/) (September 12, 2001)

al-Qaeda

al-Qaeda	
Dates of operation	August 11, 1988 – present
Leader(s)	Osama bin Laden (1988–2011) Ayman al-Zawahiri (2011 – present)
Active region(s)	Worldwide
Ideology	Sunni Islamism Strict sharia law Islamic fundamentalism[1] Takfiri[2] Pan-Islamism Worldwide Caliphate[3] [4] [5] [6] [7] Qutbism Wahhabism[8] Salafist Jihadism[9] [10]
Status	Designated as Foreign Terrorist Organization by the U.S. State Department[11] Designated as Proscribed Group by the UK Home Office[12] Designated as terrorist group by EU Common Foreign and Security Policy[13] Under the Unlawful Activities (Prevention) Act designated as terrorist organization by the Government of India[14]
Size	500–1,000 operatives

al-Qaeda (pron.: /ælˈkaɪdə/ *al-KY-də*; Arabic: القاعدة *al-qāʿidah*, Arabic: [ælqɑːʕɪdɐ], translation: "The Base" and alternatively spelled **al-Qaida** and sometimes **al-Qa'ida**) is a global militant Islamist organization founded by Osama bin Laden at some point between August 1988[15] and late 1989,[16] with its origins being traceable to the Soviet War in Afghanistan.[17] It operates as a network comprising both a multinational, stateless army[18] and a radical Sunni Muslim movement calling for global Jihad and a strict interpretation of sharia law. It has been designated as a terrorist organization by the United Nations Security Council, NATO, the European Union, the United Kingdom, the United States, and various other countries (see below). Al-Qaeda has carried out several attacks on non-Muslims,[19] [20] and other targets it considers *kafir*.[21]

Al-Qaeda has attacked civilian and military targets in various countries. For example, it carried out the September 11 attacks, 1998 U.S. embassy bombings and the 2002 Bali bombings. The U.S. government responded to the September 11 attacks by launching the War on Terror. With the loss of key leaders, culminating in the death of Osama bin Laden, al-Qaeda's operations have devolved from actions that were controlled from the top-down, to actions by franchise associated groups, to actions of lone wolf operators.

Characteristic techniques employed by Al-Qaeda include suicide attacks and simultaneous bombings of different targets.[22] Activities ascribed to it may involve members of the movement, who have taken a pledge of loyalty to Osama bin Laden, or the much more numerous "al-Qaeda-linked" individuals who have undergone training in one of its camps in Afghanistan, Pakistan, Iraq or Sudan, but who have not taken any pledge.[23] Al-Qaeda ideologues envision a complete break from all foreign influences in Muslim countries, and the creation of a new world-wide Islamic caliphate.[3] [24] [25] Among the beliefs ascribed to Al-Qaeda members is the conviction that a Christian–Jewish alliance is conspiring to destroy Islam.[26] As Salafist jihadists, they believe that the killing of civilians is religiously sanctioned, and they ignore any aspect of religious scripture which might be interpreted as forbidding the murder of civilians and internecine fighting.[9] [27] Al-Qaeda also opposes man-made laws, and wants to replace them with a strict form of sharia law.[28]

Al-Qaeda is also responsible for instigating sectarian violence among Muslims.[29] Al-Qaeda is intolerant of non-Sunni branches of Islam and denounces them by means of excommunications called "takfir". Al-Qaeda leaders regard liberal Muslims, Shias, Sufis and other sects as heretics and have attacked their mosques and gatherings.[30] Examples of sectarian attacks include the Yazidi community bombings, the Sadr City bombings, the Ashoura Massacre and the April 2007 Baghdad bombings.[31]

Organization

Al-Qaeda's management philosophy has been described as "centralization of decision and decentralization of execution."[32] It is thought that al-Qaeda's leadership, following the War on Terror, has "become geographically isolated", leading to the "emergence of decentralized leadership" of regional groups using the al-Qaeda "brand".[33][34]

Many terrorism experts do not believe that the global jihadist movement is driven at every level by al-Qaeda's leadership. Although bin Laden still held considerable ideological sway over some Muslim extremists before his death, experts argue that al-Qaeda has fragmented over the years into a variety of regional movements that have little connection with one another. Marc Sageman, a psychiatrist and former Central Intelligence Agency (CIA) officer, said that al-Qaeda is now just a "loose label for a movement that seems to target the West". "There is no umbrella organisation. We like to create a mythical entity called [al-Qaeda] in our minds, but that is not the reality we are dealing with."[35]

This view mirrors the account given by Osama bin Laden in his October 2001 interview with Tayseer Allouni:

> "...this matter isn't about any specific person and...is not about the al-Qai`dah Organization. We are the children of an Islamic Nation, with Prophet Muhammad as its leader, our Lord is one...and all the true believers [mu'mineen] are brothers. So the situation isn't like the West portrays it, that there is an 'organization' with a specific name (such as 'al-Qai`dah') and so on. That particular name is very old. It was born without any intention from us. Brother Abu Ubaida... created a military base to train the young men to fight against the vicious, arrogant, brutal, terrorizing Soviet empire... So this place was called 'The Base' ['Al-Qai`dah'], as in a training base, so this name grew and became. We aren't separated from this nation. We are the children of a nation, and we are an inseparable part of it, and from those public demonstrations which spread from the far east, from the Philippines, to Indonesia, to Malaysia, to India, to Pakistan, reaching Mauritania... and so we discuss the conscience of this nation."[36]

Others, however, see al-Qaeda as an integrated network that is strongly led from the Pakistani tribal areas and has a powerful strategic purpose. Bruce Hoffman, a terrorism expert at Georgetown University, said "It amazes me that people don't think there is a clear adversary out there, and that our adversary does not have a strategic approach."[35]

Al-Qaeda has the following direct affiliates:

- Al-Qaeda in Iraq
- Al-Qaeda Organization in the Islamic Maghreb
- Al-Shabaab (Mujahideen Youth Movement) in Somalia
- Egyptian Islamic Jihad
- Al-Qaeda in the Arabian Peninsula, which comprises
  - Al Qaeda in Saudi Arabia, and
  - Islamic Jihad of Yemen

Leadership

Pakistani journalist Hamid Mir interviewing then al-Qaeda leader Osama bin Laden in Afghanistan, in 1997

Information mostly acquired from Jamal al-Fadl provided American authorities with a rough picture of how the group was organized. While the veracity of the information provided by al-Fadl and the motivation for his cooperation are both disputed, American authorities base much of their current knowledge of al-Qaeda on his testimony.[37]

Osama bin Laden was the most historically notable emir, or commander, and Senior Operations Chief of al-Qaida prior to his assassination on May 1, 2011 by US forces. Ayman al-Zawahiri, al-Qaeda's Deputy Operations Chief prior to bin Laden's death, assumed the role of commander, according to an announcement by al-Qaida on June 16, 2011. He replaced Saif al-Adel, who had served as interim commander.[38]

Bin Laden was advised by a Shura Council, which consists of senior al-Qaeda members, estimated by Western officials to consist of 20–30 people.

Atiyah Abd al-Rahman was alleged to be second in command prior to his death on August 22, 2011.[39]

On 5 June 2012, Pakistan intelligence officials announced that al-Rahman's replacement Abu Yahya al-Libi had been killed in Pakistan.[40]

Al-Qaeda's network was built from scratch as a conspiratorial network that draws on leaders of all its regional nodes "as and when necessary to serve as an integral part of its high command."[41]

- The Military Committee is responsible for training operatives, acquiring weapons, and planning attacks.
- The Money/Business Committee funds the recruitment and training of operatives through the *hawala* banking system. U.S-led efforts to eradicate the sources of terrorist financing[42] were most successful in the year immediately following the September 11 attacks;[43] al-Qaeda continues to operate through unregulated banks, such as the 1,000 or so *hawaladars* in Pakistan, some of which can handle deals of up to $10 million.[44] It also provides air tickets and false passports, pays al-Qaeda members, and oversees profit-driven businesses.[45] In the 9/11 Commission Report, *it was estimated that al-Qaeda required $30 million-per-year to conduct its operations.*
- The Law Committee reviews Sharia law, and decides whether particular courses of action conform to it.
- The Islamic Study/*Fatwah* Committee issues religious edicts, such as an edict in 1998 telling Muslims to kill Americans.
- In the late 1990s there was a publicly known Media Committee, which ran the now-defunct newspaper *Nashrat al Akhbar (Newscast)* and handled public relations.
- In 2005, al-Qaeda formed As-Sahab, a media production house, to supply its video and audio materials.

Command structure

When asked about the possibility of al-Qaeda's connection to the July 7, 2005 London bombings in 2005, Metropolitan Police Commissioner Sir Ian Blair said: "Al-Qaeda is not an organization. Al-Qaeda is a way of working ... but this has the hallmark of that approach ... al-Qaeda clearly has the ability to provide training ... to provide expertise ... and I think that is what has occurred here."[46]

On August 13, 2005, however, *The Independent* newspaper, quoting police and MI5 investigations, reported that the July 7 bombers had acted independently of an al-Qaeda terror mastermind someplace abroad.[47]

What exactly al-Qaeda is, or was, remains in dispute. Author and journalist Adam Curtis argues that the idea of al-Qaeda as a formal organization is primarily an American invention. Curtis contends the name "al-Qaeda" was first brought to the attention of the public in the 2001 trial of bin Laden and the four men accused of the 1998 US embassy bombings in East Africa:

> The reality was that bin Laden and Ayman al-Zawahiri had become the focus of a loose association of disillusioned Islamist militants who were attracted by the new strategy. But there was no organization. These were militants who mostly planned their own operations and looked to bin Laden for funding and assistance. He was not their commander. There is also no evidence that bin Laden used the term "al-Qaeda" to refer to the name of a group until after September 11 attacks, when he realized that this was the term the Americans had given it.[48]

As a matter of law, the US Department of Justice needed to show that bin Laden was the leader of a criminal organization in order to charge him *in absentia* under the Racketeer Influenced and Corrupt Organizations Act, also known as the RICO statutes. The name of the organization and details of its structure were provided in the testimony of Jamal al-Fadl, who said he was a founding member of the organization and a former employee of bin Laden.[49] Questions about the reliability of al-Fadl's testimony have been raised by a number of sources because of his history of dishonesty, and because he was delivering it as part of a plea bargain agreement after being convicted of conspiring to attack U.S. military establishments.[37] [50] Sam Schmidt, one of his defense lawyers, said:

> There were selective portions of al-Fadl's testimony that I believe was false, to help support the picture that he helped the Americans join together. I think he lied in a number of specific testimony about a unified image of what this organization was. It made al-Qaeda the new Mafia or the new Communists. It made them identifiable as a group and therefore made it easier to prosecute any person associated with al-Qaeda for any acts or statements made by bin Laden.[48]

Field operatives

The number of individuals in the organization who have undergone proper military training, and are capable of commanding insurgent forces, is largely unknown. In 2006, it was estimated that al-Qaeda had several thousand commanders embedded in 40 different countries.[51] As of 2009, it was believed that no more than 200–300 members were still active commanders.[52]

According to the award-winning 2004 BBC documentary *The Power of Nightmares*, al-Qaeda was so weakly linked together that it was hard to say it existed apart from bin Laden and a small clique of close associates. The lack of any significant numbers of convicted al-Qaeda members, despite a large number of arrests on terrorism charges, was cited by the documentary as a reason to doubt whether a widespread entity that met the description of al-Qaeda existed.[53]

Insurgent forces

According to Robert Cassidy, al-Qaeda controls two separate forces deployed alongside insurgents in Iraq and Pakistan. The first, numbering in the tens of thousands, was "organized, trained, and equipped as insurgent combat forces" in the Soviet-Afghan war.[51] It was made up primarily of foreign *mujahideen* from Saudi Arabia and Yemen. Many went on to fight in Bosnia and Somalia for global *jihad*. Another group, approximately 10,000 strong, live in Western states and have received rudimentary combat training.[51]

Other analysts have described al-Qaeda's rank and file as being "predominantly Arab," in its first years of operation, and now also includes "other peoples" as of 2007.[54] It has been estimated that 62% of al-Qaeda members have university education.[55]

Financing

Some financing for al-Qaeda in the 1990s came from the personal wealth of Osama bin Laden.[56] By 2001 Afghanistan had become politically complex and mired. With many financial sources for al-Qaeda Bin Laden's financing role may have become comparatively minor. Sources in 2001 could also have included Jamaa Al-Islamiyya and Islamic Jihad, both associated with Afghan-based Egyptians.[57] Other sources of income in 2001 included the heroin trade and donations from supporters in Kuwait, Saudi Arabia and other Islamic countries.[56] A WikiLeaks released memo from the United States Secretary of State sent in 2009 asserted that the primary source of funding of Sunni terrorist groups worldwide was Saudi Arabia.[58]

Strategy

On March 11, 2005, *Al-Quds Al-Arabi* published extracts from Saif al-Adel's document "Al Quaeda's Strategy to the Year 2020".[59] [60] Abdel Bari Atwan summarizes this strategy as comprising five stages to rid the Ummah from all forms of oppression:

1. Provoke the United States and the West into invading a Muslim country by staging a massive attack or string of attacks on U.S. soil that results in massive civilian casualties.
2. Incite local resistance to occupying forces.
3. Expand the conflict to neighboring countries, and engage the U.S. and its allies in a long war of attrition.
4. Convert al-Qaeda into an ideology and set of operating principles that can be loosely franchised in other countries without requiring direct command and control, and via these franchises incite attacks against the U.S. and countries allied with the U.S. until they withdraw from the conflict, as happened with the 2004 Madrid train bombings, but which did not have the same effect with the July 7, 2005 London bombings.
5. The U.S. economy will finally collapse by the year 2020 under the strain of multiple engagements in numerous places, making the worldwide economic system which is dependent on the U.S. also collapse leading to global political instability, which in turn leads to a global jihad led by al-Qaeda and a Wahhabi Caliphate will then be installed across the world following the collapse of the U.S. and the rest of the Western world countries.

Atwan also noted, regarding the collapse of the U.S., "If this sounds far-fetched, it is sobering to consider that this virtually describes the downfall of the Soviet Union."[59]

Etymology

In Arabic, *al-Qaeda* has four syllables (*al-qāʿidah*, Arabic pronunciation: [ælˈqɑːʕɪdɐ] or Arabic pronunciation: [ælqɑːˈʕedæ]). However, since two of the Arabic consonants in the name (the voiceless uvular plosive [q] and the voiced pharyngeal fricative [ʕ]) are not phones found in the English language, the closest naturalized English pronunciations include /ælˈkaɪdə/, /ælˈkeɪdə/ and /ˌælkɑːˈiːdə/. al-Qaeda's name can also be transliterated as al-Qaida, al-Qa'ida, el-Qaida, or al-Qaeda.[61]

The name comes from the Arabic noun *qā'idah*, which means *foundation* or *basis*, and can also refer to a military base. The initial *al-* is the Arabic definite article *the*, hence *the base*.[62]

Bin Laden explained the origin of the term in a videotaped interview with Al Jazeera journalist Tayseer Alouni in October 2001:

> The name 'al-Qaeda' was established a long time ago by mere chance. The late Abu Ebeida El-Banashiri established the training camps for our *mujahedeen* against Russia's terrorism. We used to call the training camp al-Qaeda. The name stayed.[63]

It has been argued that two documents seized from the Sarajevo office of the Benevolence International Foundation prove that the name was not simply adopted by the *mujahid* movement and that a group called al-Qaeda was established in August 1988. Both of these documents contain minutes of meetings held to establish a new military group, and contain the term "al-Qaeda".[64]

Former British Foreign Secretary Robin Cook wrote that the word al-Qaeda should be translated as "the database", and originally referred to the computer file of the thousands of *mujahideen* militants who were recruited and trained with CIA help to defeat the Russians.[65] In April 2002, the group assumed the name *Qa'idat al-Jihad*, which means "the base of Jihad". According to Diaa Rashwan, this was "apparently as a result of the merger of the overseas branch of Egypt's al-Jihad (Egyptian Islamist Jihad, or EIJ) group, led by Ayman al-Zawahiri, with the groups Bin Laden brought under his control after his return to Afghanistan in the mid-1990s."[66]

Ideology

The radical Islamist movement in general and al-Qaeda in particular developed during the Islamic revival and Islamist movement of the last three decades of the 20th century, along with less extreme movements.

Some have argued that "without the writings" of Islamic author and thinker Sayyid Qutb, "al-Qaeda would not have existed."[67] Qutb preached that because of the lack of *sharia* law, the Muslim world was no longer Muslim, having reverted to pre-Islamic ignorance known as *jahiliyyah*.

To restore Islam, he said a vanguard movement of righteous Muslims was needed to establish "true Islamic states", implement *sharia*, and rid the Muslim world of any non-Muslim influences, such as concepts like socialism and nationalism. Enemies of Islam in Qutb's view included "treacherous Orientalists"[68] and "world Jewry", who plotted "conspiracies" and "wicked[ly]" opposed Islam.

In the words of Mohammed Jamal Khalifa, a close college friend of bin Laden:

> Islam is different from any other religion; it's a way of life. We [Khalifa and bin Laden] were trying to understand what Islam has to say about how we eat, who we marry, how we talk. We read Sayyid Qutb. He was the one who most affected our generation.[69]

Qutb had an even greater influence on bin Laden's mentor and another leading member of al-Qaeda,[70] Ayman al-Zawahiri. Zawahiri's uncle and maternal family patriarch, Mafouz Azzam, was Qutb's student, then protégé, then personal lawyer, and finally executor of his estate—one of the last people to see Qutb before his execution. "Young Ayman al-Zawahiri heard again and again from his beloved uncle Mahfouz about the purity of Qutb's character and the torment he had endured in prison."[71] Zawahiri paid homage to Qutb in his work *Knights under the Prophet's Banner*.[72]

One of the most powerful of Qutb's ideas was that many who said they were Muslims were not. Rather, they were apostates. That not only gave jihadists "a legal loophole around the prohibition of killing another Muslim," but made "it a religious obligation to execute" these self-professed Muslims. These alleged apostates included leaders of Muslim countries, since they failed to enforce *sharia* law.[73]

Religious compatibility

Abdel Bari Atwan writes that:

> While the leadership's own theological platform is essentially Salafi, the organization's umbrella is sufficiently wide to encompass various schools of thought and political leanings. Al-Qaeda counts among its members and supporters people associated with Wahhabism, Shafi'ism, Malikism, and Hanafism. There are even some whose beliefs and practices are directly at odds with Salafism, such as Yunis Khalis, one of the leaders of the Afghan mujahedin. He is a mystic who visits tombs of saints and seeks their blessings—practices inimical to bin Laden's Wahhabi-Salafi school of thought. The only exception to this pan-Islamic policy is Shi'ism. Al-Qaeda seems implacably opposed to it, as it holds Shi'ism to be heresy. In Iraq it has openly declared war on the Badr Brigades, who have fully cooperated with the US, and now considers even Shi'i civilians to be legitimate targets for acts of violence.[74]

History

Researchers have described five distinct phases in the development of al-Qaeda: the beginning in the late 1980s, the "wilderness" period in 1990–96, its "heyday" in 1996–2001, the network period of 2001–05, and a period of fragmentation from 2005 to today.[75]

Jihad in Afghanistan

The origins of al-Qaeda as a network inspiring terrorism around the world and training operatives can be traced to the Soviet War in Afghanistan (December 1979 – February 1989).[17] The U.S. viewed the conflict in Afghanistan, with the Afghan Marxists and allied Soviet troops on one side and the native Afghan *mujahideen*, some of whom were radical Islamic militants, on the other, as a blatant case of Soviet expansionism and aggression. A CIA program called Operation Cyclone channeled funds through Pakistan's Inter-Services Intelligence agency to the Afghan Mujahideen who were fighting the Soviet occupation.[76]

CIA-funded and ISI-trained Afghan mujahideen fighters crossing the Durand Line border to fight Soviet forces and the Soviet-backed Afghan government in 1985.

At the same time, a growing number of Arab *mujahideen* joined the *jihad* against the Afghan Marxist regime, facilitated by international Muslim organizations, particularly the Maktab al-Khidamat,[77] which was funded by the Saudi Arabia government as well as by individual Muslims (particularly Saudi businessmen who were approached by bin Laden). Together, these sources donated some $600 million a year to jihad.[78]

In 1984 Maktab al-Khidamat (MAK), or the "Services Office", a Muslim organization founded to raise and channel funds and recruit foreign *mujahideen* for the war against the Soviets in Afghanistan, was established in Peshawar, Pakistan, by bin Laden and Abdullah Yusuf Azzam, a Palestinian Islamic scholar and member of the Muslim Brotherhood. MAK organized guest houses in Peshawar, near the Afghan border, and gathered supplies for the construction of paramilitary training camps to prepare foreign recruits for the Afghan war front. Bin Laden became a "major financier" of the *mujahideen*, spending his own money and using his connections with "the Saudi royal family and the petro-billionaires of the Gulf" to influence public opinion about the war and raise additional funds.[79]

Omar Abdel-Rahman

From 1986, MAK began to set up a network of recruiting offices in the U.S., the hub of which was the Al Kifah Refugee Center at the Farouq Mosque on Brooklyn's Atlantic Avenue. Among notable figures at the Brooklyn center were "double agent" Ali Mohamed, whom FBI special agent Jack Cloonan called "bin Laden's first trainer,"[80] and "Blind Sheikh" Omar Abdel-Rahman, a leading recruiter of *mujahideen* for Afghanistan. Al-Qaeda evolved from MAK.

Azzam and bin Laden began to establish camps in Afghanistan in 1987.[81]

U.S. government financial support for the Afghan Islamic militants was substantial. Aid to Gulbuddin Hekmatyar, an Afghan *mujahideen* leader. and founder and leader of the Hezb-e Islami radical Islamic militant faction, alone amounted "by the most conservative estimates" to $600 million. Later, in the early 1990s, after the U.S. had withdrawn support, Hekmatyar "worked closely" with bin Laden.[82] In addition to receiving hundreds of millions of dollars in American aid, Hekmatyar was the recipient of the lion's share of Saudi aid.[83] There is evidence that the CIA supported Hekmatyar's drug trade activities by giving him immunity for his opium trafficking, which financed the operation of his militant faction.[84]

MAK and foreign *mujahideen* volunteers, or "Afghan Arabs," did not play a major role in the war. While over 250,000 Afghan *mujahideen* fought the Soviets and the communist Afghan government, it is estimated that were never more than 2,000 foreign *mujahideen* in the field at any one time.[85] Nonetheless, foreign *mujahideen* volunteers came from 43 countries, and the total number that participated in the Afghan movement between 1982 and 1992 is reported to have been 35,000.[86] Bin Laden played a central role in organizing training camps for the foreign Muslim volunteers.[87] [88]

The Soviet Union finally withdrew from Afghanistan in 1989. To the surprise of many, Mohammad Najibullah's communist Afghan government hung on for three more years, before being overrun by elements of the *mujahideen*. With *mujahideen* leaders unable to agree on a structure for governance, chaos ensued, with constantly reorganizing alliances fighting for control of ill-defined territories, leaving the country devastated.

Expanding operations

> the correlation between the words and deeds of bin Laden, his lieutenants, and their allies was close to perfect—if they said they were going to do something, they were much more than likely to try to do it. Their record in this regard puts Western leaders to shame.
>
> —Michael Scheuer, CIA Station Chief[89]

Toward the end of the Soviet military mission in Afghanistan, some *mujahideen* wanted to expand their operations to include Islamist struggles in other parts of the world, such as Israel and Kashmir. A number of overlapping and interrelated organizations were formed, to further those aspirations.

One of these was the organization that would eventually be called al-Qaeda, formed by bin Laden with an initial meeting held on August 11, 1988.[86]

Notes of a meeting of bin Laden and others on August 20, 1988, indicate al-Qaeda was a formal group by that time: "basically an organized Islamic faction, its goal is to lift the word of God, to make His religion victorious." A list of requirements for membership itemized the following: listening ability, good manners, obedience, and making a pledge (*bayat*) to follow one's superiors.[90]

According to Wright, the group's real name wasn't used in public pronouncements because "its existence was still a closely held secret."[91] His research suggests that al-Qaeda was formed at an August 11, 1988, meeting between "several senior leaders" of Egyptian Islamic Jihad, Abdullah Azzam, and bin Laden, where it was agreed to join bin

Laden's money with the expertise of the Islamic Jihad organization and take up the jihadist cause elsewhere after the Soviets withdrew from Afghanistan.[92]

Bin Laden wished to establish non-military operations in other parts of the world; Azzam, in contrast, wanted to remain focused on military campaigns. After Azzam was assassinated in 1989, the MAK split, with a significant number joining bin Laden's organization.

In November 1989, Ali Mohamed, a former special forces Sergeant stationed at Fort Bragg, North Carolina, left military service and moved to California. He traveled to Afghanistan and Pakistan and became "deeply involved with bin Laden's plans."[93]

A year later, on November 8, 1990, the FBI raided the New Jersey home of Ali Mohammed's associate El Sayyid Nosair, discovering a great deal of evidence of terrorist plots, including plans to blow up New York City skyscrapers.[94] Nosair was eventually convicted in connection to the 1993 World Trade Center bombing, and for the murder of Rabbi Meir Kahane on November 5, 1990. In 1991, Ali Mohammed is said to have helped orchestrate bin Laden's relocation to Sudan.[95]

Gulf War and the start of U.S. enmity

Following the Soviet Union's withdrawal from Afghanistan in February 1989, bin Laden returned to Saudi Arabia. The Iraqi invasion of Kuwait in August 1990 had put the Kingdom and its ruling House of Saud at risk. The world's most valuable oil fields were within easy striking distance of Iraqi forces in Kuwait, and Saddam's call to pan-Arab/Islamism could potentially rally internal dissent.

In the face of a seemingly massive Iraqi military presence, Saudi Arabia's own forces were well armed but far outnumbered. Bin Laden offered the services of his *mujahideen* to King Fahd to protect Saudi Arabia from the Iraqi army. The Saudi monarch refused bin Laden's offer, opting instead to allow U.S. and allied forces to deploy troops into Saudi territory.[96]

The deployment angered Bin Laden, as he believed the presence of foreign troops in the "land of the two mosques" (Mecca and Medina) profaned sacred soil. After speaking publicly against the Saudi government for harboring American troops, he was banished and forced to live in exile in Sudan.

Sudan

From around 1992 to 1996, al-Qaeda and bin Laden based themselves in Sudan at the invitation of Islamist theoretician Hassan al Turabi. The move followed an Islamist coup d'état in Sudan, led by Colonel Omar al-Bashir, who professed a commitment to reordering Muslim political values. During this time, bin Laden assisted the Sudanese government, bought or set up various business enterprises, and established camps where insurgents trained.

A key turning point for bin Laden, further pitting him against the Sauds, occurred in 1993 when Saudi Arabia gave support for the Oslo Accords, which set a path for peace between Israel and Palestinians.[97]

Zawahiri and the EIJ, who served as the core of al-Qaeda but also engaged in separate operations against the Egyptian government, had bad luck in Sudan. In 1993, a young schoolgirl was killed in an unsuccessful EIJ attempt on the life of the Egyptian prime minister, Atef Sedki. Egyptian public opinion turned against Islamist bombings, and the police arrested 280 of al-Jihad's members and executed 6.[98]

Due to bin Laden's continuous verbal assault on King Fahd of Saudi Arabia, on March 5, 1994 Fahd sent an emissary to Sudan demanding bin Laden's passport; bin Laden's Saudi citizenship was also revoked. His family was persuaded to cut off his monthly stipend, $7 million ($11000000 today) a year, and his Saudi assets were frozen.[99] [100] His family publicly disowned him. There is controversy over whether and to what extent he continued to garner support from members of his family and/or the Saudi government.[101]

In June 1995 an even more ill-fated attempt to assassinate Egyptian president Mubarak led to the expulsion of EIJ, and in May 1996, of bin Laden, by the Sudanese government.

According to Pakistani-American businessman Mansoor Ijaz, the Sudanese government offered the Clinton Administration numerous opportunities to arrest bin Laden. Those opportunities were met positively by Secretary of State Madeleine Albright, but spurned when Susan Rice and counter-terrorism czar Richard Clarke persuaded National Security Advisor Sandy Berger to overrule Albright. Ijaz's claims appeared in numerous Op-Ed pieces, including one in the *Los Angeles Times*[102] and one in *The Washington Post* co-written with former Ambassador to Sudan Timothy M. Carney.[103] Similar allegations have been made by *Vanity Fair* contributing editor David Rose,[104] and Richard Miniter, author of *Losing bin Laden*, in a November 2003 interview with *World*.[105]

Several sources dispute Ijaz's claim, including the National Commission on Terrorist Attacks on the U.S. (the 9–11 Commission), which concluded in part:

> Sudan's minister of defense, Fatih Erwa, has claimed that Sudan offered to hand Bin Ladin over to the U.S. The Commission has found no credible evidence that this was so. Ambassador Carney had instructions only to push the Sudanese to expel Bin Ladin. Ambassador Carney had no legal basis to ask for more from the Sudanese since, at the time, there was no indictment out-standing.[106]

Refuge in Afghanistan

After the Soviet withdrawal, Afghanistan was effectively ungoverned for seven years and plagued by constant infighting between former allies and various *mujahideen* groups.

Throughout the 1990s, a new force began to emerge. The origins of the Taliban (literally "students") lay in the children of Afghanistan, many of them orphaned by the war, and many of whom had been educated in the rapidly expanding network of Islamic schools (madrassas) either in Kandahar or in the refugee camps on the Afghan-Pakistani border.

According to Ahmed Rashid, five leaders of the Taliban were graduates of Darul Uloom Haqqania, a madrassa in the small town of Akora Khattak.[107] The town is situated near Peshawar in Pakistan, but largely attended by Afghan refugees.[107] This institution reflected Salafi beliefs in its teachings, and much of its funding came from private donations from wealthy Arabs. Bin Laden's contacts were still laundering most of these donations, using "unscrupulous" Islamic banks to transfer the money to an "array" of charities which serve as front groups for al-Qaeda, or transporting cash-filled suitcases straight into Pakistan.[108] Another four of the Taliban's leaders attended a similarly funded and influenced madrassa in Kandahar.

Many of the *mujahideen* who later joined the Taliban fought alongside Afghan warlord Mohammad Nabi Mohammadi's Harkat i Inqilabi group at the time of the Russian invasion. This group also enjoyed the loyalty of most Afghan Arab fighters.

The continuing internecine strife between various factions, and accompanying lawlessness following the Soviet withdrawal, enabled the growing and well-disciplined Taliban to expand their control over territory in Afghanistan, and it came to establish an enclave which it called the Islamic Emirate of Afghanistan. In 1994, it captured the regional center of Kandahar, and after making rapid territorial gains thereafter, conquered the capital city Kabul in September 1996.

After the Sudanese made it clear, in May 1996, that bin Laden would never be welcome to return, Taliban-controlled Afghanistan—with previously established connections between the groups, administered with a shared militancy,[109] and largely isolated from American political influence and military power—provided a perfect location for al-Qaeda to relocate its headquarters. Al-Qaeda enjoyed the Taliban's protection and a measure of legitimacy as part of their Ministry of Defense, although only Pakistan, Saudi Arabia, and the United Arab Emirates recognized the Taliban as the legitimate government of Afghanistan.

While in Afghanistan, the Taliban government tasked al-Qaeda with the training of Brigade 055, an elite part of the Taliban's army from 1997–2001. The Brigade was made up of mostly foreign fighters, many veterans from the Soviet Invasion, and all under the same basic ideology of the mujahideen. In November 2001, as Operation Enduring

Freedom had toppled the Taliban government, many Brigade 055 fighters were captured or killed, and those that survived were thought to head into Pakistan along with bin Laden.[110]

By the end of 2008, some sources reported that the Taliban had severed any remaining ties with al-Qaeda,[111] while others cast doubt on this.[112] According to senior U.S. military intelligence officials, there were fewer than 100 members of al-Qaeda remaining in Afghanistan in 2009.[113]

Call for global jihad

Around 1994, the Salafi groups waging *jihad* in Bosnia entered into a seemingly irreversible decline. As they grew less and less aggressive, groups such as EIJ began to drift away from the Salafi cause in Europe. Al-Qaeda decided to step in and assumed control of around 80% of the terrorist cells in Bosnia in late 1995.

At the same time, al-Qaeda ideologues instructed the network's recruiters to look for *Jihadi international*, Muslims who believed that *jihad* must be fought on a global level. The concept of a "global Salafi *jihad*" had been around since at least the early 1980s. Several groups had formed for the explicit purpose of driving non-Muslims out of every Muslim land, at the same time, and with maximum carnage. This was, however, a fundamentally defensive strategy.

Al-Qaeda sought to open the "offensive phase" of the global Salafi *jihad*.[114] Bosnian Islamists in 2006 called for "solidarity with Islamic causes around the world", supporting the insurgents in Kashmir and Iraq as well as the groups fighting for a Palestinian state.[115]

Fatwas

In 1996, al-Qaeda announced its *jihad* to expel foreign troops and interests from what they considered Islamic lands. Bin Laden issued a *fatwa* (binding religious edict),[116] which amounted to a public declaration of war against the U.S. and its allies, and began to refocus al-Qaeda's resources on large-scale, propagandist strikes. In June 1996, the Khobar Towers bombing took place in Khobar, Saudi Arabia, attributed by some to al-Qaeda, killing 19 Americans and one Saudi local, and wounding 372.[117]

Khobar Towers bombing

On February 23, 1998, bin Laden and Ayman al-Zawahiri, a leader of Egyptian Islamic Jihad, along with three other Islamist leaders, co-signed and issued a *fatwa* calling on Muslims to kill Americans and their allies where they can, when they can.[118] Under the banner of the World Islamic Front for Combat Against the Jews and Crusaders, they declared:

> [T]he ruling to kill the Americans and their allies—civilians and military—is an individual duty for every Muslim who can do it in any country in which it is possible to do it, in order to liberate the al-Aqsa Mosque [in Jerusalem] and the holy mosque [in Mecca] from their grip, and in order for their armies to move out of all the lands of Islam, defeated and unable to threaten any Muslim. This is in accordance with the words of Almighty Allah, 'and fight the pagans all together as they fight you all together,' and 'fight them until there is no more tumult or oppression, and there prevail justice and faith in Allah'.[119]

Neither bin Laden nor al-Zawahiri possessed the traditional Islamic scholarly qualifications to issue a *fatwa*. However, they rejected the authority of the contemporary *ulema* (which they saw as the paid servants of *jahiliyya* rulers), and took it upon themselves.[120] Former Russian FSB agent Alexander Litvinenko, who was later killed, said that the FSB trained al-Zawahiri in a camp in Dagestan eight months before the 1998 *fatwa*.[121] [122]

Iraq

Al-Qaeda is Sunni, and often attacked the Iraqi Shia majority in an attempt to incite sectarian violence and greater chaos in the country.[123] Al-Zarqawi purportedly declared an all-out war on Shiites[124] while claiming responsibility for Shiite mosque bombings.[125] The same month, a statement claiming to be by AQI rejected as "fake" a letter allegedly written by al-Zawahiri, in which he appears to question the insurgents' tactic of indiscriminately attacking Shiites in Iraq.[126] In a December 2007 video, al-Zawahiri defended the Islamic State in Iraq, but distanced himself from the attacks against civilians committed by "hypocrites and traitors existing among the ranks".[127]

U.S. and Iraqi officials accused AQI of trying to slide Iraq into a full-scale civil war between Iraq's majority Shiites and minority Sunni Arabs, with an orchestrated campaign of civilian massacres and a number of provocative attacks against high-profile religious targets.[128] With attacks such as the 2003 Imam Ali Mosque bombing, the 2004 Day of Ashura and Karbala and Najaf bombings, the 2006 first al-Askari Mosque bombing in Samarra, the deadly single-day series of bombings in which at least 215 people were killed in Baghdad's Shiite district of Sadr City, and the second al-Askari bombing in 2007, they provoked Shiite militias to unleash a wave of retaliatory attacks, resulting in death squad-style killings and spiraling further sectarian violence which escalated in 2006 and brought Iraq to the brink of violent anarchy in 2007.[129] In 2008, sectarian bombings blamed on al-Qaeda killed at least 42 people at the Imam Husayn Shrine in Karbala in March, and at least 51 people at a bus stop in Baghdad in June.

Somalia and Yemen

In Somalia, al-Qaeda agents had been collaborating closely with the its Somalian wing, which was created from the al-Shabaab group. In February 2012, al-Shabaab officially joined al-Qaeda, declaring loyalty in a joint video.[130] The Somalian al-Qaeda actively recruit children for suicide-bomber training, and export young people to participate in military actions against Americans at the AfPak border.[131]

The percentage of terrorist attacks in the West originating from the Afghanistan-Pakistan (AfPak) border declined considerably from almost 100% to 75% in 2007, and to 50% in 2010, as al-Qaeda shifted to Somalia and Yemen.[132] While al-Qaeda leaders are hiding in the tribal areas along the AfPak border, the middle-tier of the movement display heightened activity in Somalia and Yemen. "We know that South Asia is no longer their primary base," a U.S. defense agency source said. "They are looking for a hide-out in other parts of the world, and continue to expand their organization."

In January 2009, al-Qaeda's division in Saudi Arabia merged with its Yemeni wing to form al-Qaeda in the Arabian Peninsula.[133] Centered in Yemen, the group takes advantage of the country's poor economy, demography and domestic security. In August 2009, they made the first assassination attempt against a member of the Saudi royal dynasty in decades. President Obama asked his Yemen counterpart Ali Abdullah Saleh to ensure closer cooperation with the U.S. in the struggle against the growing activity of al-Qaeda in Yemen, and promised to send additional aid. Because of the wars in Iraq and Afghanistan, the U.S. was unable to pay sufficient attention to Somalia and Yemen, which could cause problems in the near future.[134] In December 2011, U.S. Secretary of Defense Leon Panetta said that the U.S. operations against al-Qaeda "are now concentrating on key groups in Yemen, Somalia and North Africa."[135] Al-Qaeda in the Arabian Peninsula claimed responsibility for the 2009 bombing attack on Northwest Airlines Flight 253 by Umar Farouk Abdulmutallab.[136] The group released photos of Abdulmutallab smiling in a white shirt and white Islamic skullcap, with the al-Qaeda in Arabian Peninsula banner in the background.

American operations

Anwar al-Awlaki

In December 1998, the Director of Central Intelligence Counterterrorist Center reported to the president that al-Qaeda was preparing for attacks in the USA, including the training of personnel to hijack aircraft.[137] On September 11, 2001, al-Qaeda attacked the United States, hijacking four airliners and deliberately crashing them.[138] The attackers killed 2,977 people.[139]

U.S. officials called Anwar al-Awlaki an "example of al-Qaeda reach into" the U.S. in 2008 after probes into his ties to the September 11 attacks hijackers. A former FBI agent identifies Awlaki as a known "senior recruiter for al-Qaeda", and a spiritual motivator.[140] Awlaki's sermons in the U.S. were attended by three of the 9/11 hijackers, as well as accused Fort Hood shooter Nidal Malik Hasan. U.S. intelligence intercepted emails from Hasan to Awlaki between December 2008 and early 2009. On his website, Awlaki has praised Hasan's actions in the Fort Hood shooting.[141]

An unnamed official claimed there was good reason to believe Awlaki "has been involved in very serious terrorist activities since leaving the U.S. [after 9/11], including plotting attacks against America and our allies."[142] In addition, "Christmas Day bomber" Umar Farouk Abdulmutallab said al-Awlaki was one of his al-Qaeda trainers, meeting with him and involved in planning or preparing the attack, and provided religious justification for it, according to unnamed U.S. intelligence officials.[143] [144] [145] In March 2010, al-Awlaki said in a videotape delivered to CNN that jihad against America was binding upon himself and every other able Muslim.[146] [147]

U.S. President Barack Obama approved the targeted killing of al-Awlaki by April 2010, making al-Awlaki the first U.S. citizen ever placed on the CIA target list. That required the consent of the U.S. National Security Council, and officials said it was appropriate for an individual who posed an imminent danger to national security.[148] [149] [150] [151] In May 2010, Faisal Shahzad, who pleaded guilty to the 2010 Times Square car bombing attempt, told interrogators he was "inspired by" al-Awlaki, and sources said Shahzad had made contact with al-Awlaki over the internet.[152] [153] [154] Representative Jane Harman called him "terrorist number one", and *Investor's Business Daily* called him "the world's most dangerous man".[155] [156] In July 2010, the U.S. Treasury Department added him to its list of Specially Designated Global Terrorists, and the UN added him to its list of individuals associated with al-Qaeda.[157] In August 2010, al-Awlaki's father initiated a lawsuit against the U.S. government with the American Civil Liberties Union, challenging its order to kill al-Awlaki.[158] In October 2010, U.S. and U.K. officials linked al-Awlaki to the 2010 cargo plane bomb plot.[159] In September 2011, he was killed in a targeted killing drone attack in Yemen.[160] It was reported on March 16, 2012 that Osama bin Laden plotted to kill United States President Barack Obama.[161]

Death of Osama bin Laden

View of Osama bin Laden's compound in Abbottabad, Pakistan, where he was killed on May 1, 2011.

On May 1, 2011 in Washington, D.C. (May 2, Pakistan Standard Time), U.S. President Barack Obama announced that Osama bin Laden had been killed by "a small team of Americans" acting under Obama's direct orders, in a covert operation in Abbottabad, Pakistan,[162] [163] about 50 km (mi) north of Islamabad.[164] According to U.S. officials a team of 20–25 US Navy SEALs under the command of the Joint Special Operations Command and working with the CIA stormed bin Laden's compound in two helicopters. Bin Laden and those with him were killed during a firefight in which U.S. forces experienced no injuries or casualties.[165] According to one US official the attack was carried out without the knowledge or consent of the Pakistani authorities.[166] In Pakistan some people were reported to be shocked at the unauthorized incursion by US armed forces.[167] The site is a few miles from the Pakistan Military Academy in Kakul.[168] In his broadcast announcement President Obama said that U.S. forces "took care to avoid civilian casualties."[169] Details soon emerged that three men and a woman were killed along with Bin Laden, the woman being killed when she was "used as a shield by a male combatant".[166] DNA from bin Laden's body, compared with DNA samples on record from his dead sister,[170] confirmed bin Laden's identity.[171] The body was recovered by the US military and was in its custody[163] until, according to one US official, his body was buried at sea according to Islamic traditions.[164] [172] One U.S. official stated that "finding a country willing to accept the remains of the world's most wanted terrorist would have been difficult."[173] U.S State Department issued a "Worldwide caution" for Americans following Bin Laden's death and U.S Diplomatic facilities everywhere were placed on high alert, a senior U.S official said.[174] Crowds gathered outside the White House and in New York City's Time Square to celebrate Bin Laden's death.[175]

Attacks

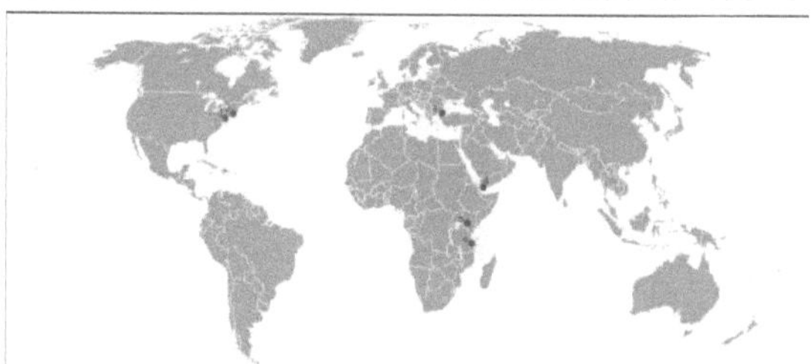

Map of major attacks attributed to al-Qaeda:
1. The Pentagon, US – Sep 11, 2001
2. World Trade Center, US – Sep 11, 2001
3. Istanbul, Turkey – Nov 15, 2003; Nov 20, 2003
4. Aden, Yemen – Oct 12, 2000
5. Nairobi, Kenya – Aug 7, 1998
6. Dar es Salaam, Tanzania – Aug 7, 1998

Al-Qaeda has carried out a total of six major terrorist attacks, four of them in its jihad against America. In each case the leadership planned the attack years in advance, arranging for the shipment of weapons and explosives and using its privatized businesses to provide operatives with safehouses and false identities.

Al-Qaeda usually does not disburse funds for attacks, and very rarely makes wire transfers.[176]

1992

On December 29, 1992, al-Qaeda's first terrorist attack took place as two bombs were detonated in Aden, Yemen. The first target was the Movenpick Hotel and the second was the parking lot of the Goldmohur Hotel.

The bombings were an attempt to eliminate American soldiers on their way to Somalia to take part in the international famine relief effort, Operation Restore Hope. Internally, al-Qaeda considered the bombing a victory that frightened the Americans away, but in the U.S. the attack was barely noticed.

No Americans were killed because the soldiers were staying in a different hotel altogether, and they went on to Somalia as scheduled. However little noticed, the attack was pivotal as it was the beginning of al-Qaeda's change in direction, from fighting armies to killing civilians.[177] Two people were killed in the bombing, an Australian tourist and a Yemeni hotel worker. Seven others, mostly Yemenis, were severely injured.

Two fatwas are said to have been appointed by the most theologically knowledgeable of al-Qaeda's members, Mamdouh Mahmud Salim, to justify the killings according to Islamic law. Salim referred to a famous fatwa appointed by Ibn Taymiyyah, a 13th-century scholar much admired by Wahhabis, which sanctioned resistance by any means during the Mongol invasions.[178]

1993 World Trade Center bombing

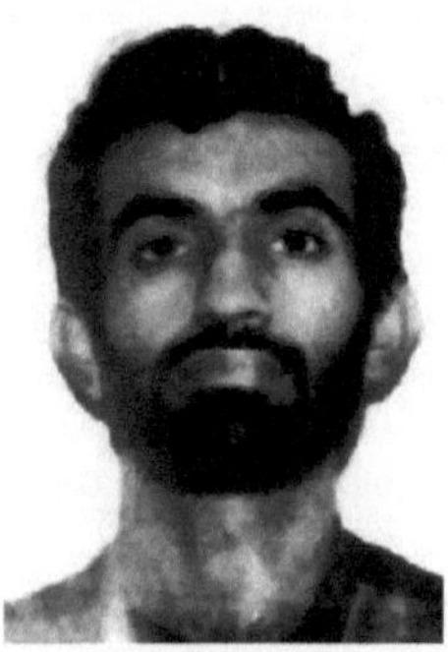
Ramzi Yousef

In 1993, Ramzi Yousef used a truck bomb to attack the World Trade Center in New York City. The attack was intended to break the foundation of Tower One knocking it into Tower Two, bringing the entire complex down.

Yousef hoped this would kill 250,000 people. The towers shook and swayed but the foundation held and he succeeded in killing only six people (although he injured 1,042 others and caused nearly $300 million in property damage).[179][180]

After the attack, Yousef fled to Pakistan and later moved to Manila. There he began developing the Bojinka Plot plans to implode a dozen American airliners simultaneously, to assassinate Pope John Paul II and President Bill Clinton, and to crash a private plane into CIA headquarters. He was later captured in Pakistan.[179]

None of the U.S. government's indictments against bin Laden have suggested that he had any connection with this bombing, but Ramzi Yousef is known to have attended a terrorist training camp in Afghanistan. After his capture, Yousef declared that his primary justification for the attack was to punish the U.S. for its support for the Israeli occupation of Palestinian territories and made no mention of any religious motivations.[180]

Late 1990s

1998 Nairobi embassy bombing

In 1996, bin Laden personally engineered a plot to assassinate Clinton while the president was in Manila for the Asia Pacific Economic Cooperation. However, intelligence agents intercepted a message just minutes before the motorcade was to leave, and alerted the U.S. Secret Service. Agents later discovered a bomb planted under a bridge.[181]

The 1998 U.S. embassy bombings in East Africa, resulting in upward of 300 deaths, mostly locals. A barrage of cruise missiles launched by the U.S. military in response devastated an al-Qaeda base in Khost, Afghanistan, but the network's capacity was unharmed.

In October 2000, al-Qaeda militants in Yemen bombed the missile destroyer *U.S.S. Cole* in a suicide attack, killing 17 U.S. servicemen and damaging the vessel while it lay offshore. Inspired by the success of such a brazen attack, al-Qaeda's command core began to prepare for an attack on the U.S. itself.

September 11 attacks

Aftermath of the September 11 attacks

The September 11, 2001 attacks were the most devastating terrorist acts in American history, killing approximately 3,000 people. Two commercial airliners were deliberately flown into the World Trade Center towers, a third into The Pentagon, and a fourth, originally intended to target the United States Capitol, crashed in a field in Shanksville, Pennsylvania.

The attacks were conducted by al-Qaeda, acting in accord with the 1998 *fatwa* issued against the U.S. and its allies by military forces under the command of bin Laden, al-Zawahiri, and others.[182] Evidence points to suicide squads led by al-Qaeda military commander Mohamed Atta as the culprits of the attacks, with bin Laden, Ayman al-Zawahiri, Khalid Shaikh Mohammed, and Hambali as the key planners and part of the political and military command.

Messages issued by bin Laden after September 11, 2001 praised the attacks, and explained their motivation while denying any involvement.[183] Bin Laden legitimized the attacks by identifying grievances felt by both mainstream and Islamist Muslims, such as the general perception that the U.S. was actively oppressing Muslims.[184]

Bin Laden asserted that America was massacring Muslims in 'Palestine, Chechnya, Kashmir and Iraq' and that Muslims should retain the 'right to attack in reprisal'. He also claimed the 9/11 attacks were not targeted at women and children, but 'America's icons of military and economic power'.[185]

Evidence has since come to light that the original targets for the attack may have been nuclear power stations on the east coast of the U.S. The targets were later altered by al-Qaeda, as it was feared that such an attack "might get out of hand".[186] [187]

Designation as terrorist organization

Al-Qaeda has been designated a *terrorist organization* by the following countries and international organizations:

- Australia[188]
- Canada[189]
- European Union[190]
- France[191]
- Germany
- India[192]
- Ireland
- Israel[193]
- Japan[194]
- Netherlands[195]
- NATO[196] [197]
- Philippines
- Russia[198]
- South Korea[199]
- Sweden[200]
- Switzerland[201]
- Turkey[202]
- United Kingdom[203]
- United Nations Security Council[204]
- United States[205]

War on Terrorism

In the immediate aftermath of the attacks, the U.S. government decided to respond militarily, and began to prepare its armed forces to overthrow the Taliban regime it believed was harboring al-Qaeda. Before the U.S. attacked, it offered Taliban leader Mullah Omar a chance to surrender bin Laden and his top associates. The first forces to be inserted into Afghanistan were Paramilitary Officers from the CIA's elite Special Activities Division (SAD).

U.S. troops in Afghanistan

The Taliban offered to turn over bin Laden to a neutral country for trial if the U.S. would provide evidence of bin Laden's complicity in the attacks. U.S. President George W. Bush responded by saying: "We know he's guilty. Turn him over",[206] and British Prime Minister Tony Blair warned the Taliban regime: "Surrender bin Laden, or surrender power".[207]

Soon thereafter the U.S. and its allies invaded Afghanistan, and together with the Afghan Northern Alliance removed the Taliban government in the war in Afghanistan.

As a result of the U.S. using its special forces and providing air support for the Northern Alliance ground forces, both Taliban and al-Qaeda training camps were destroyed, and much of the operating structure of al-Qaeda is believed to have been disrupted. After being driven from their key positions in the Tora Bora area of Afghanistan, many al-Qaeda fighters tried to regroup in the rugged Gardez region of the nation.

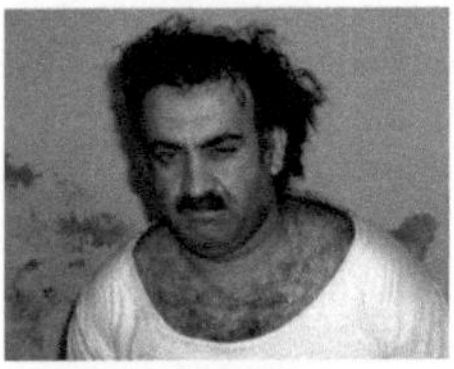
Khalid Sheikh Mohammed after his arrest in Rawalpindi, Pakistan, in March 2003.

Again, under the cover of intense aerial bombardment, U.S. infantry and local Afghan forces attacked, shattering the al-Qaeda position and killing or capturing many of the militants. By early 2002, al-Qaeda had been dealt a serious blow to its operational capacity, and the Afghan invasion appeared an initial success. Nevertheless, a significant Taliban insurgency remains in Afghanistan, and al-Qaeda's top two leaders, bin Laden and al-Zawahiri, evaded capture.

Debate raged about the exact nature of al-Qaeda's role in the 9/11 attacks, and after the U.S. invasion began, the U.S. State Department also released a videotape showing bin Laden speaking with a small group of associates somewhere in Afghanistan shortly before the Taliban was removed from power.[208] Although its authenticity has been questioned by some,[209] the tape appears to implicate bin Laden and al-Qaeda in the September 11 attacks and was aired on many television channels all over the world, with an accompanying English translation [210] provided by the U.S. Defense Department.

In September 2004, the US government 9/11 Commission investigating the September 11 attacks officially concluded that the attacks were conceived and implemented by al-Qaeda operatives.[211] In October 2004, bin Laden appeared to claim responsibility for the attacks in a videotape released through Al Jazeera, saying he was inspired by Israeli attacks on high-rises in the 1982 invasion of Lebanon: "As I looked at those demolished towers in Lebanon, it entered my mind that we should punish the oppressor in kind and that we should destroy towers in America in order that they taste some of what we tasted and so that they be deterred from killing our women and children."[212]

By the end of 2004, the U.S. government proclaimed that two-thirds of the most senior al-Qaeda figures from 2001 had been captured and interrogated by the CIA: Abu Zubaydah, Ramzi bin al-Shibh and Abd al-Rahim al-Nashiri in 2002;[213] Khalid Sheikh Mohammed in 2003; and Saif al Islam el Masry in 2004. Mohammed Atef and several others were killed.

Activities

Africa

Al-Qaeda involvement in Africa has included a number of bombing attacks in North Africa, as well as supporting parties in civil wars in Eritrea and Somalia. From 1991 to 1996, bin Laden and other al-Qaeda leaders were based in Sudan.

Islamist rebels in the Sahara calling themselves al-Qaeda in the Islamic Maghreb have stepped up their violence in recent years.[215] French officials say the rebels have no real links to the al-Qaeda leadership, but this is a matter of some dispute in the international press and amongst security analysts. It seems likely that bin Laden approved the group's name in late 2006, and the rebels "took on the al Qaeda franchise label", almost a year before the violence began to escalate.[216]

Front page of *The Guardian Weekly* on the eighth anniversary of the September 11 attacks. The article claimed that al-Qaeda's activity is "increasingly dispersed to 'affiliates' or 'franchises' in Yemen and North Africa."[214]

Europe

In 2003, Islamists carried out a series of bombings in Istanbul killing fifty-seven people and injuring seven hundred. Seventy-four people were charged by the Turkish authorities. Some had previously met Bin Laden, and though they specifically declined to pledge allegiance to al-Qaeda they asked for its blessing and help.[217] [218]

In 2009, three Londoners, Tanvir Hussain, Assad Sarwar and Ahmed Abdullah Ali, were convicted of conspiring to detonate bombs disguised as soft drinks on seven airplanes bound for Canada and the U.S. The massively complex police and MI5 investigation of the plot involved more than a year of surveillance work conducted by over two hundred officers.[219] [220] British and U.S. officials said the plan—unlike many recent homegrown European terrorist plots—was directly linked to al-Qaeda and guided by senior Islamic militants in Pakistan.[221] [222]

In 2012, Russian Intelligence indicated that al-Qaeda had give a call for "forest jihad" and has been starting massive forest fires as part of a strategy of "thousand cuts".[223]

Arab world

Following Yemeni unification in 1990, Wahhabi networks began moving missionaries into the country in an effort to subvert the capitalist north. Although it is unlikely bin Laden or Saudi al-Qaeda were directly involved, the personal connections they made would be established over the next decade and used in the *USS Cole* bombing.[224] Concerns grow over Al Qaeda's group in Yemen.[225]

USS *Cole* after the October 2000 attack

In Iraq, al-Qaeda forces loosely associated with the leadership were embedded in the Jama'at al-Tawhid wal-Jihad organization commanded by Abu Musab al-Zarqawi. Specializing in suicide operations, they have been a "key driver" of the Sunni insurgency.[226] Although they played a small part in the overall insurgency, between 30% and 42% of all suicide bombings which took place in the early years were claimed by Zarqawi's organization.[227] Reports have indicated that oversights such as the failure to control access to the Qa'qaa munitions factory in Yusufiyah have allowed large quantities of munitions to fall into the hands of al-Qaida.[228] In November 2010, the Islamic State of Iraq militant group, which is linked to al-Qaeda in Iraq, threatened to "exterminate Iraqi Christians".[229] [230]

Significantly, it was not until the late 1990s that al-Qaeda began training Palestinians. This is not to suggest that resistance fighters are underrepresented in the network as a number of Palestinians, mostly coming from Jordan, wanted to join and have risen to serve high-profile roles in Afghanistan.[231] Rather, large groups such as Hamas and Palestinian Islamic Jihad—which cooperate with al-Qaeda in many respects—have had difficulties accepting a strategic alliance, fearing that al-Qaeda will co-opt their smaller cells. This may have changed recently, as Israeli security and intelligence services believe al-Qaeda has managed to infiltrate operatives from the Occupied Territories into Israel, and is waiting for the right time to mount an attack.[231]

Kashmir

Bin Laden and Ayman al-Zawahiri consider India to be a part of the 'Crusader-Zionist-Hindu' conspiracy against the Islamic world.[232] According to the 2005 report 'Al Qaeda: Profile and Threat Assessment' by Congressional Research Service, bin Laden was involved in training militants for Jihad in Kashmir while living in Sudan in the early nineties. By 2001 Kashmiri militant group Harkat-ul-Mujahideen had become a part of the al-Qaeda coalition.[233] According to the United Nations High Commissioner for Refugees al-Qaeda was thought to have established bases in Pakistan-administered Kashmir (in Azad Kashmir, and to some extent in Gilgit–Baltistan) during the 1999 Kargil War and continued to operate there with tacit approval of Pakistan's Intelligence services.[234]

Many of the militants active in Kashmir were trained in the same Madrasahs as Taliban and al-Qaeda. Fazlur Rehman Khalil of Kashmiri militant group Harkat-ul-Mujahideen was a signatory of al-Qaeda's 1998 declaration of Jihad against America and its allies.[235] In a 'Letter to American People' written by bin Laden in 2002 he stated that one of the reasons he was fighting America is because of her support to India on the Kashmir issue.[236] [237] In November 2001 Kathmandu airport went on high alert after threats that Bin Laden planned to hijack a plane from there and crash it into a target in New Delhi.[238] In 2002 U.S. Secretary of Defense Donald Rumsfeld, on a trip to Delhi, suggested that al-Qaeda was active in Kashmir though he did not have any hard evidence.[239] [240] He proposed hi tech ground sensors along the line of control to prevent militants from infiltrating into Indian administered Kashmir.[240] An investigation in 2002 unearthed evidence that al-Qaeda and its affiliates were prospering in Pakistan-administered Kashmir with tacit approval of Pakistan's National Intelligence agency Inter-Services Intelligence[241] In 2002 a special team of Special Air Service and Delta Force was sent into Indian Administered Kashmir to hunt for Bin Laden after reports that he was being sheltered by Kashmiri militant group Harkat-ul-Mujahideen which had previously been responsible for 1995 Kidnapping of western tourists in Kashmir.[242] Britain's highest ranking al-Qaeda operative Rangzieb Ahmed had previously fought in Kashmir with the group Harkat-ul-Mujahideen and spent time in Indian prison after being captured in Kashmir.[243]

U.S. officials believe that al-Qaeda was helping organize a campaign of terror in Kashmir in order to provoke conflict between India and Pakistan.[244] Their strategy was to force Pakistan to move its troops to the border with India, thereby relieving pressure on al-Qaeda elements hiding in northwestern Pakistan.[245] In 2006 al-Qaeda claimed they had established a wing in Kashmir; this has worried the Indian government.[235] [246] However the Indian Army Lt. Gen. H.S. Panag, GOC-in-C Northern Command, said to reporters that the army has ruled out the presence of al-Qaeda in Indian-administered Jammu and Kashmir; furthermore he said that there is nothing that can verify reports from the media of al-Qaeda presence in the state. He however stated that al-Qaeda had strong ties with Kashmiri militant groups Lashkar-e-Taiba and Jaish-e-Mohammed based in Pakistan.[247] It has been noted that Waziristan has now become the new battlefield for Kashmiri militants fighting NATO in support of al-Qaeda and Taliban.[248] [249] [250] Dhiren Barot, who wrote the *Army of Madinah In Kashmir*[251] and was an al-Qaeda operative convicted for involvement in the 2004 financial buildings plot, had received training in weapons and explosives at a militant training camp in Kashmir.[252]

Maulana Masood Azhar, the founder of another Kashmiri group Jaish-e-Mohammed, is believed to have met bin Laden several times and received funding from him.[235] In 2002 Jaish-e-Mohammed organized the kidnapping and murder of Daniel Pearl in an operation run in conjunction with al-Qaeda and funded by Bin Laden.[253] According to American counter-terrorism expert Bruce Riedel, al-Qaeda and Taliban were closely involved in the 1999 hijacking of Indian Airlines Flight 814 to Kandahar which led to the release of Maulana Masood Azhar & Ahmed Omar Saeed Sheikh from an Indian prison in exchange for the passengers. This hijacking, Riedel stated, was rightly described by then Indian Foreign minister Jaswant Singh as a 'dress rehearsal' for September 11 attacks.[254] Bin laden personally welcomed Azhar and threw a lavish party in his honor after his release, according to Abu Jandal, bodyguard of Bin Laden.[255] [256] Ahmed Omar Saeed Sheikh, who had been in Indian prison for his role in 1994 kidnappings of Western tourists in India, went on to murder Daniel Pearl and was sentenced to death by Pakistan. Al-Qaeda operative Rashid Rauf, who was one of the accused in 2006 transatlantic aircraft plot, was related to Maulana Masood Azhar by marriage.[257]

Lashkar-e-Taiba, a Kashmiri militant group which is thought to be behind 2008 Mumbai attacks, is also known to have strong ties to senior al-Qaeda leaders living in Pakistan.[258] In Late 2002 top al-Qaeda operative Abu Zubaydah was arrested while being sheltered by Lashkar-e-Taiba in a safe house in Faisalabad.[259] The FBI believes that al-Qaeda and Lashkar have been 'intertwined' for a long time while the CIA has said that al-Qaeda funds Lashkar-e-Taiba.[259] French investigating magistrate Jean-Louis Bruguière, who was the top French counter-terrorism official, told Reuters in 2009 that 'Lashkar-e-Taiba is no longer a Pakistani movement with only a Kashmir political or military agenda. Lashkar-e-Taiba is a member of al-Qaeda.'[260] [261]

In a video released in 2008, senior al-Qaeda operative American-born Adam Yahiye Gadahn stated that "victory in Kashmir has been delayed for years; it is the liberation of the jihad there from this interference which, Allah willing, will be the first step towards victory over the Hindu occupiers of that Islam land."[262]

In September 2009 a U.S. Drone strike reportedly killed Ilyas Kashmiri who was the chief of Harkat-ul-Jihad al-Islami, a Kashmiri militant group associated with al-Qaeda.[263] Kashmiri was described by Bruce Riedel as a 'prominent' al-Qaeda member[264] while others have described him as head of military operations for al-Qaeda.[265] [266] Kashmiri was also charged by the U.S. in a plot against Jyllands-Posten, the Danish newspaper which was at the center of Jyllands-Posten Muhammad cartoons controversy.[267] U.S. officials also believe that Kashmiri was involved in the Camp Chapman attack against the CIA.[268] In January 2010 Indian authorities notified Britain of an al-Qaeda plot to hijack an Indian airlines or Air India plane and crash it into a British city. This information was uncovered from interrogation of Amjad Khwaja, an operative of Harkat-ul-Jihad al-Islami, who had been arrested in India.[269]

In January 2010 U.S. Defense secretary Robert Gates, while on a visit to Pakistan, stated that al-Qaeda was seeking to destabilize the region and planning to provoke a nuclear war between India and Pakistan.[270]

Internet

Timothy L. Thomas claims that in the wake of its evacuation from Afghanistan, al-Qaeda and its successors have migrated online to escape detection in an atmosphere of increased international vigilance. As a result, the organization's use of the Internet has grown more sophisticated, encompassing financing, recruitment, networking, mobilization, publicity, as well as information dissemination, gathering and sharing.[271]

Abu Ayyub al-Masri

Abu Ayyub al-Masri's al-Qaeda movement in Iraq regularly releases short videos glorifying the activity of jihadist suicide bombers. In addition, both before and after the death of Abu Musab al-Zarqawi (the former leader of al-Qaeda in Iraq), the umbrella organization to which al-Qaeda in Iraq belongs, the Mujahideen Shura Council, has a regular presence on the Web. The range of multimedia content includes guerrilla training clips, stills of victims about to be murdered, testimonials of suicide bombers, and videos that show participation in jihad through stylized portraits of mosques and musical scores. A website associated with al-Qaeda posted a video of captured American entrepreneur Nick Berg being decapitated in Iraq. Other decapitation videos and pictures, including those of Paul Johnson, Kim Sun-il, and Daniel Pearl, were first posted on jihadist websites.

In December 2004 an audio message claiming to be from Bin Laden was posted directly to a website, rather than sending a copy to al Jazeera as he had done in the past.

Al-Qaeda turned to the Internet for release of its videos in order to be certain it would be available unedited, rather than risk the possibility of al Jazeera editors editing the videos and cutting out anything critical of the Saudi royal family.[272] Bin Laden's December 2004 message was much more vehement than usual in this speech, lasting over an hour.

In the past, Alneda.com and Jehad.net were perhaps the most significant al-Qaeda websites. Alneda was initially taken down by American Jon Messner, but the operators resisted by shifting the site to various servers and strategically shifting content.

The U.S. is currently attempting to extradite a British information technology specialist, Babar Ahmad, on charges of operating a network of English-language al-Qaeda websites, such as Azzam.com.[273] [274] Ahmad's extradition is opposed by various British Muslim organizations, such as the Muslim Association of Britain.

Aviation network

Al-Qaeda is believed to be operating a clandestine aviation network including "several Boeing 727 aircraft", turboprops and executive jets, according to a Reuters story. Based on a U.S. Department of Homeland Security report, the story said that al-Qaeda is possibly using aircraft to transport drugs and weapons from South America to various unstable countries in West Africa. A Boeing 727 can carry up to 10 tons of cargo. The drugs eventually are smuggled to Europe for distribution and sale, and the weapons are used in conflicts in Africa and possibly elsewhere. Gunmen with links to al-Qaeda have been increasingly kidnapping some Europeans for ransom. The profits from the drug and weapon sales, and kidnappings can, in turn, fund more militant activities.[275]

Alleged CIA involvement

Experts debate whether or not the al-Qaeda attacks were blowback from the American CIA's "Operation Cyclone" program to help the Afghan mujahideen. Robin Cook, British Foreign Secretary from 1997 to 2001, has written that al-Qaeda and Bin Laden were "a product of a monumental miscalculation by western security agencies", and that "Al-Qaida, literally 'the database', was originally the computer file of the thousands of mujahideen who were recruited and trained with help from the CIA to defeat the Russians."[276]

Munir Akram, Permanent Representative of Pakistan to the United Nations from 2002 to 2008, wrote in a letter published in the New York Times on January 19, 2008:

> The strategy to support the Afghans against Soviet military intervention was evolved by several intelligence agencies, including the C.I.A. and Inter-Services Intelligence, or ISI. After the Soviet withdrawal, the Western powers walked away from the region, leaving behind 40,000 militants imported from several countries to wage the anti-Soviet jihad. Pakistan was left to face the blowback of extremism, drugs and guns.[277]

A variety of sources—CNN journalist Peter Bergen, Pakistani ISI Brigadier Mohammad Yousaf, and CIA operatives involved in the Afghan program, such as Vincent Cannistraro—deny that the CIA or other American officials had contact with the foreign mujahideen or Bin Laden, let alone armed, trained, coached or indoctrinated them.

Bergen and others argue that there was no need to recruit foreigners unfamiliar with the local language, customs or lay of the land since there were a quarter of a million local Afghans willing to fight;[278] that foreign mujahideen themselves had no need for American funds since they received several hundred million dollars a year from non-American, Muslim sources; that Americans could not have trained mujahideen because Pakistani officials would not allow more than a handful of them to operate in Pakistan and none in Afghanistan; and that the Afghan Arabs were almost invariably militant Islamists reflexively hostile to Westerners whether or not the Westerners were helping the Muslim Afghans.

According to Bergen, known for conducting the first television interview with bin Laden in 1997, the idea that "the CIA funded bin Laden or trained bin Laden ...[is] a folk myth. There's no evidence of this. ... Bin Laden had his own money, he was anti-American and he was operating secretly and independently. ... The real story here is the CIA didn't really have a clue about who this guy was until 1996 when they set up a unit to really start tracking him."[279] But Bergen conceded that, in one "strange incident," the CIA appeared to give visa help to mujahideen-recruiter Omar Abdel-Rahman.[280]

In his widely praised account of al-Qaeda, English journalist Jason Burke wrote:

> It is often said that bin Laden was funded by the CIA. This is not true and, indeed, would have been impossible given the structure of funding that General Zia ul-Haq, who had taken power in Pakistan in 1977, had set up. A condition of Zia's cooperation with the American plan to turn Afghanistan into the Soviets' 'Vietnam' was that all American funding to the Afghan resistance had to be channelled through the Pakistani government, which in effect meant the Afghan bureau of the Inter Services Intelligence (ISI), the military spy agency. The American funding, which went exclusively to the Afghan mujahideen groups, not the Arab volunteers, was supplemented by Saudi government money and huge funds raised from mosques, non-governmental charitable institutions and private donors throughout the Islamic world.[281]

Broader influence

Anders Breivik was inspired by al-Qaeda, calling it "the most successful revolutionary movement in the world." While admitting different aims, he sought to "create a European version of al-Qaida."[282] [283]

Criticism

According to a number of sources there has been a "wave of revulsion" against al-Qaeda and its affiliates by "religious scholars, former fighters and militants" alarmed by al-Qaeda's takfir and killing of Muslims in Muslim countries, especially Iraq.[284]

An Al Qaida recruit dreams about Osama bin Laden

Noman Benotman, a former Afghan Arab and militant of the Libyan Islamic Fighting Group, went public with an open letter of criticism to Ayman al-Zawahiri in November 2007 after persuading imprisoned senior leadership of his former group to enter into peace negotiations with the Libyan regime. While Ayman al-Zawahiri announced the affiliation of the group with al-Qaeda in November 2007, the Libyan government released 90 members of the group from prison several months later after "they were said to have renounced violence."[285]

In 2007, around the sixth anniversary of the September 11 attacks and a couple of months before *Rationalizing Jihad* first appeared in the newspapers,[92] the Saudi sheikh Salman al-Ouda delivered a personal rebuke to bin Laden. Al-Ouda, a religious scholar and one of the fathers of the Sahwa, the fundamentalist awakening movement that swept through Saudi Arabia in the 1980s, is a widely respected critic of jihadism. Al-Ouda addressed al-Qaeda's leader on television asking him

> My brother Osama, how much blood has been spilt? How many innocent people, children, elderly, and women have been killed ... in the name of al-Qaeda? Will you be happy to meet God Almighty carrying the burden of these hundreds of thousands or millions [of victims] on your back?[286]

According to Pew polls, support for al-Qaeda has been slightly dropped for parts of the Muslim world in the years before 2008.[287] The numbers supporting suicide bombings in Indonesia, Lebanon, and Bangladesh, for instance, have dropped by half or more in the last five years. In Saudi Arabia, only 10 percent now have a favorable view of al-Qaeda, according to a December poll by Terror Free Tomorrow, a Washington-based think tank.[288]

In 2007, the imprisoned Sayyed Imam Al-Sharif, an influential Afghan Arab, "ideological godfather of al-Qaeda", and former supporter of takfir, sensationally withdrew his support from al-Qaeda with a book *Wathiqat Tarshid Al-'Aml Al-Jihadi fi Misr w'Al-'Alam* (*Rationalizing Jihad in Egypt and the World*).

Although once associated with al-Qaeda, in September 2009 LIFG completed a new "code" for jihad, a 417-page religious document entitled "Corrective Studies". Given its credibility and the fact that several other prominent Jihadists in the Middle East have turned against al-Qaeda, the LIFG's about face may be an important step toward staunching al-Qaeda's recruitment.[289]

See also

- Al Qaeda Network Exord
- Bin Laden Issue Station (former CIA unit for tracking Bin Laden)
- Bosnian mujahideen
- Fatawā of Osama bin Laden
- List of designated terrorist organizations
- Muttahida Majlis-e-Amal
- Operation Cannonball
- Pakistan and state terrorism
- Psychological warfare
- Religious terrorism
- Steven Emerson
- Takfir wal-Hijra
- Terrorist organizations as destructive cults
- Videos of Osama bin Laden

Publications:

- *Al Qaeda Handbook*
- *Management of Savagery*

References

[1] Atwan 2006, p. 40

[2] http://www.fas.org/irp/crs/RS21745.pdf

[3] "The Future of Terrorism: What al-Qaida Really Wants – SPIEGEL ONLINE – News – International" (http://www.spiegel.de/international/0,1518,369448,00.html). *Der Spiegel*. September 11, 2001. . Retrieved October 18, 2011.

[4] *The Daily Telegraph* (London). http://my.telegraph.co.uk/riteman/riteway/16309030/al-qaeda-seeks-global-dominance/.

[5] "Jihadists Want Global Caliphate" (http://www.thepolitic.com/archives/2005/07/27/jihadists-want-global-caliphate/). ThePolitic.com. July 27, 2005. . Retrieved October 18, 2011.

[6] John Pike. "Al-Qaida" (http://www.globalsecurity.org/military/world/para/al-qaida.htm). Globalsecurity.org. . Retrieved October 18, 2011.

[7] Burke, Jason (March 21, 2004). "What exactly does al-Qaeda want?" (http://www.guardian.co.uk/world/2004/mar/21/alqaida.terrorism). *The Guardian* (London). .

[8] Saudi Arabia, Wahhabism and the Spread of Sunni Theofascism (http://www.globalpolitician.com/23661-saudi#_ftn15) retrieved 3 June 2012

[9] Moghadam, Assaf (2008). *The Globalization of Martyrdom: Al Qaeda, Salafi Jihad, and the Diffusion of Suicide Attacks*. Johns Hopkins University. pp. 48. ISBN 978-0-8018-9055-0.

[10] Livesey, Bruce (January 25, 2005). "Special Reports – The Salafist Movement: Al Qaeda's New Front" (http://www.pbs.org/wgbh/pages/frontline/shows/front/special/sala.html). *PBS Frontline*. WGBH educational foundation. . Retrieved October 18, 2011.

[11] "Foreign Terrorist Organizations List" (http://web.archive.org/web/20070712193357/http://www.state.gov/s/ct/rls/fs/2002/12535.htm). United States Department of State. Archived from the original (http://www.state.gov/s/ct/rls/fs/2002/12535.htm) on July 12, 2007. . Retrieved August 3, 2007. – USSD Foreign Terrorist Organization

[12] "Terrorism Act 2000" (http://www.opsi.gov.uk/acts/acts2000/20000011.htm). Home Office. Archived (http://web.archive.org/web/20070811053906/http://www.opsi.gov.uk/ACTS/acts2000/20000011.htm) from the original on August 11, 2007. . Retrieved August 14, 2007. – Terrorism Act 2000

[13] "Council Decision" (http://web.archive.org/web/20070714142855/http://www.state.gov/s/ct/rls/45394.htm). Council of the European Union. Archived from the original (http://www.state.gov/s/ct/rls/45394.htm) on July 14, 2007. . Retrieved August 14, 2007.

[14] "Terrorism Act 2000" (http://www.mha.nic.in/uniquepage.asp?Id_Pk=292). Ministry of Home Affairs (India). . Retrieved May 20, 2012.

[15] Bergen 2006, p. 75.

[16] *United States v. Usama bin Laden et al.*, S (7) 98 Cr. 1023 (http://cryptome.org/usa-v-ubl-02.htm), Testimony of Jamal Ahmed Mohamed al-Fadl (S.D.N.Y. February 6, 2001).

[17] Cooley, John K. (Spring 2003). "Unholy Wars: Afghanistan, America and International Terrorism" (http://findarticles.com/p/articles/mi_qa3996/is_200304/ai_n9199132) (reprint). *Demokratizatsiya*. .

[18] Gunaratna 2002, pp. 95–96. "Al-Qaeda's global network, as we know it today, was created while it was based in Khartoum, from December 1991 till May 1996. To coordinate its overt and covert operations as al-Qaeda's ambitions and resources increased, it developed a decentralized, regional structure. [...] As a global multinational, al-Qaeda makes its constituent nationalities and ethnic groups, of which there are several dozen, responsible for a particular geographic region. Although its *modus operandi* is cellular, familial relationships play a key role."

See also:

- Naím, Moisés (January/February 2003). "The Five Wars of Globalization". *Foreign Policy* (134): 28–37.

[19] Bosnian Model of Al Qaeda Terrorism - Page 41, Darko Trifunovic & Jill Starr

[20] Homeland Security in the UK: Future Preparedness for Terrorist, Paul Wilkinson - 2007

[21] Jihadi Terrorism and the Radicalisation Challenge: p.219, Rik Coolsaet - 2011

[22] Wright 2006, pp. 107–108, 185, 270–271

[23] Wright 2006, p. 270.

[24] "al Qaida's Ideology" (https://www.mi5.gov.uk/output/al-qaidas-ideology.html). MI5. . Retrieved May 19, 2012.

[25] "Dreaming of a caliphate" (http://www.economist.com/node/21525400). *The Economist.* Aug 6, 2011. . Retrieved May 19, 2012.

[26] Fu'ad Husayn `Al-Zarqawi, "The Second Generation of al-Qa'ida, Part Fourteen," *Al-Quds al-Arabi*, July 13, 2005

[27] Ranstorp, Magnus (2009). *Unconventional Weapons and International Terrorism.* Routledge. pp. 33. ISBN 978-0-415-48439-8.

[28] Lawrence Wright (2006). *The Looming Tower: Al-Qaeda and the Road to 9/11.* Knopf. pp. 246. ISBN 0-375-41486-X.

[29] Dragons and Tigers: A Geography of South, East, and Southeast Asia – (2011) – Barbara A. Weightman

[30] Security strategy and transatlantic relations (2006) Roland Dannreuther

[31] Jihad and Just War in the War on Terror (2011) Alia Brahimi

[32] al-Hammadi, Khalid, "The Inside Story of al-Qa'ida", part 4, *Al-Quds al-Arabi*, March 22, 2005

[33] Aug 13, 2004 (August 13, 2004). "Evolution of the al-Qaeda brand name" (http://www.atimes.com/atimes/Middle_East/FH13Ak05.html). Atimes.com. Archived (http://web.archive.org/web/20100411010338/http://www.atimes.com/atimes/Middle_East/FH13Ak05.html) from the original on April 11, 2010. . Retrieved March 22, 2010.

[34] "The Moral Logic and Growth of Suicide Terrorism – SPRING 2006" (http://www.sitemaker.umich.edu/satran/files/twq06spring_atran.pdf) (PDF). . Retrieved March 22, 2010.

[35] Blitz, James (January 19, 2010). "A threat transformed" (http://www.ft.com/cms/s/0/af31e344-0499-11df-8603-00144feabdc0.html). *Financial Times.* . Retrieved January 23, 2010.

[36] "A Discussion on the New Crusader Wars: Tayseer Allouni with Usamah bin Laden" (http://www.islamicawakening.com/viewarticle.php?articleID=977&pageID=64). *IslamicAwakening.com.* .

[37] McGeary 2001.

[38] ""Al-Qaida Says Al-Zawahri Has Succeeded Bin Laden", June 16, 2011, Associated Press" (http://www.nytimes.com/aponline/2011/06/16/world/middleeast/AP-ML-Al-Qaida-Zawahri.html?hp). *The New York Times.* June 16, 2011. . Retrieved June 6, 2011.

[39] Balz, Dan (August 27, 2011). *The Washington Post.* http://www.washingtonpost.com/politics/us-official-al-qaidas-no-2-leader-atiyah-abd-al-rahman-killed-in-pakistan/2011/08/27/gIQABVppiJ_story.html.

[40] [Al-Qaida No. 2 dies in US drone strike in Pakistan "Al-Qaida No. 2 dies in US drone strike in Pakistan"]. 6 June 2012. Al-Qaida No. 2 dies in US drone strike in Pakistan.

[41] Gunaratna 2002, p. 54.

[42] White House 2003.

[43] Basile 2004, p. 177.

[44] Wechsler 2001, p. 135; cited in Gunaratna 2002, p. 63.

[45] Businesses are run from below, with the council only being consulted on new proposals and collecting funds.

See:

- Hoffman 2002.

[46] "Cops: London Attacks Were Homicide Blasts" (http://www.foxnews.com/story/0,2933,162476,00.html). Fox News. July 15, 2005. . Retrieved June 15, 2008.

[47] Bennetto, Jason; Ian Herbert (August 13, 2005). "London bombings: the truth emerges" (http://news.independent.co.uk/uk/crime/article305547.ece). *The Independent* (UK). Archived (http://web.archive.org/web/20061026100045/http://news.independent.co.uk/uk/crime/article305547.ece) from the original on October 26, 2006. . Retrieved December 3, 2006.

[48] *The Power of Nightmares* (http://www.archive.org/details/ThePowerOfNightmares), BBC Documentary.

[49] McCloud, Kimberly; Osborne, Matthew (March 7, 2001). "WMD Terrorism and Usama bin Laden" (http://cns.miis.edu/reports/binladen.htm). *CNS Reports.* James Martin Center for Nonproliferation Studies. Archived (http://web.archive.org/web/20110506145417/http://cns.miis.edu/reports/binladen.htm) from the original on May 06 2011. . Retrieved May 4, 2011.

[50] "Witness: Bin Laden planned attack on U.S. embassy in Saudi Arabia" (http://edition.cnn.com/2001/LAW/02/13/Embassy.bombings.trial/). CNN. February 13, 2001. . Retrieved June 12, 2007.

[51] Cassidy 2006, p. 9.

[52] Noah, Timothy (February 25, 2009). "The Terrorists-Are-Dumb Theory: Don't mistake these guys for criminal masterminds" (http://www.slate.com/id/2211994/). Slate. .

[53] Gerges, Fawaz A (September 5, 2005). *The Far Enemy: Why Jihad Went Global.* Cambridge University Press. ISBN 0-521-79140-5.

[54] Jihad's New Leaders (http://www.meforum.org/article/1710) by Daveed Gartenstein-Ross and Kyle Dabruzzi, *Middle East Quarterly*, Summer 2007
[55] July 3, 2007 (July 3, 2007). "Today's jihadists: educated, wealthy and bent on killing?" (http://www.canada.com/nationalpost/news/story.html?id=c841b52c-b2e7-4e41-b27e-33d10245b935&k=0). Canada.com. . Retrieved March 22, 2010.
[56] Who is Bin Laden? (http://www.forbes.com/2001/09/14/0914whoisobl.html) retrieved May 5, 2011
[57] Cost of bin Laden (http://www.forbes.com/2001/09/14/0914ladenmoney.html) retrieved May 5, 2011
[58] Eric Lichtbau and Eric Schmitt Cash Flow to Terrorists Evades U.S. Efforts (http://www.nytimes.com/2010/12/06/world/middleeast/06wikileaks-financing.html) *The New York Times*, December 5, 2010
[59] *Abdel Bari Atwan. "The Secret History of Al Qaeda", p. 221. University of California Press, 2006. ISBN 0-520-24974-7* (http://books.google.com/books?id=ypICzykNXiAC&pg=PA221&lpg=PA221&dq=makkawi+strategy+2020&source=bl&ots=1iFik9U91G&sig=vF2xfwHZ76rhNWD74T2nSmonrdc&hl=en&ei=GJo7TOLyJYH88AbAu4GPBw&sa=X&oi=book_result&ct=result&resnum=5&ved=0CCMQ6AEwBA#v=onepage&q=makkawi strategy 2020&f=false). Google Books. March 11, 2005. Archived (http://web.archive.org/web/20110512042028/http://books.google.com/books?id=ypICzykNXiAC&pg=PA221&lpg=PA221&dq=makkawi+strategy+2020&source=bl&ots=1iFik9U91G&sig=vF2xfwHZ76rhNWD74T2nSmonrdc&hl=en&ei=GJo7TOLyJYH88AbAu4GPBw&sa=X&oi=book_result&ct=result&resnum=5&ved=0CCMQ6AEwBA) from the original on May 12, 2011. . Retrieved May 8, 2011.
[60] [tt_news]=181&tx_ttnews[backPid]=238&no_cache=1 |title=Al-Qaeda's Strategy Until 2020, by Stephen Ulph, May 5, 2005 (http://www.jamestown.org/programs/gta/single/?tx_ttnews), The Jamestown Foundation
[61] Listen to the U.S. pronunciation (http://ibb7.ibb.gov/pronunciations/sounds/2930.ra) (RealPlayer).
[62] Arabic Computer Dictionary: English-Arabic, Arabic-English By Ernest Kay, Multi-lingual International Publishers, 1986.
[63] "Transcript of Bin Laden's October interview" (http://archives.cnn.com/2002/WORLD/asiapcf/south/02/05/binladen.transcript/index.html). CNN. February 5, 2002. Archived (http://web.archive.org/web/20061206081331/http://archives.cnn.com/2002/WORLD/asiapcf/south/02/05/binladen.transcript/index.html) from the original on December 06 2006. . Retrieved October 22, 2006.
[64] Bergen 2006, p. 75. Wright indirectly quotes one of the documents, based on an exhibit from the "Tareek Osama" document presented in *United States v. Enaam M. Arnaout* (http://fl1.findlaw.com/news.findlaw.com/hdocs/docs/bif/usarnaout10603prof.pdf).
[65] Robin Cook (July 8, 2005). "Robin Cook: The struggle against terrorism cannot be won by military means" (http://www.guardian.co.uk/uk/2005/jul/08/july7.development). *The Guardian* (UK). Archived (http://web.archive.org/web/20110514051125/http://www.guardian.co.uk/uk/2005/jul/08/july7.development) from the original on May 14, 2011. . Retrieved May 8, 2011.
[66] " After Mombassa (http://weekly.ahram.org.eg/2003/619/op13.htm)", *Al-Ahram Weekly Online*, January 2–8, 2003 (Issue No. 619). Retrieved September 3, 2006.
[67] Wright 2006, p. 332.
[68] Qutb 2003, pp. 63, 69.
[69] Wright 2006, p. 79.
[70] "How Did Sayyid Qutb Influence Osama bin Laden?" (http://gemsofislamism.tripod.com/qutb_milest_influence_obl.html). Gemsofislamism.tripod.com. . Retrieved March 22, 2010.
[71] Mafouz Azzam; cited in Wright 2006, p. 36.
[72] "Sayyid Qutb's Milestones (footnote 24)" (http://gemsofislamism.tripod.com/milestones_qutb.html#footnote_24). Gemsofislamism.tripod.com. . Retrieved March 22, 2010.
[73] Qutbism: An Ideology of Islamic-Fascism (http://www.carlisle.army.mil/usawc/Parameters/07spring/eikmeier.htm) DALE C. EIKMEIER From *Parameters*, Spring 2007, pp. 85–98.
[74] Abdel Bari Atwan. "The Secret History of Al Qaeda", p. 233. University of California Press, 2006. ISBN 0-520-24974-7.
[75] Jason Burke and Paddy Allen (September 11 attacks, 2009). "The five ages of al-Qaida" (http://www.guardian.co.uk/world/interactive/2009/sep/10/al-qaida-five-ages-terror-attacks). *The Guardian* (UK). . Retrieved March 22, 2010.
[76] "1986–1992: CIA and British Recruit and Train Militants Worldwide to Help Fight Afghan War" (http://www.cooperativeresearch.org/context.jsp?item=a86operationcyclone). Cooperative Research History Commons. . Retrieved January 9, 2007.
[77] "Maktab al-Khidamat" (http://www.globalsecurity.org/security/profiles/maktab_al-khidamat.htm). GlobalSecurity.org. January 11, 2006. Archived (http://web.archive.org/web/20070214205625/http://www.globalsecurity.org/security/profiles/maktab_al-khidamat.htm) from the original on February 14, 2007. . Retrieved February 11, 2007.
[78] Wright 2006.
[79] Gunaratna 2002, p. 19. Quotes taken from Riedel 2008, p. 42 and Wright 2006, p. 103.
[80] Cloonan *Frontline* interview (http://www.pbs.org/wgbh/pages/frontline//////torture/interviews/cloonan.html), PBS, July 13, 2005.
[81] Sageman 2004, p. 35.
[82] Bergen, Peter L., *Holy war, Inc.: Inside the Secret World of Osama bin Laden*, New York: Free Press, c2001., p.70-1
[83] Bergen, Peter L., *Holy war, Inc.: inside the secret world of Osama bin Laden*, New York: Free Press, c2001., p. 69
[84] Interview with Alfred McCoy, November 9, 1991 by Paul DeRienzo (http://www.bearcave.com/bookrev/nugan_hand.html)
[85] Wright 2006, p. 137.
[86] "The War on Terror and the Politics of Violence in Pakistan" (http://web.archive.org/web/20061208102951/http://www.jamestown.org/news_details.php?news_id=54). The Jamestown Foundation. July 2, 2004. Archived from the original (http://www.jamestown.org/news_details.php?news_id=54) on December 8, 2006. . Retrieved January 9, 2007.
[87] " Who Is Osama Bin Laden? (http://www.forbes.com/2001/09/14/0914whoisobl.html)". *Forbes*. September 14, 2001.

[88] " Frankenstein the CIA created (http://www.guardian.co.uk/world/1999/jan/17/yemen.islam)". January 17, 1999. *The Guardian*.

[89] Scehuer, Michael. "Marching Towards Hell", 2008. p. 10

[90] Wright 2006, pp. 133–134.

[91] Wright 2006, p. 260.

[92] Wright 2008.

[93] Wright 2006, p. 181.

[94] "Legal case profiles – USA v. Omar Ahmad Ali Abdel-Rahman et al: 93-CR-181-KTD" (http://classic-web.archive.org/web/20080302005340/http://www.tkb.org/CaseHome.jsp?caseid=332). MIPT Terrorism Knowledge Base. . Retrieved March 22, 2010.

[95] "Osama bin Laden: The Past" (http://www.iacsp.com/itobli3.html). Archived (http://web.archive.org/web/20070218211532/http://www.iacsp.com/itobli3.html) from the original on February 18, 2007. . Retrieved January 12, 2007.

[96] Jehl, Douglas (December 27, 2001). "A Nation Challenged: Holy war lured Saudis as rulers looked away" (http://www.nytimes.com/2001/12/27/world/a-nation-challenged-saudi-arabia-holy-war-lured-saudis-as-rulers-looked-away.html). *The New York Times*: pp. A1, B4. . Retrieved September 5, 2009.

[97] Riedel 2008, p. 52.

[98] Wright 2006, p. 186.

[99] Wright 2006, p. 195.

[100] "Osama bin Laden: A Chronology of His Political Life" (http://www.pbs.org/wgbh/pages/frontline///shows/binladen/etc/cron.html). PBS. Archived (http://web.archive.org/web/20061205100703/http://www.pbs.org/wgbh/pages/frontline/shows/binladen/etc/cron.html) from the original on December 05 2006. . Retrieved January 12, 2007.

[101] "Context of 'Shortly After April 1994'" (http://www.cooperativeresearch.org/context.jsp?item=a94familydisown). Cooperative Research History Commons. . Retrieved January 12, 2007.

[102] Ijaz, Mansoor (December 5, 2001). "Clinton Let Bin Laden Slip Away and Metastasize" (http://articles.latimes.com/2001/dec/05/opinion/oe-ijaz05). *Los Angeles Times*. . Retrieved March 22, 2010.

[103] Carney, Timothy; Mansoor Ijaz (June 30, 2002). "Intelligence Failure? Let's Go Back to Sudan" (http://www.washingtonpost.com/ac2/wp-dyn/A64828-2002Jun29). *The Washington Post*. . Retrieved December 1, 2008.

[104] Rose, David (January, 2002). "The Osama Files" (http://www.vanityfair.com/politics/features/2002/01/osama200201?currentPage=1). *Vanity Fair*. Archived (http://web.archive.org/web/20081205155110/http://www.vanityfair.com/politics/features/2002/01/osama200201?currentPage=1) from the original on December 05 2008. . Retrieved December 1, 2008.

[105] Belz, Mindy (November 1, 2003). "'Clinton did not have the will to respond'" (http://www.worldmag.com/articles/8206). *World*. Archived (http://web.archive.org/web/20081206014541/http://www.worldmag.com/articles/8206) from the original on December 06 2008. . Retrieved December 1, 2008.

[106] "National Commission on Terrorist Attacks Upon the United States" (http://govinfo.library.unt.edu/911/report/911Report_Ch4.htm). Govinfo.library.unt.edu. . Retrieved March 22, 2010.

[107] Rashid 2002.

[108] Napoleoni 2003, pp. 121–123; Akacem 2005 ("[Napoleoni] does a decent job of covering al-Qaida and presents some numbers and estimates that are of value [to terrorism scholars].").

[109] Kronstadt & Katzman 2008.

[110] Eisenberg, Daniel (October 28, 2001). "Secrets Of Brigade 055" (http://www.time.com/time/magazine/article/0,9171,1101011105-181591,00.html). *Time*. .

[111] Nic Robertson. "CNN.com" (http://edition.cnn.com/2008/WORLD/asiapcf/10/06/afghan.saudi.talks/?iref=mpstoryview). CNN. . Retrieved March 22, 2010.

[112] Roggio, Bill Taliban have not split from al Qaeda: sources (http://www.longwarjournal.org/archives/2008/10/taliban_have_not_spl.php) October 7, 2008 *The Long War Journal*

[113] Partlow, Joshua. In Afghanistan, Taliban surpasses al-Qaeda" (http://www.washingtonpost.com/wp-dyn/content/article/2009/11/10/AR2009111019644.html) November 11, 2009

[114] Sageman 2004, pp. 48, 51.

[115] Trofimov 2006, p. 282.

[116] "Bin Laden's Fatwa" (http://www.pbs.org/newshour/terrorism/international/fatwa_1996.html). Al Quds Al Arabi. August 1996. Archived (http://web.archive.org/web/20070108175653/http://www.pbs.org/newshour/terrorism/international/fatwa_1996.html) from the original on January 08 2007. . Retrieved January 9, 2007.

[117] Abdel Bari Atwan (2006). *The Secret History of Al Qaeda*. University of California Press. pp. 168–169. ISBN 0-520-24974-7.

[118] Summary taken from bin Laden's May 26, 1998 interview with American journalist John Miller. Most recently broadcast in the documentary *Age of Terror*, part 4, with translations checked by Barry Purkis (archive researcher).

[119] "Text of Fatwah Urging Jihad Against Americans" (http://web.archive.org/web/20060422210853/http://www.ict.org.il/articles/fatwah.htm). Archived from the original (http://www.ict.org.il/articles/fatwah.htm) on April 22, 2006. . Retrieved May 15, 2006.

[120] Benjamin & Simon 2002, p. 117. "By issuing fatwas, bin Laden and his followers are acting out a kind of self-appointment as alim: they are asserting their rights as interpreters of Islamic law."

[121] Nyquist, J.R. (August 13, 2005). "Is Al Qaeda a Kremlin Proxy?" (http://www.jrnyquist.com/nyquist_2005_0813.htm). jrnyquist.com. Archived (http://web.archive.org/web/20080417001945/http://www.jrnyquist.com/nyquist_2005_0813.htm) from the original on April

17, 2008. . Retrieved April 17, 2008.

[122] "Obituary: Alexander Litvinenko" (http://news.bbc.co.uk/2/hi/uk_news/6163502.stm). BBC News. November 24, 2006. Archived (http://web.archive.org/web/20080307180315/http://news.bbc.co.uk/2/hi/uk_news/6163502.stm) from the original on March 07 2008. . Retrieved April 16, 2008.

[123] Al Qaeda's hand in tipping Iraq toward civil war (http://www.csmonitor.com/2006/0320/p09s01-coop.html), *The Christian Science Monitor/Al-Quds Al-Arabi*, March 20, 2006

[124] " Another wave of bombings hit Iraq (http://www.iht.com/articles/2005/09/15/africa/web.0915iraq.php)", Associated Press/*International Herald Tribune*, September 15, 2005

[125] " 20 die as insurgents in Iraq target Shiites (http://www.iht.com/articles/2005/09/16/news/iraq.php)", *International Herald Tribune/The New York Times*, September 17, 2005

[126] " Al-Qaeda disowns 'fake letter' (http://news.bbc.co.uk/2/hi/middle_east/4339912.stm)", CNN, October 13, 2005

[127] "British 'fleeing' claims al-Qaeda" (http://www.adnkronos.com/AKI/English/Security/?id=1.0.1678520501). Adnkronos.com. April 7, 2003. Archived (http://web.archive.org/web/20110512042036/http://www.adnkronos.com/AKI/English/Security/?id=1.0.1678520501) from the original on May 12, 2011. . Retrieved May 8, 2011.

[128] " Al Qaeda leader in Iraq 'killed by insurgents' (http://www.abc.net.au/news/newsitems/200705/s1911670.htm)", ABC News, May 1, 2007

[129] DeYoung, Karen/Pincus, Walter. " Al-Qaeda in Iraq May Not Be Threat Here (http://www.washingtonpost.com/wp-dyn/content/article/2007/03/17/AR2007031701373.html)", *The Washington Post*, March 18, 2007

[130] "Somalia's al-Shabab join al-Qaeda" (http://www.bbc.co.uk/news/world-africa-16979440). *BBC*. February 10, 2012. .

[131] "Al-Shabaab joining al Qaeda, monitor group says" (http://www.cnn.com/2012/02/09/world/africa/somalia-shabaab-qaeda/). CNN. February 9, 2012. . Retrieved February 9, 2012.

[132] Johnston, Philip (September 17, 2010). "Anwar al Awlaki: the new Osama bin Laden?" (http://www.telegraph.co.uk/news/uknews/terrorism-in-the-uk/8009819/Anwar-al-Awlaki-the-new-Osama-bin-Laden.html). *The Daily Telegraph* (London). .

[133] "NEWS.BBC.co.uk" (http://news.bbc.co.uk/2/hi/middle_east/8437724.stm). NEWS.BBC.co.uk. January 3, 2010. Archived (http://web.archive.org/web/20100323005536/http://news.bbc.co.uk/2/hi/middle_east/8437724.stm) from the original on March 23, 2010. . Retrieved March 22, 2010.

[134] "Al-Qaeda Slowly Makes Its Way to Somalia and Yemen" (http://english.pravda.ru/hotspots/terror/109278-0/). Pravda.ru. . Retrieved September 23, 2009.

[135] http://worldnews.msnbc.msn.com/_news/2011/12/13/9410244-hunt-for-terrorists-shifts-to-dangerous-north-africa-panetta-says

[136] Font size Print E-mail Share 1 Comment (December 28, 2009). "CBS News Dec. 28, 2009 Al Qaeda: We Planned Flight 253 Bombing Terrorist Group Says It Was In Retaliation for U.S. Operation in Yemen; Obama Orders Reviews of Watchlist and Air Safety" (http://www.cbsnews.com/stories/2009/12/28/cbsnews_investigates/main6031295.shtml). Cbsnews.com. . Retrieved March 22, 2010.

[137] "Bin Ladin Preparing to Hijack US Aircraft and Other Attacks" (http://www.foia.cia.gov/docs/DOC_0001110635/0001110635_0001.gif). Director of Central Intelligence. December 4, 1998. . Retrieved April 18, 2010.

[138] National Commission on Terrorist Attacks (July 22, 2004 (first edition)). *The 9/11 Commission Report* (http://govinfo.library.unt.edu/911/report/911Report_Ch9.pdf). W. W. Norton & Company. ISBN 0-393-32671-3. .

[139] "Lost lives remembered during 9/11 ceremony" (http://www.theonlinerocket.com/news/lost-lives-remembered-during-9-11-ceremony-1.2333384). The Online Rocket. September 12, 2008. . Retrieved 2012-04-04.

[140] Chucmach, Megan, and Ross, Brian, "Al Qaeda Recruiter New Focus in Fort Hood Killings Investigation Army Major Nidal Hasan Was In Contact With Imam Anwar Awlaki, Officials Say," *[[ABC News* (http://abcnews.go.com/Blotter/al-qaeda-recruiter-focus-fort-hood-killings-investigation/story?id=9045492)], *November 10, 2009, accessed November 12, 2009]*

[141] Esposito, Richard, Cole, Matthew, and Ross, Brian, "Officials: U.S. Army Told of Hasan's Contacts with al Qaeda; Army Major in Fort Hood Massacre Used 'Electronic Means' to Connect with Terrorists," *[[ABC News* (http://abcnews.go.com/Blotter/fort-hood-shooter-contact-al-qaeda-terrorists-officials/story?id=9030873)], *November 9, 2009, accessed November 12, 2009]*

[142] WashingtonPost.com (http://www.washingtonpost.com/wp-dyn/content/article/2008/02/26/AR2008022603267_pf.html) February 27, 2008 Imam From Va. Mosque Now Thought to Have Aided Al-Qaeda.

[143] Bennett, Chuck (January 3, 2010). "Ft. Hood link in 'crotch' case" (http://www.nypost.com/p/news/international/jet_terrorist_met_evil_imam_yemen_tvUzCVFH5WxPhHgr3UtQIM). *The New York Post*. . Retrieved January 21, 2010.

[144] CBS News staff (December 29, 2009). "Did Abdulmutallab Talk to Radical Cleric?" (http://www.cbsnews.com/stories/2009/12/29/world/main6034880.shtml). CBS News. Archived (http://web.archive.org/web/20100124073713/http://www.cbsnews.com/stories/2009/12/29/world/main6034880.shtml) from the original on January 24, 2010. . Retrieved January 4, 2010.

[145] Meyer, Josh (December 31, 2009). "U.S.-born cleric linked to airline bombing plot" (http://www.latimes.com/news/nation-and-world/la-na-terror-intel31-2009dec31,0,4146208,full.story). *Los Angeles Times*. . Retrieved May 7, 2010.

[146] Newton, Paula (March 10, 2010). "Purported al-Awlaki message calls for jihad against U.S." (http://www.cnn.com/2010/WORLD/europe/03/17/al.awlaki.message/index.html). CNN. Archived (http://web.archive.org/web/20100419233413/http://www.cnn.com/2010/WORLD/europe/03/17/al.awlaki.message/index.html) from the original on April 19, 2010. . Retrieved May 7, 2010.

[147] Newton, Paula (March 10, 2010). "CNN Report: A Message From Anwar Al-Awlaki" (http://www.youtube.com/watch?v=Gr8TUEfNrBk). YouTube. . Retrieved May 7, 2010.

[148] Miller, Greg (April 6, 2010). "Muslim cleric Aulaqi is 1st U.S. citizen on list of those CIA is allowed to kill" (http://www.washingtonpost.com/wp-dyn/content/article/2010/04/06/AR2010040604121.html). *Washington Post*. . Retrieved April 9, 2010.

[149] Shane, Scott (April 6, 2010). "U.S. Approves Targeted Killing of American Cleric" (http://www.nytimes.com/2010/04/07/world/middleeast/07yemen.html). *New York Times*. Archived (http://web.archive.org/web/20100408031248/http://www.nytimes.com//2010//04//07//world//middleeast//07yemen.html) from the original on April 08 2010. . Retrieved April 6, 2010.

[150] Leonard, Tom (April 7, 2010). "Barack Obama orders killing of US cleric Anwar al-Awlaki" (http://www.telegraph.co.uk/news/worldnews/northamerica/usa/barackobama/7564581/Barack-Obama-orders-killing-of-US-cleric-Anwar-al-Awlaki.html). London: Telegraph (UK). Archived (http://web.archive.org/web/20100411042550/http://www.telegraph.co.uk/news/worldnews/northamerica/usa/barackobama/7564581/Barack-Obama-orders-killing-of-US-cleric-Anwar-al-Awlaki.html) from the original on April 11, 2010. . Retrieved April 8, 2010.

[151] Fox News staff (May 1, 2010). "Times Square Suspect Contacted Radical Cleric" (http://www.myfoxdetroit.com/dpps/news/times-square-suspect-contacted-radical-cleric-dpgonc-km-20100506_7411371). NewsCore. MyFoxDetroit.com. . Retrieved May 7, 2010.

[152] Dreazen, Yochi J.; Perez, Evan (May 6, 2010). "Suspect Cites Radical Imam's Writings" (http://online.wsj.com/article/SB10001424052748704370704575228150116907566.html). *The Wall Street Journal*. Archived (http://web.archive.org/web/20100509120618/http://online.wsj.com/article/SB10001424052748704370704575228150116907566.html?) from the original on May 09 2010. . Retrieved May 6, 2010.

[153] Herridge, Catherine (May 6, 2010). "Times Square Bomb Suspect a 'Fan' of Prominent Radical Cleric, Sources Say" (http://www.foxnews.com/us/2010/05/06/times-square-bomb-suspect-fan-prominent-radical-cleric-sources-say/). Fox News. Archived (http://web.archive.org/web/20100507101825/http://www.foxnews.com/us/2010/05/06/times-square-bomb-suspect-fan-prominent-radical-cleric-sources-say/) from the original on May 07 2010. . Retrieved May 7, 2010.

[154] Esposito, Richard; Chris Vlasto and Chris Cuomo (May 6, 2010). "Faisal Shahzad Had Contact With Anwar Awlaki, Taliban, and Mumbai Massacre Mastermind, Officials Say" (http://abcnews.go.com/Blotter/faisal-shahzad-contact-awlaki-taliban-mumbai-massacre-mastermind/story?id=10575061). *The Blotter from Brian Ross* (ABC News). Archived (http://web.archive.org/web/20100509013338/http://abcnews.go.com/Blotter/faisal-shahzad-contact-awlaki-taliban-mumbai-massacre-mastermind/story?id=10575061) from the original on May 09 2010. . Retrieved May 7, 2010.

[155] "Awlaki Not Among FBI's 'Most Wanted' Terrorists; No Reward Offered for His Capture May 24, 2010, by Patrick Goodenough" (http://www.cnsnews.com/news/article/66496). Cnsnews.com. May 24, 2010. Archived (http://web.archive.org/web/20100812062811/http://www.cnsnews.com/news/article/66496) from the original on August 12, 2010. . Retrieved July 17, 2010.

[156] May 10, 2010 editorial in the Investor's Business Daily

[157] "Awlaki lands on al-Qaida suspect list" (http://www.upi.com/Top_News/Special/2010/07/21/Awlaki-lands-on-al-Qaida-suspect-list/UPI-55521279731808/). UPI.com. Archived (http://web.archive.org/web/20101023142601/http://www.upi.com/Top_News/Special/2010/07/21/Awlaki-lands-on-al-Qaida-suspect-list/UPI-55521279731808/) from the original on October 23, 2010. . Retrieved October 30, 2010.

[158] Mark Wilson (August 5, 2010). "CIA on the verge of lawsuit" (http://seerpress.com/cia-on-the-verge-of-lawsuit/3341/). Seer Press News. .

[159] Rayment, Sean; Hennessy, Patrick; Barrett, David (October 30, 2010). "Yemen cargo bomb plot may have been targeted at Britain" (http://www.telegraph.co.uk/news/uknews/terrorism-in-the-uk/8099345/Yemen-cargo-bomb-plot-may-have-been-targeted-at-Britain.html). London: Telegraph. Archived (http://web.archive.org/web/20101101031903/http://www.telegraph.co.uk/news/uknews/terrorism-in-the-uk/8099345/Yemen-cargo-bomb-plot-may-have-been-targeted-at-Britain.html) from the original on November 01 2010. . Retrieved October 31, 2010.

[160] http://blogs.cfr.org/zenko/2011/09/30/targeted-killings-the-death-of-anwar-al-awlaki/

[161] BBC staff (17 March 2012). "Osama Bin Laden 'plotted to kill Obama' before death" (http://www.bbc.co.uk/news/world-asia-17413055). BBC. .

[162] Adams, Richard; Walsh, Declan; MacAskill, Ewen (May 1, 2011). "Osama bin Laden is dead, Obama announces" (http://www.guardian.co.uk/world/2011/may/02/osama-bin-laden-dead-obama). *The Guardian* (London). .

[163] "Osama Bin Laden Killed by US Strike" (http://abcnews.go.com/Blotter/osama-bin-laden-killed/story?id=13505703). ABC News. May 1, 2011. .

[164] the CNN Wire Staff (2). "How U.S. forces killed Osama bin Laden" (http://edition.cnn.com/2011/WORLD/asiapcf/05/02/bin.laden.raid/). Cable News Network. . Retrieved May 2, 2011.

[165] "Osama Bin Laden Killed By Navy Seals in Firefight – ABC News" (http://abcnews.go.com/Blotter/osama-bin-laden-killed-navy-seals-firefight/story?id=13505792). Abcnews.go.com. Archived (http://web.archive.org/web/20110604035658/http://abcnews.go.com/Blotter/osama-bin-laden-killed-navy-seals-firefight/story?id=13505792) from the original on June 04 2011. . Retrieved May 2, 2011.

[166] Balz, Dan (May 2, 2011). "Osama bin Laden is killed by U.S. forces in Pakistan" (http://www.washingtonpost.com/politics/osama-bin-laden-is-killed-by-us-forces-in-pakistan/2011/05/01/AFXMZyVF_story.html). *The Washington Post*. .

[167] "Chitralis bewildered at OBL episode" (http://www.chitralnews.com/News- AweStruckt-02-Mayr-11.htm). Chitralnews.com. May 2, 2011. . Retrieved May 8, 2011.

[168] "Osama bin Laden, the face of terror, killed in Pakistan" (http://edition.cnn.com/2011/WORLD/asiapcf/05/01/bin.laden.obit/index.html). CNN. May 2, 2011. Archived (http://web.archive.org/web/20110506084553/http://edition.cnn.com/2011/WORLD/asiapcf/05/01/bin.laden.obit/index.html) from the original on May 06 2011. . Retrieved May 2, 2011.

[169] "Osama Bin Laden Dead: Obama Speech Video And Transcript" (http://www.huffingtonpost.com/2011/05/02/osama-bin-laden-dead-obama-speech-video-transcript_n_856122.html) Huffington Post, May 2, 2011

[170] POSTED: 1:11 am EDT May 2, 2011 (February 5, 2011). "Report: DNA At Mass. General Confirms bin Laden's Death" (http://www.thebostonchannel.com/news/27739824/detail.html). Thebostonchannel.com. Archived (http://web.archive.org/web/20110518182150/http://www.thebostonchannel.com/news/27739824/detail.html) from the original on May 18, 2011. . Retrieved May 2, 2011.

[171] "Osama bin Laden Killed; ID Confirmed by DNA Testing" (http://abcnews.go.com/Blotter/osama-bin-laden-killed/story?id=13505703). ABC News. May 1, 2011. .

[172] "US forces kill Osama bin Laden in Pakistan" (http://www.msnbc.msn.com/id/42852700/ns/world_news-south_and_central_asia). MSN. .

[173] "Official: Bin Laden buried at sea" (http://news.yahoo.com/s/ap/us_bin_laden_burial). News.yahoo.com. May 2, 2011. . Retrieved May 8, 2011.

[174] "U.S forces kill elusive terror figure Osama Bin Laden in Pakistan" (http://www.cnn.com/2011/WORLD/asiapcf/05/02/bin.laden.dead/index.html?hpt=T1). CNN World. May 2, 2011. .

[175] "Crowds celebrate Bin Laden's death" (http://www.euronews.net/2011/05/02/crowds-celebrate-bin-laden-s-death/). Euronews. May 2, 2011. .

[176] Eichenwald, Kurt (December 10, 2001). "A Nation Challenged: The Money; Terror Money Hard to Block, Officials Find" (http://www.nytimes.com/2001/12/10/world/a-nation-challenged-the-money-terror-money-hard-to-block-officials-find.html?scp=1&sq=A+NATION+CHALLENGED:+THE+MONEY;+Terror+Money+Hard+to+Block,+Officials+Find&st=cse&pagewanted=all). *The New York Times*. . Retrieved May 4, 2011.

[177] Wright 2006, p. 174.

[178] Jansen 1997.

[179] Wright 2006, p. 178; Reeve 1999.

[180] "February 1993 Bombing of the World Trade Center in New York City" (http://web.archive.org/web/20061207130217/http://cns.miis.edu/pubs/reports/wtc93.htm). Center for Nonproliferation Studies. November 12, 2001. Archived from the original (http://www.cns.miis.edu/pubs/reports/wtc93.htm) on December 7, 2006. . Retrieved January 9, 2007.

[181] Tom Leonard (December 25, 2009). "Osama bin Laden came within minutes of killing Bill Clinton" (http://www.telegraph.co.uk/news/worldnews/asia/philippines/6867331/Osama-bin-Laden-came-within-minutes-of-killing-Bill-Clinton.html). *The Daily Telegraph* (London). Archived (http://web.archive.org/web/20091225150252/http://www.telegraph.co.uk/news/worldnews/asia/philippines/6867331/Osama-bin-Laden-came-within-minutes-of-killing-Bill-Clinton.html) from the original on December 25, 2009. . Retrieved December 25, 2009.

[182] "Jihad Against Jews and Crusaders" (http://www.fas.org/irp/world/para/docs/980223-fatwa.htm). February 23, 1998. . Retrieved June 16, 2010.

[183] "Bin Laden says he wasn't behind attacks" (http://archives.cnn.com/2001/US/09/16/inv.binladen.denial/). CNN. September 17, 2001. Archived (http://web.archive.org/web/20060705161654/http://archives.cnn.com/2001/US/09/16/inv.binladen.denial/) from the original on July 05 2006. . Retrieved July 6, 2006.

[184] Esposito 2002, p. 22.

[185] Hamid Miir 'Osama claims he has nukes: If U.S. uses N-arms it will get the same response' "Dawn: the Internet Edition" November 10, 2001

[186] Tremlett, Giles (September 9, 2002). "Al-Qaida leaders say nuclear power stations were original targets" (http://www.guardian.co.uk/afghanistan/story/0,1284,788431,00.html). *The Guardian* (UK). Archived (http://web.archive.org/web/20070122160702/http://www.guardian.co.uk/afghanistan/story/0,1284,788431,00.html) from the original on January 22, 2007. . Retrieved January 11, 2007.

[187] "Al Qaeda Scaled Back 10-Plane Plot" (http://www.washingtonpost.com/wp-dyn/articles/A45853-2004Jun16_2.html). *Washington Post*. June 17, 2004. . Retrieved January 11, 2007.

[188] Australian Government. "Listing of Terrorist Organisations" (http://www.nationalsecurity.gov.au/agd/www/nationalsecurity.nsf/AllDocs/95FB057CA3DECF30CA256FAB001F7FBD?OpenDocument). . Retrieved July 3, 2006.

[189] Public Safety and Emergency Preparedness Canada. "Entities list" (http://www.psepc.gc.ca/prg/ns/le/cle-en.asp). . Retrieved July 3, 2006.

[190] Commission of the European Communities (October 20, 2004). "Communication from the Commission to the Council and the European Parliament" (http://eur-lex.europa.eu/LexUriServ/site/en/com/2004/com2004_0700en01.doc) (DOC). Archived (http://web.archive.org/web/20070614032134/http://eur-lex.europa.eu/LexUriServ/site/en/com/2004/com2004_0700en01.doc) from the original on June 14, 2007. . Retrieved June 11, 2007.

[191] "La France face au terrorisme" (http://lesrapports.ladocumentationfrancaise.fr/cgi-bin/brp/telestats.cgi?brp_ref=064000275&brp_file=0000.pdf) (in (French)). Secrétariat général de la défense nationale (France). . Retrieved August 6, 2009.

[192] "The Hindu : Centre bans Al-Qaeda" (http://www.hinduonnet.com/2002/04/09/stories/2002040903651100.htm). Hinduonnet.com. April 9, 2002. . Retrieved March 22, 2010.

[193] "Summary of indictments against Al-Qaeda terrorists in Samaria" (http://www.mfa.gov.il/MFA/Terrorism-+Obstacle+to+Peace/Terrorism+and+Islamic+Fundamentalism-/Summary+of+indictments+against+Al-Qaeda+terrorists+in+Samaria+21-Mar-2006.htm). Israel Ministry of Foreign Affairs. March 21, 2006. . Retrieved May 4, 2011.

[194] Diplomatic Bluebook (2002). "B. Terrorist Attacks in the United States and the Fight Against Terrorism" (http://www.mofa.go.jp/policy/other/bluebook/2002/chap1-b.pdf) (PDF). Archived (http://web.archive.org/web/20070614032134/http://www.mofa.go.jp/policy/other/bluebook/2002/chap1-b.pdf) from the original on June 14, 2007. . Retrieved June 11, 2007.

[195] General Intelligence and Security Service. "Annual Report 2004" (http://ftp.fas.org/irp/world/netherlands/aivd2004-eng.pdf) (PDF). Archived (http://web.archive.org/web/20070614032136/http://ftp.fas.org/irp/world/netherlands/aivd2004-eng.pdf) from the original on June 14, 2007. . Retrieved June 11, 2007.

[196] NATO. "Press Conference with NATO Secretary General, Lord Robertson" (http://www.nato.int/docu/speech/2001/s011122b.htm). Archived (http://web.archive.org/web/20061026040125/http://www.nato.int/docu/speech/2001/s011122b.htm) from the original on October 26, 2006. . Retrieved October 23, 2006.

[197] NATO Library (2005). "AL QAEDA" (http://www.nato.int/structur/library/bibref/them0305.pdf) (PDF). Archived (http://web.archive.org/web/20070614032137/http://www.nato.int/structur/library/bibref/them0305.pdf) from the original on June 14, 2007. . Retrieved June 11, 2007.

[198] "Russia Outlaws 17 Terror Groups; Hamas, Hezbollah Not Included" (http://web.archive.org/web/20061114154904/http://www.mosnews.com/news/2006/07/28/russiaterrorlist.shtml). Archived from the original (http://www.mosnews.com/news/2006/07/28/russiaterrorlist.shtml) on November 14, 2006. .

[199] Korean Foreign Ministry (August 14, 2007). "Seoul confirms release of two Korean hostages in Afghanistan" (http://web.archive.org/web/20071215153000/http://www.korea.net/news/news/NewsView.asp?serial_no=20070813015&part=102). Archived from the original (http://www.korea.net/news/news/NewsView.asp?serial_no=20070813015&part=102) on December 15, 2007. . Retrieved September 16, 2007.

[200] Ministry for Foreign Affairs Sweden (March – June 2006). "Radical Islamist Movements in the Middle East" (http://www.sweden.gov.se/content/1/c6/06/12/67/01b99143.pdf) (PDF). Archived (http://web.archive.org/web/20070614032136/http://www.sweden.gov.se/content/1/c6/06/12/67/01b99143.pdf) from the original on June 14, 2007. . Retrieved June 11, 2007.

[201] "Report on counter-terrorism submitted by Switzerland to the Security Council Committee established pursuant to resolution 1373 (2001)" (http://www.efv.admin.ch/d/dokumentation/downloads/themen/finanzkrimi/CH_UNTerror_1201_e.pdf) (PDF). December 20, 2001. Archived (http://web.archive.org/web/20070609153811/http://www.efv.admin.ch/d/dokumentation/downloads/themen/finanzkrimi/CH_UNTerror_1201_e.pdf) from the original on June 09 2007. . Retrieved June 11, 2007.

[202] ""Türkiye'de halen faaliyetlerine devam eden başlıca terör örgütleri listesi" (Emniyet Genel Müdürlüğü)" (http://www.egm.gov.tr/temuh/terorgrup1.html). Egm.gov.tr. . Retrieved March 22, 2010.

[203] United Kingdom Home Office. "Proscribed terrorist groups" (http://www.homeoffice.gov.uk/security/terrorism-and-the-law/terrorism-act/proscribed-groups). Archived (http://web.archive.org/web/20060630031633/http://www.homeoffice.gov.uk/security/terrorism-and-the-law/terrorism-act/proscribed-groups) from the original on June 30, 2006. . Retrieved July 3, 2006.

[204] "Security Council Resolutions Related to the Work of the Committee Established Pursuant to Resolution 1267 (1999) Concerning Al-Qaida and the Taliban and Associated Individuals and Entities" (http://www.un.org/Docs/sc/committees/1267/1267ResEng.htm). United Nations Security Council. Archived (http://web.archive.org/web/20070112115326/http://www.un.org/Docs/sc/committees/1267/1267ResEng.htm) from the original on January 12, 2007. . Retrieved January 9, 2007.

[205] United States Department of State. "Foreign Terrorist Organizations (FTOs)" (http://web.archive.org/web/20060628111524/http://www.state.gov/s/ct/rls/fs/37191.htm). Archived from the original (http://www.state.gov/s/ct/rls/fs/37191.htm) on June 28, 2006. . Retrieved July 3, 2006.

[206] "U.S. Jets Pound Targets Around Kabul" (http://www.seacoastonline.com/articles/20011015-NEWS-310159983). *The Portsmouth Herald*. October 15, 2001. . Retrieved July 25, 2012.

[207] "Blair to Taliban: Surrender bin Laden or surrender power" (http://www.cbc.ca/world/story/2001/10/02/blair_war011002.html). Canadian Broadcasting Corporation. October 3, 2001. . Retrieved March 22, 2010.

[208] "U.S. Releases Videotape of Osama Bin Laden" (http://web.archive.org/web/20060625223529/http://www.defenselink.mil/releases/2001/b12132001_bt630-01.html). December 13, 2001. Archived from the original (http://www.defenselink.mil/releases/2001/b12132001_bt630-01.html) on June 25, 2006. . Retrieved July 4, 2006.

[209] Morris, Steven (December 15, 2001). "US urged to detail origin of tape" (http://www.guardian.co.uk/september11/story/0,11209,619188,00.html). *The Guardian* (UK). . Retrieved July 11, 2006.

[210] http://archives.cnn.com/2001/US/12/13/tape.transcript/

[211] "National Commission on Terrorist Attacks Upon the United States" (http://www.9-11commission.gov/). September 20, 2004. Archived (http://web.archive.org/web/20060427220632/http://www.9-11commission.gov/) from the original on April 27, 2006. . Retrieved April 27, 2006.

[212] "Full transcript of bin Ladin's speech" (http://english.aljazeera.net/NR/exeres/79C6AF22-98FB-4A1C-B21F-2BC36E87F61F.htm). Al Jazeera. November 1, 2004. Archived (http://web.archive.org/web/20060711055450/http://english.aljazeera.net/NR/exeres/79C6AF22-98FB-4A1C-B21F-2BC36E87F61F.htm) from the original on July 11, 2006. . Retrieved July 12, 2006.

[213] Shane, Scott (June 22, 2008). "Inside the interrogation of a 9/11 mastermind" (http://www.nytimes.com/2008/06/22/washington/22ksm.html). *The New York Times*: pp. A1, A12–A13. . Retrieved September 5, 2009.

[214] Black, Ian; Norton-Taylor, Richard (September 18, 2009). "Al-Qaida weakened by global recruitment woes" (http://www.guardian.co.uk/world/2009/sep/10/al-qaida-recruitment-crisis). *The Guardian Weekly* (London) **181** (14): pp. 1–2. . Retrieved September 19, 2009.

- Trofimov, Yaroslav (August 15, 2009). "Islamic rebels gain strength in the Sahara" (http://online.wsj.com/article/SB125030117348933737.html). *The Wall Street Journal* **254** (39): p. A9. . Retrieved September 15, 2009.
- Trofimov, Yaroslav (August 17, 2009). "Islamic rebels gain strength in the Sahara". *The Wall Street Journal Europe* **27** (136): p. 12.
- Trofimov, Yaroslav (August 18, 2009). "Islamic rebels gain in the Sahara". *The Wall Street Journal Asia* **33** (245): p. 12.

[216] Riedel 2008, p. 126.

[217] View all comments that have been posted about this article.. "Washington Post – Al-Qaeda's Hand In Istanbul Plot" (http://www.washingtonpost.com/wp-dyn/content/article/2007/02/12/AR2007021201715_pf.html). *The Washington Post.* . Retrieved March 22, 2010.

[218] 8:51 am ET (December 17, 2003). "Msn News – Bin Laden allegedly planned attack in Turkey – Stymied by tight security at U.S. bases, militants switched targets" (http://www.msnbc.msn.com/id/3735645/). MSNBC. . Retrieved March 22, 2010.

- Gardham, Duncan (September 8, 2009). "Gang is brought to justice by most complex operation since the war" (http://www.telegraph.co.uk/news/uknews/terrorism-in-the-uk/6152185/Airline-bomb-plot-investigation-one-of-biggest-since-WW2.html). *The Daily Telegraph* (London): p. 2. . Retrieved September 15, 2009.
- Gardham, Duncan (September 16, 2009). "Complex operation brings gang to justice". *The Weekly Telegraph* (Australian edition, issue 947): p. 9.

[220] Milmo, Cahal (September 8, 2009). "Police watched the plot unfold, then pounced" (http://www.independent.co.uk/news/uk/crime/police-watched-the-plot-unfold-then-pounced-1783388.html). *The Independent* (London): pp. 2–4. . Retrieved September 15, 2009.

[221] Press, Associated (September 7, 2009). "UK court convicts 3 of plot to blow up airliners | International | Jerusalem Post" (http://fr.jpost.com/servlet/Satellite?cid=1251804512098&pagename=JPost/JPArticle/ShowFull). Fr.jpost.com. Archived (http://web.archive.org/web/20110512043550/http://fr.jpost.com/servlet/Satellite?cid=1251804512098&pagename=JPost/JPArticle/ShowFull) from the original on May 12, 2011. . Retrieved May 8, 2011.

[222] Sandford, Daniel (September 7, 2009). "UK | Airline plot: Al-Qaeda connection" (http://news.bbc.co.uk/2/hi/uk_news/8221375.stm). BBC News. . Retrieved March 22, 2010.

[223] Elder, Miriam (3 Oct 2012). "Russia accuses al-Qaida of 'forest jihad' in Europe" (http://www.guardian.co.uk/world/2012/oct/03/russia-al-qaida-forest-jihad). *The Guardian.* . Retrieved 6 November 2012.

[224] Weir, Shelagh (July/September 1997). *A Clash of Fundamentalisms: Wahhabism in Yemen* (http://www.merip.org/mer/mer204/weir.htm). Middle East Report. Middle East Research and Information Project. . Retrieved January 19, 2009.; cited in Burke, Jason (2003). *Al-Qaeda: Casting a Shadow of Terror*. New York: I.B. Tauris. pp. 128–129. ISBN 1-85043-396-8.

[225] " Yemen: The Next Front Line Against al Qaeda (http://www.cbsnews.com/stories/2010/10/30/earlyshow/saturday/main7006282.shtml)". CBS News. October 30, 2010

[226] Riedel 2008, p. 100.

[227] See the following works cited in Riedel 2008, p. 101:

- Hafez 2007, pp. 97–98
- Al-Shishani, Murad Batal (November 17, 2005). "Al-Zarqawi's Rise to Power: Analyzing Tactics and Targets". *Jamestown Foundation Terrorism Monitor* **3** (22).

[228] Streatfeild, Dominic (January 7, 2011). "How the US let al-Qaida get its hands on an Iraqi weapons factory" (http://www.guardian.co.uk/world/2011/jan/07/iraq-weapons-factory-al-qaida-us-failure). *The Guardian* (UK). Archived (http://web.archive.org/web/20110107134826/http://www.guardian.co.uk/world/2011/jan/07/iraq-weapons-factory-al-qaida-us-failure) from the original on January 07 2011. . Retrieved January 7, 2011.

[229] " CSI Urges Obama to Protect Iraq's Endangered Christian Community (http://news.yahoo.com/s/usnw/20101102/pl_usnw/DC92946)". Yahoo News. November 1, 2010.

[230] " Iraqi Christians Mourn 58 Dead in Church Siege (http://www.cbsnews.com/stories/2010/11/01/world/main7011759.shtml)". CBS News. November 1, 2010.

[231] Gunaratna 2002, p. 150.

[232] Terrorism in India and the Global Jihad (http://www.brookings.edu/articles/2008/1130_india_terrorism_riedel.aspx), Brookings Institution, November 30, 2008

[233] Al Qaeda: Profile and Threat Assessment (http://www.fas.org/irp/crs/RS22049.pdf), Congressional Research Service, February 10, 2005

[234] United Nations High Commissioner for Refugees (July 2, 2008). "Freedom in the World 2008 – Kashmir Pakistan, 2 July 2008" (http://www.unhcr.org/refworld/topic,463af2212,469f2dcf2,487ca21a2a,0.html). Unhcr.org. Archived (http://web.archive.org/web/20110512172350/http://www.unhcr.org/refworld/topic,463af2212,469f2dcf2,487ca21a2a,0.html) from the original on May 12, 2011. . Retrieved May 8, 2011.

[235] Kashmir Militant Extremists (http://www.cfr.org/publication/9135/), Council on Foreign Relations, July 9, 2009

[236] Osama bin Laden "letter to the American people" (http://www.globalsecurity.org/security/library/report/2002/021120-ubl.htm), GlobalSecurity.org, November 20, 2002

[237] Full text: bin Laden's 'letter to America' (http://www.guardian.co.uk/world/2002/nov/24/theobserver), The Guardian, November 24, 2002

[238] Osama men plan to target Delhi: Kathmandu receives threat (http://www.dawn.com/2001/11/10/int3.htm), Dawn, November 10, 2001

[239] Analysis: Is al-Qaeda in Kashmir? (http://news.bbc.co.uk/2/hi/2043800.stm), BBC, June 13, 2002

[240] Rumsfeld offers US technology to guard Kashmir border (http://www.smh.com.au/articles/2002/06/13/1023864326179.html), The Sydney Morning Herald, June 14, 2002

[241] Al Qaeda thriving in Pakistani Kashmir (http://www.csmonitor.com/2002/0702/p01s02-wosc.html), The Christian Science Monitor, July 2, 2002

[242] SAS joins Kashmir hunt for bin Laden (http://www.telegraph.co.uk/news/worldnews/asia/india/1385795/SAS-joins-Kashmir-hunt-for-bin-Laden.html), The Telegraph, February 23, 2002

[243] Al-Qaeda terror trial: Rangzieb Ahmed was highest ranking al-Qaeda operative in Britain (http://www.telegraph.co.uk/news/uknews/3832366/Al-Qaeda-terror-trial-Rangzieb-Ahmed-was-highest-ranking-al-Qaeda-operative-in-Britain.html), The Telegraph. December 18, 2008

[244] Bin Laden's finger on Kashmir trigger? (http://edition.cnn.com/2002/WORLD/asiapcf/south/06/12/kashmir.qaeda/), CNN, June 12, 2002

[245] Taliban, al-Qaeda linked to Kashmir (http://www.usatoday.com/news/world/2002/05/29/taliban-kashmir.htm), USA Today, May 29, 2002

[246] Al Qaeda claim of Kashmir link worries India (http://www.nytimes.com/2006/07/13/world/asia/13iht-india.2194572.html), The New York Times,2006-07-13

[247] No Al Qaeda presence in Kashmir: Army (http://www.hindu.com/2007/06/18/stories/2007061801191400.htm), The Hindu,2007-06-18

[248] Ilyas Kashmiri had planned to attack COAS (http://www.thenews.com.pk/daily_detail.asp?id=199076), The News International, September 18, 2009

[249] Waziristan new battlefield for Kashmiri militants (http://www.thenews.com.pk/daily_detail.asp?id=148529), The News International, November 24, 2008

[250] Kashmiri militants move to Waziristan, open training camps (http://www.indianexpress.com/news/kashmiri-militants-move-to-waziristan-open/390569/) The Indian Express, November 26, 2008

[251] Army of Madinah in Kashmir (http://www.nefafoundation.org/miscellaneous/Barot/ArmyMadinahinKashmir.pdf), Nine Eleven Finding Answers Foundation

[252] How radical Islam turned a schoolboy into a terrorist (http://www.timesonline.co.uk/tol/news/uk/article628497.ece), The Times, November 7, 2006

[253] The Long Hunt for Osama (http://www.theatlantic.com/doc/200410/bergen), The Atlantic, October 2004

[254] Al Qaeda Strikes Back (http://www.brookings.edu/articles/2007/05terrorism_riedel.aspx), Brookings Institution

[255] Al-Qaeda involved in Indian plane hijack plot (http://www.hindu.com/2006/09/18/stories/2006091814740100.htm), The Hindu, September 18, 2006

[256] Osama threw lavish party for Azhar after hijack (http://www.indianexpress.com/news/osama-threw-lavish-party-for-azhar-after-hi/12921/), The Indian Express, September 18, 2006

[257] Rashid Rauf: profile of a terror mastermind (http://www.telegraph.co.uk/news/worldnews/asia/pakistan/3500661/Rashid-Rauf-profile-of-a-terror-mastermind.html), The Telegraph, November 22, 2008

[258] LeT, which is based at Muridke, near Lahore in Pakistan, has networks throughout India and its leadership has close links with core al-Qaeda figures living in Pakistan Focus on Westerners suggests al-Qaeda was pulling strings (http://www.timesonline.co.uk/tol/news/world/asia/article5248563.ece), The Times, November 28, 2008

[259] Lashkar-e-Taiba Served as Gateway for Western Converts Turning to Jihad (http://online.wsj.com/article/SB122834970727777709.html), The Wall Street Journal, December 4, 2008

[260] Lashkar-e-Taiba threat revived after Chicago arrest (http://www.dawn.com/wps/wcm/connect/dawn-content-library/dawn/news/world/03-lashkar-e-taiba-threat-revived-after-chicago-arrest-ss-09), Dawn, November 20, 2009

[261] INTERVIEW-French magistrate details Lashkar's global role (http://www.reuters.com/article/idUSLC383495), Reuters, November 13, 2009

[262] 'Azzam the American' releases video focusing on Pakistan (http://www.cnn.com/2008/WORLD/asiapcf/10/04/gadahn.video/index.html), CNN, October 4, 2008

[263] US drones killed two terrorist leaders in Pak (http://www.dawn.com/wps/wcm/connect/dawn-content-library/dawn/news/world/12-us+drones+killed+two+terrorist+leaders+in+pak--bi-10), Dawn, September 17, 2009

[264] Al Qaeda's American Mole (http://www.brookings.edu/opinions/2009/1215_terrorism_riedel.aspx), Brookings Institution, December 15, 2009

[265] Ilyas Kashmiri alive, lays out future terror strategy (http://www.dailytimes.com.pk/default.asp?page=2009\10\15\story_15-10-2009_pg1_11), Daily Times (Pakistan), October 15, 2009

[266] United States of America vs Tahawwur Hussain Rana (http://www.chicagotribune.com/news/chi-102709-rana-headley,0,5879769.htmlpage), *Chicago Tribune*

[267] US charges Ilyas Kashmiri in Danish newspaper plot (http://www.dawn.com/wps/wcm/connect/dawn-content-library/dawn/news/world/04-ilyas-kashmiri-danish-plot-qs-04), Dawn, January 15, 2010

[268] US seeks Harkat chief for Khost CIA attack (http://www.thenews.com.pk/daily_detail.asp?id=217152), The News International, January 6, 2010

[269] Indian hijack plot caused new UK terror alert (http://www.timesonline.co.uk/tol/news/politics/article6999959.ece), The Times, January 24, 2010

[270] Al Qaeda could provoke new India-Pakistan war: Gates (http://www.dawn.com/wps/wcm/connect/dawn-content-library/dawn/news/world/03-al-qaeda-could-provoke-new-india-pakistan-war-gates-ss-02), Dawn, January 20, 2010

[271] Timothy Thomas, "Al Qaeda and the Internet: The Danger of Cyberplanning" (http://www.carlisle.army.mil/usawc/Parameters/03spring/thomas.pdf). Retrieved February 14, 2007.

[272] Scheuer, Michael (January 2008). "Bin Laden Identifies Saudi Arabia as the Enemy of Mujahideen Unity" (http://jamestown.org/terrorism/news/article.php?articleid=2373884). *Terrorism Focus*. Jamestown Foundation. .

[273] Whitlock, Craig (August 8, 2005). "Briton Used Internet As His Bully Pulpit" (http://www.washingtonpost.com/wp-dyn/content/article/2005/08/07/AR2005080700890.html). *The Washington Post*: p. A1. . Retrieved September 4, 2009.

[274] "Babar Ahmad Indicted on Terrorism Charges" (http://web.archive.org/web/20060526115318/http://www.usdoj.gov/usao/ct/Press2004/20041006.html). United States Attorney's Office District of Connecticut. October 6, 2004. Archived from the original (http://www.usdoj.gov/usao/ct/Press2004/20041006.html) on May 26, 2006. . Retrieved May 29, 2006.

[275] Gaynor, Tim (January 13, 2010). "Al Qaeda linked to rogue aviation network" (http://www.reuters.com/article/idUSTRE60C3E820100113?type=politicsNews). Reuters. . Retrieved May 8, 2011.

[276] Cook, Robin (July 8, 2005). "The struggle against terrorism cannot be won by military means" (http://www.guardian.co.uk/terrorism/story/0,12780,1523838,00.html). *The Guardian* (UK). Archived (http://web.archive.org/web/20050710025703/http://www.guardian.co.uk/terrorism/story/0,12780,1523838,00.html) from the original on July 10, 2005. . Retrieved July 8, 2005.

[277] Akram, Munir (January 19, 2008). "Pakistan, Terrorism and Drugs" (http://www.nytimes.com/2008/01/19/opinion/lweb22pakistan.html?_r=1&ref=opinion). *New York Times*. . Retrieved October 17, 2009.

[278] Coll 2005, pp. 145–146, 155–156.

[279] Bergen, Peter. "Bergen: Bin Laden, CIA links hogwash" (http://www.cnn.com/2006/WORLD/asiapcf/08/15/bergen.answers/index.html). CNN. Archived (http://web.archive.org/web/20060821221916/http://www.cnn.com/2006/WORLD/asiapcf/08/15/bergen.answers/index.html) from the original on August 21, 2006. . Retrieved August 15, 2006.

[280] Bergen 2001, pp. 72–73.

[281] Burke, Jason (2004) [2003]. *Al-Qaeda: The True Story of Radical Islam* (revised ed.). London: Penguin. p. 59.

[282] Ritter, Karl (April 20, 2012). "Breivik Studied al-Qaeda Attacks" (http://www.time.com/time/world/article/0,8599,2112701,00.html#ixzz1uHIlb0DH). *Time Magazine*. . Retrieved May 8, 2012.

[283] "Norway: Militant Studied Al Qaeda" (http://www.nytimes.com/2012/04/21/world/europe/norway-anders-breivik-studied-al-qaeda.html). *New York Times*. April 20, 2012. . Retrieved May 8, 2012.

[284] Bergen & Cruickshank 2008; Wright 2008. Quotes taken from Riedel 2008, pp. 106–107 and Bergen & Cruickshank 2008.

[285] "Libya releases scores of prisoners APRIL 9, 2008" (http://english.aljazeera.net/NR/exeres/AA64F530-BCA8-40B0-8226-22154CCD1032.htm). English.aljazeera.net. April 9, 2008. . Retrieved March 22, 2010.

[286] Bergen & Cruickshank 2008.

[287] "Taking Stock of the War on Terror" (http://www.realclearpolitics.com/articles/2008/05/taking_stock_in_the_war_on_ter.html). Realclearpolitics.com. May 22, 2008. . Retrieved March 22, 2010.

[288] "December 18, 2007 Poll: Most Saudis oppose al Qaeda" (http://www.cnn.com/2007/WORLD/meast/12/17/saudi.poll/index.html). CNN. December 18, 2007. . Retrieved March 22, 2010.

[289] New jihad code threatens al Qaeda (http://www.cnn.com/2009/WORLD/africa/11/09/libya.jihadi.code/), Nic Robertson and Paul Cruickshank, *CNN*, November 10, 2009

Further reading

Bibliography

- Atran, Scott (2010). *Talking to the Enemy: Faith, Brotherhood, and the (un)making of Terrorists*. New York: Ecco Press. ISBN 978-0-06-134490-9.
- Atwan, Abdel Bari (2006). *The Secret History of al Qaeda*. Berkeley, CA: University of California Press. ISBN 978-0-520-24974-5.
- Basile, Mark (May 2004). "Going to the Source: Why Al Qaeda's Financial Network Is Likely to Withstand the Current War on Terrorist Financing". *Studies in Conflict and Terrorism* **27** (3): 169–185. doi:10.1080/10576100490438237.

- Benjamin, Daniel; Simon, Steven (2002). *The Age of Sacred Terror* (1st ed.). New York: Random House. ISBN 0-375-50859-7.
- Bergen, Peter (2001). *Holy War, Inc.: Inside the Secret World of Osama bin Laden* (1st ed.). New York: Free Press. ISBN 0-7432-3495-2.
- Bergen, Peter (2006). *The Osama bin Laden I Know: An Oral History of al Qaeda's Leader* (2nd ed.). New York: Free Press. ISBN 0-7432-7892-5.
- Bergen, Peter; Cruickshank, Paul (June 11, 2008). "The Unraveling: The jihadist revolt against bin Laden" (http://www.tnr.com/article/the-unraveling). *The New Republic* **238** (10): pp. 16–21. Retrieved May 4, 2011.
- Bergen, Peter (2011). *The Longest War: The Enduring Conflict between America and al-Qaeda*. New York: Free Press. ISBN 978-0-7432-7893-5.
- Bin Laden, Osama (2005). Lawrence, Bruce. ed. *Messages to the World: The Statements of Osama bin Laden*. London: Verso. ISBN 1-84467-045-7.
- Cassidy, Robert M. (2006). *Counterinsurgency and the Global War on Terror: Military Culture and Irregular War*. Westport, CT: Praeger Security International. ISBN 0-275-98990-9.
- Coll, Steve (2005). *Ghost Wars: The Secret History of the CIA, Afghanistan, and Bin Laden, from the Soviet Invasion to September 10, 2001* (2nd ed.). New York: Penguin Books. ISBN 0-14-303466-9.
- Esposito, John L. (2002). *Unholy War: Terror in the Name of Islam*. New York: Oxford University Press. ISBN 0-19-515435-5.
- Gunaratna, Rohan (2002). *Inside Al Qaeda* (1st ed.). London: C. Hurst & Co.. ISBN 1-85065-671-1.
- Hafez, Mohammed M. (March 2007). "Martyrdom Mythology in Iraq: How Jihadists Frame Suicide Terrorism in Videos and Biographies". *Terrorism and Political Violence* **19** (1): 95–115. doi:10.1080/09546550601054873.
- Hoffman, Bruce (2002). "The Emergence of the New Terrorism". In Tan, Andrew; Ramakrishna, Kumar. *The New Terrorism: Anatomy, Trends, and Counter-Strategies*. Singapore: Eastern Universities Press. pp. 30–49. ISBN 981-210-210-8.
- Jansen, Johannes J.G. (1997). *The Dual Nature of Islamic Fundamentalism*. Ithaca, NY: Cornell University Press. ISBN 0-8014-3338-X.
- McGeary, Johanna (February 19, 2001). "A Traitor's Tale" (http://www.time.com/time/magazine/article/0,9171,999237,00.html). *TIME* **157** (7): pp. 36–37. Retrieved September 15, 2009.
- Napoleoni, Loretta (2003). *Modern Jihad: Tracing the Dollars Behind the Terror Networks*. London: Pluto Press. ISBN 0-7453-2117-8.
- Qutb, Sayyid (2003). *Milestones*. Chicago: Kazi Publications. ISBN 0-911119-42-6.
- Rashid, Ahmed (2002) [2000]. *Taliban: Militant Islam, Oil and Fundamentalism in Central Asia*. New Haven: Yale University Press. ISBN 1-86064-830-4.
- Reeve, Simon (1999). *The New Jackals: Ramzi Yousef, Osama Bin Laden and the Future of Terrorism*. Boston: Northeastern University Press. ISBN 1-55553-407-4.
- Riedel, Bruce (2008). *The Search for al Qaeda: Its Leadership, Ideology, and Future*. Washington, D.C.: Brookings Institution Press. ISBN 978-0-8157-7414-3.
- Sageman, Marc (2004). *Understanding Terror Networks*. Philadelphia: University of Pennsylvania Press. ISBN 0-8122-3808-7.
- Trofimov, Yaroslav (2006). *Faith At War: A Journey On the Frontlines of Islam, From Baghdad to Timbuktu*. New York: Picador. ISBN 978-0-8050-7754-4.
- Wechsler, William F. (2001). "Strangling The Hydra: Targeting Al Qaeda's Finances". In Hoge, James; Rose, Gideon. *How Did This Happen? Terrorism and the New War*. New York: PublicAffairs. pp. 129–143. ISBN 1-58648-130-4.
- Wright, Lawrence (2006). *The Looming Tower: Al-Qaeda and the Road to 9/11*. New York: Knopf. ISBN 0-375-41486-X.

- Wright, Lawrence (June 2, 2008). "The Rebellion Within" (http://www.newyorker.com/reporting/2008/06/02/080602fa_fact_wright?currentPage=all). *The New Yorker* **84** (16): pp. 36–53. Retrieved September 15, 2009.

Reviews

- Akacem, Mohammed (August 2005). "Review: Modern Jihad: Tracing the Dollars behind the Terror Networks". *International Journal of Middle East Studies* **37** (3): 444–445. doi:10.1017/S0020743805362143.
- Bale, Jeffrey M. (October 2006). "Deciphering Islamism and Terrorism". *Middle East Journal* **60** (4): 777–788.

Government reports

- Kronstadt, K. Allen; Katzman, Kenneth (November 2008). "Islamist Militancy in the Pakistan-Afghanistan Border Region and U.S. Policy" (http://fpc.state.gov/documents/organization/113202.pdf) (PDF). U.S. Congressional Research Service.
- White House (September 2003). "Progress Report on the Global War on Terrorism" (http://web.archive.org/web/20030922090723/http://www.state.gov/s/ct/rls/rpt/24087.htm). United States Department of State. Archived from the original (http://www.state.gov/s/ct/rls/rpt/24087.htm) on September 22, 2003.

External links

- U.S. Dept. of Justice, Al Qaeda Training Manual (http://web.archive.org/web/20050331091340/http://www.usdoj.gov/ag/trainingmanual.htm)
- Al-Qaeda in Oxford Islamic Studies Online (http://www.oxfordislamicstudies.com/article/opr/t125/e1909?_hi=27&_pos=4)
- Decade After 9/11, Al-Qaeda in Shambles (http://www.onislam.net/english/news/global/453845-decade-after-911-al-qaeda-in-shambles.html)
- 17 de-classified documents captured during the Abbottabad raid and released to the Combating Terrorism Center (http://www.ctc.usma.edu/posts/letters-from-abbottabad-bin-ladin-sidelined)
- Bin Laden documents at a glance (http://www.cbsnews.com/8301-501704_162-57427765/bin-laden-documents-at-a-glance/)

Media

- Peter Taylor. (2007). " War on the West (http://news.bbc.co.uk/2/hi/programmes/age_of_terror/7306413.stm)". *Age of Terror*, No. 4, series 1. BBC.
- Investigating Al-Qaeda (http://news.bbc.co.uk/2/hi/in_depth/world/2001/war_on_terror/default.stm), *BBC News*
- Adam Curtis (2004). *The Power of Nightmares*. BBC.
- Al Qaeda's New Front (http://www.pbs.org/wgbh/pages/frontline/shows/front) from *PBS Frontline*, January 2005
- Inside al Qaeda (http://www.youtube.com/watch_popup?v=zUUUS_wRKLo&vq=medium) – video report by *National Geographic*
- Al-Qaida (http://www.guardian.co.uk/world/al-qaida) collected news and commentary at *The Guardian*
- al-Qaeda (http://topics.nytimes.com/top/reference/timestopics/organizations/a/al_qaeda/index.html) collected news and commentary at *The New York Times*

Abdulaziz_al-Omari

Abdulaziz al-Omari	

Born	Abdulaziz al-Omari (in Arabic: عبد العزيز العمري)May 28, 1979Asir, Saudi Arabia
Died	September 11, 2001 (aged 22)Manhattan, New York, United States

An airport security guard and Imam, **Abdulaziz al-Omari** (Arabic: عبد العزيز العمري, *'Abd al-'Azīz al-'Umarī*, also transliterated as **Alomari** or **al-Umari**) (May 28, 1979[1] – September 11, 2001) was one of five hijackers of American Airlines Flight 11 as part of the September 11 attacks.

A Saudi, Omari arrived in the United States in June 2001, on a tourist visa, obtained through the Visa Express program.

On September 11, 2001, Omari boarded American Airlines Flight 11 and assisted in the hijacking of the plane, which was crashed into the North Tower of the World Trade Center, as part of the coordinated attacks.

History

Little is known about Omari's life, and it is unclear whether some information refers to Omari or another person by that name. He has used birth dates of December 24, 1972 and May 28, 1979.

Omari came from Asir Province, a poor region in southwestern Saudi Arabia that borders Yemen, and graduated with honours from high school, attained a degree from the Imam Muhammad Ibn Saud Islamic University, was married, and had a daughter.[2]

He is alleged to have often served as an Imam at his mosque in Saudi Arabia and is believed by American Authorities to have been a student of a Saudi cleric named Sulayman al Alwan, whose mosque is located in Al-Qassim Province.

According to Tawfiq bin Attash, Omari was one of a group of future hijackers who provided security at Kandahar airport after their basic training at an al-Qaeda camp. During the 2000 Al Qaeda Summit in Kuala Lumpur, American authorities claim that immigration records show that a person named Abdulaziz al-Omari was visiting the country, although they say they are not sure that this was the same person.

In the autumn of 2001, after the September 11 attacks, al Jazeera television broadcast a tape they claim was made by Omari. The speaker made a farewell suicide video. In it, he reads "I am writing this with my full conscience and I am writing this in expectation of the end, which is near. . . God praise everybody who trained and helped me, namely the leader Sheikh Osama bin Laden."

According to FBI director Robert Mueller and the 9/11 Commission, Omari entered the United States through a Dubai flight on June 29, 2001, with Salem al-Hazmi, landing in New York.[3] He had used the controversial Visa Express program to gain entry. He apparently stayed with several other hijackers in Paterson, New Jersey, before moving to his own place at 4032 57th Terrace, Vero Beach, Florida. On his rental agreement form for that house, Omari gave two license-plates authorized to park in his space, one of which was registered to Atta.[4] Omari occasionally trained on simulators at the FlightSafety Academy in Vero Beach, Florida together with Mohand al-Shehri and Saeed al-Ghamdi.[5]

Omari obtained a fake United States ID card from All Services Plus in Passaic County, New Jersey, which was in the business of selling fake documents, including another to Khalid al-Mihdhar.[6]

Attacks

On September 10, 2001, Mohamed Atta picked up Omari from the Milner Hotel in Boston, Massachusetts, and the two drove their rented Nissan to a Comfort Inn in South Portland, Maine, where they spent the night in room 232. It was initially reported that Adnan and Ameer Bukhari were the two hijackers who had rented and driven the car.[7]

Atta (blue shirt) and Omari at Portland International Jetport on 9/11

Wrongly accused Abdul Rahman al-Omari

In the early hours of September 11, they boarded a commuter flight back to Boston to connect to American Airlines Flight 11. American 11 was hijacked 15 minutes after the flight departed by Omari and four other hijackers, which allowed trained pilot Mohamed Atta to crash the Boeing 767 into the North Tower of the World Trade Center as part of an attack that killed thousands of people.

Mistaken identity

Controversy over Omari's identity erupted shortly after the attacks. At first, the FBI had named Abdul Rahman al-Omari, a pilot for Saudi Arabian Airlines, as the pilot of Flight 11. It was quickly shown that this person was still alive, and the FBI issued an apology. It was also quickly determined that Mohamed Atta was the pilot among the hijackers. The FBI then named Abdulaziz al-Omari as a hijacker.

A man with the same name as those given by the FBI turned up alive in Saudi Arabia, saying that he had studied at the University of Denver and his passport was stolen there in 1995. The name, origin, birth date, and occupation were released by the FBI, but the picture was not of him. "I couldn't believe it when the FBI put me on their list", he said. "They gave my name and my date of birth, but I am not a suicide bomber. I am here. I am alive. I have no idea how to fly a plane. I had nothing to do with this."[8] [9] [10]

See also

- PENTTBOM

References

[1] John J. Lumpkin. "Abdul Aziz al Omari" (http://www.globalsecurity.org/security/profiles/abdul_aziz_al_omari.htm). Globalsecurity.org. . Retrieved 2012-01-06.

[2] CIA analytic report,"The Plot and the Plotters," June 1, 2003, p. 25.

[3] "Statement of Robert S. Mueller: Joint Investigation Into September 11: (published September 26, 2002)" (http://www.fas.org/irp/congress/2002_hr/092602mueller.html). Fas.org. . Retrieved 2012-01-06.

[4] FBI Affidavit: Page 11 (http://www.abc.net.au/4corners/atta/resources/documents/fbiaffidavit11.htm) ABC

[5] "Piecing together the shadowy lives of the hijackers" (http://www.telegraph.co.uk/news/worldnews/northamerica/usa/1341136/Piecing-together-the-shadowy-lives-of-the-hijackers.html). *The Daily Telegraph* (London). September 20, 2001. Archived (http://web.archive.org/web/20100410205817/http://www.telegraph.co.uk/news/worldnews/northamerica/usa/1341136/Piecing-together-the-shadowy-lives-of-the-hijackers.html) from the original on 10 April 2010. . Retrieved April 23, 2010.

[6] Miller, Jonathan (2003-03-08). "A Plea Deal, Then Freedom, in Terror Case Where Prosecutors Kept Evidence a Secret" (http://query.nytimes.com/gst/fullpage.html?res=9803E0DD1E3FF93BA35750C0A9659C8B63&sec=&spon=&pagewanted=all). The New York Times. . Retrieved 2008-09-29.

[7] Two Brothers among Hijackers: CNN Report (http://english.peopledaily.com.cn/200109/13/eng20010913_80131.html) People's Daily

[8] Sack, Kevin (September 16, 2001). "AFTER THE ATTACKS: MISSED CUES; Saudi May Have Been Suspected in Error, Officials Say" (http://query.nytimes.com/gst/abstract.html?res=F2071FF63D5F0C758DDDA00894D9404482). *The New York Times*. .

[9] (http://news.independent.co.uk/world/middle_east/story.jsp?story=94438)

[10] "Middle East | Hijack 'suspects' alive and well" (http://news.bbc.co.uk/1/hi/world/middle_east/1559151.stm). BBC News. 2001-09-23. . Retrieved 2012-01-06.

External links

- The Final 9/11 Commission Report (http://www.gpoaccess.gov/911/index.html)
- portal.telegraph.co.uk (http://www.portal.telegraph.co.uk/news/main.jhtml?xml=/news/2001/09/23/widen23.xml) (Article which reports that the Saudi Arabian Airlines pilot named Omari was not involved with the terrorist attacks)

Abu_Zubaydah

Abu Zubaydah	

Born	March 12, 1971Saudi Arabia
Detained at	CIA black sites, Guantanamo
Alternate name	Abu Zubaydah زين العابدين محمد حسين
ISN	10016
Charge(s)	No charge
Status	Indefinitely detained without trial

Abu Zubaydah (Arabic: ابو زبيدة, *Abū Zubaydah*; born March 12, 1971 as **Zayn al-Abidin Muhammad Husayn**) is a Saudi Arabian citizen currently held in U.S. custody in Guantanamo Bay detention camps, in Cuba.[1]

In April 2001, months prior to 9/11, Abu Zubaydah called his brother Hesham (who was living in Oregon) 3 times. At this time the NSA and other agencies had stated that Abu Zubaydah was under the radar and radio silent. Arrested in Pakistan in March 2002, Abu Zubaydah has been in US custody for more than nine years, four-and-a-half of them as a ghost detainee in the CIA secret prison network. The CIA transferred him between prisons in various countries as part of their extraordinary rendition program. During interrogation in that time he was water-boarded 83 times and has been subjected to numerous other controversial interrogation techniques including forced nudity, sleep deprivation, confinement in small dark boxes, deprivation of solid food, stress positions, and physical assaults. Videotapes of some of his interrogations are amongst those destroyed by the CIA in 2005.[2] [3] [4] [5]

He was transferred to Guantanamo in September 2006 where he and other former CIA detainees are held in Camp 7 where conditions are the most isolating. He has never been charged, despite being accused for years by US authorities.[6] [7] [8] [9]

Biography and his early years in Afghanistan

According to his younger brother Hesham they had eight siblings.[10] According to Jason Leopold, Lead Investigative Reporter of *Truthout*, Hesham remembers his older brother *"as a happy-go-lucky guy, and something of a womanizer."*

Born in Saudi Arabia, Abu Zubaydah moved to the West Bank as a teenager where he joined in Palestinian demonstrations against the Israelis.[11] [12]

Abu Zubaydah is reported to have studied Computer Science in Pune, India prior to his travel to Afghanistan/Pakistan in 1991.[13]

Abu Zubaydah moved to Afghanistan in 1991 to fight alongside the mujahideen in the Afghan civil war.[14] In 1992 Abu Zubaydah was injured from a mortar shell blast which left shrapnel in his head and caused severe memory loss, as well as the loss of his ability to speak for over one year.[15] [16] [17] Abu Zubaydah eventually became involved in the jihad training camp known as the Khalden Camp where he oversaw the flow of recruits into and out of the camp. He obtained passports and paperwork for men transferring to other training camps or home.[18]

The Khalden Camp has been described by the U.S. Government as an al-Qaeda training facility—an assertion that has been utilized as evidence of Abu Zubaydah's, and over 50 other Guatanamo detainees' alleged connection to al-Qaeda.[15] This allegation has been contested, however, by multiple detainees, the 9/11 Commission Report, and Brynjar Lia, head of the international terrorism and global jihadism at the Norwegian Defence Research Establishment.[15] [19] [20] [21] [22] Abu Zubaydah testified in his Combatant Status Review Tribunal that the Khalden Camp was at such odds with al-Qaeda and Bin Laden that it was closed by the Taliban in 2001, at al-Qaeda's request.[15] This account was corroborated by two other detainees, Noor Uthman Muhammed, who was alleged by the U.S. Government to have been the emir, or leader, of the Khalden Camp, and a close friend of Abu Zubaydah, Khalid Sulayman Jaydh Al Hubayshi.[19] [20] In addition, Noor Uthman Muhamed's charge sheet references the closing of the Khalden camp at the request of terrorist leaders.[23] Brynjar Lia states in his book that there was an ideological conflict between the leaders of the Khalden Camp on one side, and the Taliban and al-Qaeda on the other, and that this led to the closing of the Khalden Camp.[22] Abu Zubaydah, Khalid Sulayman Jaydh Al Hubayshi, and Noor Uthman Muhammed confirmed this divide in their CSRT testimony.[15] [19] [20] Of the 57 detainees the U.S. Government claims are associated with the Khalden Camp, 27 have been released, including Abu Zubaydah's good friend Khalid Sulayman Jaydh Al Hubayshi.[24]

On September 12, 2009, Colin Freeze, writing in the *Globe and Mail*, reported on recent interviews with Mohamad Kamal Elzahabi, a Lebanese citizen, US resident, and green card holder, who has been held without charge in the USA since 2003.[25] Elzahabi told the *Globe and Mail* that Abu Zubaydah had served under him when he was a squad commander during the war against the Soviets. Elzahabi said that after the Soviet ouster Elzahabi worked as an instructor at Khalden.

Early activities

By 1999, the U.S. Government was attempting to run surveillance on Abu Zubaydah.[26] By March 2000, United States officials were reporting that Abu Zubaydah was a "senior bin Laden official", the "former head of Egypt-based Islamic Jihad", a "trusted aide" to bin Laden with "growing power," who had "played a key role in the East Africa embassy attacks."[27] None of these allegations have been corroborated at this point, however.

Internationally Abu Zubaydah was convicted in absentia and sentenced to death[28] by a Jordanian court for his role in plots to bomb U.S. and Israeli targets in Jordan.[29] A senior Middle East security official stated Abu Zubaydah had directed the Jordanian cell and was part of "bin Laden's inner circle."[30]

In August 2001, a classified FBI report entitled "Bin Laden determined to strike in U.S.", which would not become public until much later, stated that the foiled millennium bomber, Ahmed Ressam, had confessed that Abu Zubaydah had not only encouraged him to blow up the Los Angeles airport, but had facilitated his mission.[31] The report also claims Abu Zubaydah was planning his own attack on the U.S.[31] An unclassified FBI report also stated that Ahmed Ressam attempted to buy a laptop for Abu Zubaydah.[32] Despite all of these supposed connections, when Ahmed Ressam went to trial in December 2001 federal prosecutors did not attempt to link him to Abu Zubaydah, and did not use any of this supposed evidence in its case.[32] He also recanted his confessions after the trial saying he was coerced into giving them.

Capture

On March 28, 2002, CIA and FBI agents, in conjunction with Pakistani intelligence services, raided several safe houses in Pakistan searching for him.[33] [34] [35] [36] Abu Zubaydah was apprehended from one of the targeted safe houses in Faisalabad, Pakistan.[33] [34] [35] [36] [37] During his apprehension he was shot in the thigh, the testicle, and the stomach with rounds from an AK-47 assault rifle.[11] [33] [38] [39] [40] He was not recognised at first, and simply thrown into a pick-up truck along with other prisoners by the Pakistani forces, until a senior FBI agent identified him as Abu Zubaydah.[41] He was taken by the FBI to a Pakistani hospital nearby and treated for his wounds, where the attending doctor admitted to John Kiriakou, the co-leader of the CIA group that apprehended Abu Zubaydah, he had never before seen a patient survive such severe wounds.[38] The FBI and CIA flew in a doctor from Johns Hopkins University to ensure Abu Zubaydah would not succumb to his wounds during transit out of Pakistan.[42]

His pocket litter supposedly contained two bank cards which showed he had access to Saudi and Kuwaiti bank accounts, which was considered rare since most al-Qaeda members used the preferred untraceable hawala banking.[41] According to James Risen, "It is not clear whether an investigation of the cards simply fell through the cracks, or whether they were ignored because no one wanted to know the answers about connections between al Qaeda and important figures in the Middle East – particularly in Saudi Arabia." One of Risen's sources chalks up the failure to investigate the cards to incompetence rather than foul play, "The cards were sent back to Washington and were never fully exploited. I think nobody ever looked at them because of incompetence."[41] When American investigators finally did get around to looking into the cards, they worked with "a Muslim financier with a questionable past, and with connections to the Afghan Taliban, al Qaeda, and Saudi intelligence."[41] He reported back that "Saudi intelligence officials had seized all of the records related to the card from the Saudi financial institution in question; the records then disappeared. There was no longer any way to trace the money that had gone into the account."[41]

A search of the safehouse turned up his personal 10,000 page diaries, in which he recorded his thoughts in seemingly split personalities of a young boy, old man, and himself. What appears to be split personalities is actually how Zubaydah was piecing his memories together after his 1992 shrapnel head wound; as part of his therapy to regain his memories, he began recording the diary that detailed his life, emotions, and what people were telling him. He split information into categories, such as what he knew about himself and what people told him about himself, and listed them under different names to distinguish one set from the other. This was later incorrectly interpreted by some analysts reviewing the diary as symptoms of split personality disorder.[11] [43]

Abu Zubaydah was turned over to the CIA,[44] [45] and was allegedly transferred to secret CIA-operated prisons in Pakistan, Thailand, Afghanistan, Poland, Northern Africa, and Diego Garcia.[46] [47] [48] [49] [50] [51] Historically renditions to countries which commit torture have been illegal, however, a memo written by John Yoo and Jay Bybee days before Abu Zubaydah's capture, provided legal cover for renditions to places such as Thailand.[52] In March 2009, the U.S. Senate Intelligence Committee launched a year-long study on how the CIA operated the secret prisons around the world.[53]

Interrogation of Abu Zubaydah

Abu Zubaydah was interrogated by two separate interrogation teams, one from the FBI and one from the CIA.

The interrogation of Abu Zubaydah

Abu Zubaydah's treatment at the hands of the CIA has been called torture by Ali Soufan, the FBI interrogator who witnessed part of Abu Zubaydah's CIA interrogation, multiple U.S. officials including President Obama, and by the International Committee of the Red Cross.[54] [55] [56] [57]

The CIA subjected Abu Zubaydah to various forms of increasingly harsh interrogation techniques including temperature extremes, music played at debilitating volumes, and sexual humiliation.[16] [48] [57] [58] Abu Zubaydah was also subjected to beatings, isolation, waterboarding, long-time standing, continuous cramped confinement, and sleep deprivation.[11] [47] [48] [57] [59] [60] [61] During Abu Zubaydah's interrogation President Bush learned he was on painkillers for the wounds he suffered during his capture and was therefore difficult to get information from.[62] President Bush exclaimed to then CIA director George Tenet "[w]ho authorized putting him on pain medication?"[62] It would later be reported that Abu Zubaydah was denied painkillers during his interrogation.[11] [63] [64] [65] [66] [67] [68] [69] [70]

Many of the interrogation techniques used against Abu Zubaydah, including waterboarding, cold cell, long-time standing, and sleep deprivation were previously considered illegal under U.S. and international law and treaties at the time of Abu Zubaydah's capture.[51] [71] In fact, the United States had prosecuted Japanese military officials after World War II and American soldiers after the Vietnam War for waterboarding.[71] Since 1930, the United States had defined sleep deprivation as an illegal form of torture.[11] Many other techniques developed by the CIA constitute inhuman and degrading treatment and torture under the United Nations Convention against Torture and Article 3 of the European Convention on Human Rights.[51]

At least one detainee claims to have seen pictures of the injuries caused to Abu Zubaydah during his torture.[72]

International Committee of the Red Cross report

The International Committee of the Red Cross concluded a report on the treatment of 14 high-value detainees in February 2007.[57] The report would be made public April 7, 2009.[73] The report is composed of interviews with the detainees, although the ICRC stated "[t]he ICRC wishes to underscore that the consistency of the detailed allegations provided separately by each of the fourteen [detainees] adds particular weight to the information provided."[57] The ICRC outlined twelve interrogation techniques suffocation by water (waterboarding), prolonged stress standing position, beatings by use of a collar, beating and kicking, confinement in a box, prolonged nudity, sleep deprivation, exposure to cold temperature, prolonged shackling, threats of ill-treatment, forced shaving, and deprivation/restricted provision of solid food.[57] Of the 12 interrogation techniques, only Abu Zubaydah was subjected to all of them.[57]

Abu Zubaydah was the only one of the fourteen detainees to be put into close confinement.[57] This would be corroborated, in part, by one of the Torture Memos written by the Department of Justice Office of Legal Counsel, which noted "[i]n OMS's view, however, cramped confinement "ha[s] not proved particularly effective" because it provides "a safehaven offering respite from interrogation."[74] In Abu Zubaydah's case, however, this is debatable since the smaller of the close confinement boxes caused his wound to reopen and begin to bleed again.[57] He was placed in two boxes, a 'large' one in which he could stand upright, and a much smaller one, in which he had to crouch down.[57]

Waterboarding

Abu Zubaydah was one of three high value detainees to be waterboarded.[57] [75] Originally the Government's position was that Abu Zubaydah had only been waterboarded once.[38] [39] [76] [77] [78] [79] Indeed, John Kiriakou, a CIA officer who had seen the cables regarding Abu Zubaydah's interrogation publicly stated Abu Zubaydah was waterboarded only once for 35 seconds before he started talking.[80] [81] [82] Even before the May 30, 2005 Torture Memo surfaced, Government officials familiar with Abu Zubaydah's detention questioned this account of events.[11] [16] In fact, intelligence sources claimed as early as 2008 that Abu Zubaydah had been waterboarded no less than ten times in the span of one week.[11] In reality, Abu Zubaydah was waterboarded 83 times within the month of August 2002, which happens to be the same month waterboarding was approved for use on him.[74] [75] [83] [84] [85] [86] In January 2010, Kiriakou, in a memoir, said "Now we know that Zubaydah was waterboarded eighty-three times in a single month, raising questions about how much useful information he actually supplied."[87]

The May 10, 2005 torture memo stated that no more than 60 applications of water could be conducted in a 30 day period. However, that number applied only to those applications of "ten seconds or more."[88] A U.S. official with knowledge of the interrogation program reported to Fox News that many of the applications lasted "a matter of seconds" and it was these less than ten second applications "that created the huge numbers." He further stated that "[a]ll of those individual pours were scrupulously counted by the CIA, according to the memos, to abide by the procedures set up for the waterboardings."[89]

However, the May 10, 2005 torture memo reported that the Department of Justice, Inspector General report noted that

> [T]he waterboard technique… was different from the technique described in the DoJ opinion and used in the SERE training. The difference was in the manner in which the detainee's breathing was obstructed. At the SERE school and in the DoJ opinion the subject's airflow is disrupted by the firm application of a damp cloth over the air passage; the interrogator applies a small amount of water to the cloth in a controlled manner. By contrast, the Agency interrogator… applied large volumes of water to a cloth that covered the detainee's mouth and nose. One of the psychologists/interrogators acknowledged that the Agency's use of the technique is different from that used in SERE training because it is 'for real' and is 'more poignant and convincing.'" The Inspector General further reported that "OMS contends that the expertise of the SERE psychologist/interrogator on the waterboard was probably misrepresented at the time, as the SERE waterboard experience is so different from the subsequent Agency usage as to make it almost irrelevant. Consequently, according to OMS, there was no a priori reason to believe that applying the waterboard with the frequency and intensity with which it was used by the psychologist/interrogators was either efficacious or medically safe.[88]

The Inspector General also noted that the use of waterboarding was discontinued in every armed services branch except the Navy SERE training "because of its dramatic effect on the students who were subjects."[88] The CIA Office of Medical Services contradicted this conclusion, however, stating that "[w]hile SERE trainers believe that trainees are unable to maintain psychological resistance to the waterboard our experience was otherwise. Some subjects unquestionably can withstand a large number of applications, with no immediately discernible cumulative impact beyond their strong aversion to the experience."[88] The memo also noted that at a Senate Judiciary Committee hearing, Douglas Johnson, Executive Director of the Center for Victims of Torture, testified that some U.S. military personnel who have undergone waterboard training have apparently stated "that it's taken them 15 years of therapy to get over it", although his claim has not substantiated.[88] Despite the conflicting information the Department of Justice Office of Legal Counsel approved all of the tactics listed above.[88]

Several people who have subjected themselves to controlled waterboarding have stated that it is torture.[90] [91] [92] Conservative Chicago radio host, Erich "Mancow" Muller agreed to be waterboarded so he could "prove it wasn't torture."[93] Instead Muller stated "It is such an odd feeling to have water poured down your nose with your head back... It was instantaneous... and I don't want to say this: absolutely torture."[93]

Malcolm Wrightson Nance, an instructor at the Navy's SERE school in California testified before congress that waterboarding "is an overwhelming experience that induces horror and triggers frantic survival instincts. As the event unfolded, I was fully conscious of what was happening: I was being tortured."[90] Nance further testified that waterboarding is "[N]ot simulated anything. It's slow motion suffocation with enough time to contemplate the inevitability of blackout and expiration."[90] Richard E. Mezo, who was waterboarded during training to become a Navy flight crewmember described waterboarding as an act that is "indeed torture" which is "real drowning that simulates death."[91] The experience was so horrific, Mezo stated that "Pulling out my fingernails or even cutting off a finger would have been preferable."[91] Henri Alleg, who was waterboarded by the French during the Algerian War, wrote that "a terrible agony, that of death itself, took possession of me."[92]

Abu Zubaydah said of his experience:

> I struggled against the straps, trying to breathe, but it was hopeless. I thought I was going to die.[57]

During his waterboarding Abu Zubaydah lost control of his bladder and would later admit to the ICRC that "[s]ince then I still lose control of my urine when under stress."[57]

Some politicians and administration officials, including Richard Armitage, the former Deputy Secretary of State, Mike McConnell, the former National Intelligence Directorate, Tom Ridge, the former Homeland Security Secretary, and former Republican Presidential Candidate John McCain, have also declared it unequivocal torture.[11] [94] Senator John McCain stated "It is not a complicated procedure. It is torture."[11]

Top U.S. officials approved enhanced interrogation techniques

In the Spring of 2002, immediately following Abu Zubaydah's capture, top US Government officials including Dick Cheney, Colin Powell, George Tenet, Condoleezza Rice, Donald Rumsfeld, and John Ashcroft discussed at length whether or not the CIA could legally use harsh techniques against Abu Zubaydah.[95] [96] Condoleezza Rice specifically mentioned the SERE program during the meeting stating "I recall being told that U.S. military personnel were subjected to training to certain physical and psychological interrogation techniques…"[95] In addition, in 2002 and 2003, several Democratic congressional leaders were briefed on the proposed "enhanced interrogation techniques."[97] These congressional leaders included Nancy Pelosi, the future Speaker of the House, and Representative Jane Harman.[97] Congressional officials have stated that the attitude in the briefings ranged from "quiet acquiescence, if not downright support."[97] The documents show that top U.S. Officials were intimately involved in the discussion and approval of the harsher interrogation techniques used on Abu Zubaydah.[95] Condoleezza Rice ultimately told the CIA the harsher interrogation tactics were acceptable,[98] [99] and Dick Cheney stated "I signed off on it; so did others."[99] [100] During the discussions John Ashcroft is reported as saying "Why are we talking about this in the White House? History will not judge this kindly."[96]

The torture memos

In 2009, President Obama released four Justice Department memos which outlined the procedures CIA operatives wished to use on Abu Zubaydah.[101] [102] [103] The memos were written at the request of Dick Cheney and the Bush White House and were written under extreme pressure from the Office of the Vice President.[104] In e-mails to former U.S. Attorney Chuck Rosenberg, Jim Comey, former Deputy Attorney General, stated that the Attorney General, Alberto Gonzalez, had informed him he was "under great pressure from the Vice President to complete both [May] memos."[104]

Opposition within the Bush Administration

The memos were opposed by members of the White House and Congress. Within the White House Philip Zelikow, a top advisor to Condoleezza Rice, opposed the new, harsher interrogation techniques.[105] Upon reading the August 1, 2002, memo which justified the torture, Zelikow authored his own memo contesting the Justice Department's conclusions, since he believed they were legally incorrect.[105] The Bush Administration attempted to collect all of the copies of Zelikow's memo and destroy them.[55] [105] In addition, military advisors to the CIA declared the proposed tactics as torture thirteen times within two pages in a memo regarding the techniques.[106] The memo was written one month before the torture memos were drafted and the interrogation techniques were used against Abu Zubaydah.[106]

In Congress, House Representative Jane Harman, who was briefed on the proposed enhanced interrogation techniques in 2002, was the sole member of Congress to object.[107]

Former CIA Inspector General, John Helgerson, drafted a report in 2004 which reportedly condemns not only the legality of the torture memos, but also the effectiveness of the torture techniques used.[108] The report was so appalling it prompted then director of the Office of Legal Counsel, Jack Goldsmith, to repudiate the John Yoo, Jay Bybee, and Stephen Bradbury memos.[108]

In addition, a federal judge in San Francisco has ruled that José Padilla has standing to sue John Yoo for his detention and torture, due to John Yoo's role in drafting the torture memos.[109] Dick Cheney's order to keep secret from Congress a CIA program recently killed by CIA Director Leon Panetta may also come under legal fire.[110] [111] Under the Intelligence Authorization Act, the office of the President is required to keep the Congress completely informed of the operations of the intelligence community.[110]

August 1, 2002 memo

In August 2002, the Department of Justice's Office of Legal Counsel, Jay Bybee and John Yoo drafted the first Torture Memo.[112] Addressed to CIA acting General Counsel John A. Rizzo, the purpose of the memo was to gain approval for harsh interrogation tactics to be used on Abu Zubaydah.[83] [112] Although some believe the harsh tactics were already in effect before the memo granting authority to use them was written,[11] [95] [112] [113] [114] and was used to provide after-the-fact legal support for harsh interrogation techniques.[115] In fact, a Department of Justice report regarding prisoner abuses reportedly stated the preparation of the memos occurred one month after Abu Zubaydah was subjected to the techniques allowed in the August 1, 2002, memo.[116] John Kiriakou stated in July 2009, that Abu Zubaydah was waterboarded in the early summer of 2002, months before the August 1, 2002 memo was written.[117] [118] Alberto Gonzales would later testify before Congress that the opinion was sought after the detention of Abu Zubaydah.[119] Questions by CIA officers over which tactics could be used on Abu Zubaydah had spurred the torture memo's existence,[115] which is reflected in the language of the memo; "You have asked for this advice in the course of conducting interrogations of Abu Zubaydah."[120] The memo's author, John Yoo, acknowledged the memo was the basis for Abu Zubaydah's interrogation.[121] Yoo even told an interviewer "there was an urgency to decide so that valuable intelligence could be acquired from Abu Zubaydah, before further attacks could occur."[121] John Yoo would later be harshly criticized by the Department of Justice for failing to cite legal precedent and existing case law when drafting his memos.[116] In particular, the report chastises Yoo for failing to cite Youngstown Sheet & Tube Co. v. Sawyer, a seminal case on the powers of the Executive in times of war.[116] In addition, Jack Goldsmith, the former head of the Office of Legal Counsel who withdrew the torture memos before resigning, stated he was "astonished" by the "deeply flawed" and "sloppily reasoned" legal analysis in the memos.[116] [122] David Luban, a law professor at Georgetown Law School, testified before Congress on May 13, 2009, that the memos were "an ethical train wreck" and had been drafted to "reverse engineer" a defense for illegal actions already committed.[123]

The memo contemplated ten techniques the interrogators wanted to use: "(1) attention grasp, (2) walling, (3) facial hold, (4) facial slap (insult slap), (5) cramped confinement, (6) wall standing, (7) stress positions, (8) sleep deprivation, (9) insects placed in a confinement box, and (10) the waterboard."[120] Many of the techniques were, until then, generally considered illegal.[11] [51] [71] [94] [95] [112] Many other techniques developed by the CIA constituted inhumane and degrading treatment and torture under the United Nations Convention against Torture and Article 3 of the European Convention on Human Rights.[51]

According to a psychological evaluation conducted of Abu Zubaydah upon his capture, the memo alleges that Abu Zubaydah:

- Quickly rose from very low level mujahedin to third or fourth man in al Qaeda.
- Served as Osama Bin Laden's senior lieutenant.
- Managed a network of training camps.
- Was instrumental in the training of operatives for al Qaeda, the Egyptian Islamic Jihad, and other terrorist elements inside Pakistan and Afghanistan.
- Acted as the Deputy Camp Commander for al Qaeda training camp in Afghanistan, personally approving entry and graduation of all trainees during 1999–2000.
- Approved all individuals going in and out of Afghanistan to the training camps from 1996–1999.
- No one went in and out of Peshawar, Pakistan without his knowledge and approval.
- Acted as al Qaeda's coordinator of external contacts and foreign communications.
- Acted as al Qaeda's counter-intelligence officer and had been trusted to find spies within the organization.
- Was involved in every major terrorist operation carried out by al Qaeda.
- Was a planner for the Millennium plot to attack U.S. and Israeli targets during the Millennium celebrations in Jordan.
- Served as a planner for the Paris Embassy plot in 2001.
- Was one of the planners of 9/11.
- Engaged in planning future terrorist attacks against U.S. interests.
- Wrote al Qaeda's manual on resistance techniques,[120]

May 10, 2005 memo

Another subsequent memo addressed the legality of additional interrogation techniques such as nudity, dietary manipulation, abdominal slap, water dousing, and water flicking.[88] It also expanded on the techniques of walling, stress positions, and sleep deprivation, allowing for an additional stress position and extended sleep deprivation up to 180 consecutive hours.[88] The memo also outlined the amount of waterboarding applications a detainee could be subjected to.[88]

> The waterboard can only be used with a given detainee during one 30-day period. During that 30-day period the waterboard can be used no more than 5 days. In any given day that waterboarding occurs interrogators may use no more than two "sessions", with a "session" defined as the time that the detainee is strapped to the board, and that a session can last no more than 2 hours. During any session no more than six applications of water of 10 seconds or more can be used. The total cumulative time of all water applications in a 24 hour period may not exceed 12 minutes.[88]

May 10, 2005 memo (combined interrogation techniques)

Another memo penned on May 10, 2005 authorized the use of the above outlined individual techniques in conjunction with one another, but stressed the importance of constant vigilance on the part of medical observers to ensure the techniques did not cause "severe physical or mental pain."[124]

May 30, 2005 memo

The final memo mentioned Abu Zubaydah several times and claimed that due to the increased interrogation techniques Zubaydah also "provided significant information on two operatives, [including] José Padilla[,] who planned to build and detonate a 'dirty bomb' in the Washington DC area."[74] This claim is heavily disputed, however, by Ali Soufan the FBI interrogator who first interrogated Abu Zubaydah following his capture, and other intelligence officials.[11] [54] [55] [56] [125] Ali Soufan, when asked by Senator Sheldon Whitehouse during a Congressional hearing if the memo was incorrect, testified that it was.[126] In fact, the memo itself noted that not all of the waterboarding sessions were necessary for Abu Zubaydah, since the on-scene interrogation team determined he had stopped producing actionable intelligence.[74] The memo reads:

> This is not to say that the interrogation program has worked perfectly. According to the IG Report, the CIA, at least initially, could not always distinguish detainees who had information but were successfully resisting interrogation from those who did not actually have the information. See IG Report at 83–85. On at least one occasion, this may have resulted in what might be deemed in retrospect to have been the unnecessary use of enhanced techniques. On that occasion, although the on-scene interrogation team judged Abu Zubaydah to be compliant, elements within CIA Headquarters still believed he was withholding information. See id at 84. At the direction of CIA Headquarters interrogators therefore used the waterboard one more time on Zubaydah.[74]

U.S. officials defend interrogation techniques

Dennis C. Blair, President Obama's new director of national intelligence wrote in a memo to his staff in April 2009 that "high value information came from interrogations in which [enhanced interrogation] methods were used."[127] Former Vice-President, Dick Cheney claims there are memos in existence which show the success of the enhanced interrogation techniques.[128] Cheney stated:

> I haven't talked about it, but I know specifically of reports that I read, that I saw, that lay out what we learned through the interrogation process and what the consequences were for the country I've now formally asked the CIA to take steps to declassify those memos so we can lay them out there and the American people have a chance to see what we obtained and what we learned and how good the intelligence was.[128]

Additionally, four successive CIA directors echoed these sentiments, with the most recent, Michael V. Hayden, stating that he believed the methods "got the maximum amount of information" from prisoners, specifically Abu Zubaydah.[127] As noted, however, the belief that enhanced interrogation techniques were necessary to get actionable intel from Abu Zubaydah has been heavily contested. John McLaughlin, former acting CIA director, stated in 2006 "I totally disagree with the view that the capture of Abu Zubaydah was unimportant. Abu Zubaydah was woven through all of the intelligence prior to 9/11 that signaled a major attack was coming, and his capture yielded a great deal of important information."[129] In his memoir, former CIA Director George Tenet writes:

> A published report in 2006 contended that Abu Zubaydah was mentally unstable and that the administration had overstated his importance. Baloney. Abu Zubaydah had been at the crossroads of many al-Qa'ida operations and was in position to – and did – share critical information with his interrogators. Apparently, the source of the rumor that Abu Zubaydah was unbalanced was his personal diary, in which he adopted various personas. From that shaky perch, some junior Freudians leapt to the conclusion that Zubaydah had multiple personalities. In fact, Agency psychiatrists eventually

determined that in his diary he was using a sophisticated literary device to express himself. And, boy, did he express himself.[130]

The changing depiction of Abu Zubaydah

When Abu Zubaydah was captured, the Bush Administration believed he was an unparalleled source of intelligence on al-Qaeda and terrorism plots. Abu Zubaydah was touted as the biggest catch of the War on Terror until the capture of Khalid Sheikh Mohammed.[131] Immediately after Abu Zubaydah's capture the director of the FBI stated Abu Zubaydah's capture would help deter future attacks.[132] Also, White House spokesman Ari Fleischer stated Abu Zubaydah could provide a treasure-trove of information about al-Qaeda.[133] Donald Rumsfeld echoed these sentiments claiming Abu Zubaydah was "a man who knows of additional attacks", who has "trained people to do this", and was a big fish who had a fountain of knowledge.[134]

U.S. government accounts of Abu Zubaydah's importance

After Abu Zubaydah's capture, officials from the U.S. Government spoke out publicly about Abu Zubaydah's supposed role in al-Qaeda. However, as would later be reported in 2009, the U.S. Government's depiction of Abu Zubaydah was overly inflated and he was, as Justice Department Officials stated, "[t]he above ground support... To make him the mastermind of anything is ridiculous",[14] a "personnel clerk",[135] a "logistics chief",[136] and a "travel agent."[17] In fact, the CIA reportedly told Abu Zubaydah during his interrogation that they discovered he was not an al-Qaeda fighter, partner, or even a member.[137]

The allegations included:

- Abu Zubaydah was "sinister" and "[t]here is evidence that he is a planner and a manager as well. I think he's a major player." – Former State Department director of counter-terrorism, Michael Sheehan[138]
- Abu Zubaydah was "extremely dangerous" and a planner of 9/11. – John B. Bellinger III in a June 2007 briefing on Guantanamo Bay.[139]
- Abu Zubaydah was a trainer, a recruiter, understood bomb-making, was a forger, a logistician, and someone who made things happen, and made "al-Qaeda function." – Former CIA station chief, Bob Grenier[140]
- "I don't think there's any doubt but a man named Abu Zubaydah is a close associate of UBL's, and if not the number two, very close to the number two person in the organization. I think that's well established." -Donald Rumsfeld[141]
- Abu Zubaydah was "a very senior al Qaeda official who has been intimately involved in a range of activities for the al Qaeda." – Donald Rumsfeld[142]
- Abu Zubaydah was a "very senior al Qaeda operative." – Donald Rumsfeld[134]
- Abu Zubaydah was a "key terrorist recruiter and operational planner and member of Osama bin Laden's inner circle." – White House spokesman Ari Fleischer[143]
- The capture of Abu Zubaydah was a "very serious blow" to al-Qaeda and that one of al-Qaeda's "many tentacles" was "cut off." – White House spokesman Ari Fleischer[133]
- Abu Zubaydah was "one of the top operatives plotting and planning death and destruction on the United States." – Former President George W. Bush[144]
- Abu Zubaydah was "one of al-Qaeda's top leaders" who was "spending a lot of time as one of the top operating officials of al Qaeda, plotting and planning murder." – Former President George W. Bush[145]
- Abu Zubaydah was "al Qaeda's chief of operations." – President George W. Bush[146]
- "Abu Zubaydah was one of the top three leaders" in al-Qaeda. – President George W. Bush[147]
- Abu Zubaydah's interrogation "led to reliable information", that Abu Zubaydah was a "prolific producer" of information,[148] and that roughly 25 percent of the information on al Qaeda that came from human sources originated from Abu Zubaydah. – Michael Hayden[149]

- Abu Zubaydah was one of three individuals "best positioned to know about impending terrorist atrocities." – Michael Hayden[150]
- Abu Zubaydah is someone who was "carefully trained in techniques of disinformation." – Richard C. Shelby[151]
- Nancy Pelosi, the senior Democrat on the House Intelligence Committee would describe Abu Zubaydah as being "very skilled at avoiding interrogation. He is an agent of disinformation."[152]

Numerous anonymous U.S. officials have also made allegations against Abu Zubaydah in the press, including:

- "One Abu Zubaydah is worth a ton of guys at Gitmo."[153]
- He was a "senior bin Laden official" and the "former head of Egypt-based Islamic Jihad."[27]
- He "played a key role in the East Africa embassy attacks."[27]
- He was listed as a "trusted aide" to bin Laden with "growing power."[27]
- Western officials believed al-Qaeda may have been under the control of Abu Zubaydah.[154]
- Abu Zubaydah was an aide of bin Laden who ran training camps in Afghanistan and "coordinated terror cells in Europe and North America."[154]
- Abu Zubaydah was a "key terrorist recruiter, operational planner, and member of Osama Bin Laden's inner circle."[155]
- Abu Zubaydah was "bin Laden's CEO",[156] "a central figure in Al Qaeda",[157] and a "bin Laden lieutenant."[158]
- Abu Zubaydah is Bin Laden's "travel planner."[159]
- Abu Zubaydah is one of bin Laden's "confidants."[160]
- Abu Zubaydah is "one of a handful of men entrusted with running the terrorism network in the event of Osama bin Laden's death or capture."[161]
- Abu Zubaydah was a senior bin Laden lieutenant who was believed "to be organizing al Qaida resources to carry out attacks on American targets."[162]
- Abu Zubaydah was the fourth ranking member of al Qaida behind Bin Laden, Ayman al-Zawahiri, and Muhamed Atef.[162]
- Abu Zubaydah was the "successor as chief of operations for al Qaida" after Muhamed Atef's death.[162]
- Abu Zubaydah knows the identities of "thousands" of terrorists that passed through al Qaida training camps in Afghanistan.[162]
- Abu Zubaydah briefed Richard Reid, the shoe-bomber.[143]
- Abu Zubaydah was one of bin Laden's top planners of terrorist operations who knew of al Qaida plots and cells.[163]
- Abu Zubaydah stated al-Qaeda knew how to smuggle a dirty-bomb into the United States.[164] Abu Zubaydah's information was further proof al-Qaeda was attempting to acquire weapons of mass destruction.[164]
- Abu Zubaydah is the "connection between bin Laden and many of al-Qaida's operational cells."[165]
- Abu Zubaydah is linked to plots to detonate apartment buildings.[166]
- Abu Zubaydah confirmed the fourth September 11 flight, UA 93, was intended to hit the White House.[167] [168] (His claim is contradicted by Khalid Sheikh Mohamed and Ramzi Bin al Shibh though, who stated UA 93 was on its way to the capitol, not the White House.)[169]
- Osama bin Laden wrote a handwritten note to Abu Zubaydah in December 2001.[170] The letter allegedly tells Abu Zubaydah to continue fighting the United States if bin Laden passes away.[170] [171]

While Abu Zubaydah would provide important intelligence on the War on Terror, his value as an intel source was greatly inflated, much like his role in the global terror network.

Despite all of these allegations the U.S. Government has not officially charged Abu Zubaydah with any crimes.[172]

Exploitation of Abu Zubaydah's perceived value

President Bush personally used Abu Zubaydah's perceived "value" as a detainee to justify the use of the CIA's harsher interrogation techniques[173] as well as Abu Zubaydah's detention in secret CIA prisons around the world.[174]

In a speech in 2006, President Bush claimed that Abu Zubaydah initially revealed useful intelligence, including information that allegedly helped foil a terrorist attack on American soil, but that Abu Zubaydah became uncooperative.[173] It was only then, he reported, that an "alternative set of procedures" was used on Abu Zubaydah in order to gain valuable intelligence and were "safe and lawful."[173] He also stated that Abu Zubaydah had received training in how to resist interrogation, and thus more aggressive techniques were mandated.[173] These claims directly conflict with the reports of the original F.B.I. agents tasked with interrogating Abu Zubaydah who had been receiving crucial pieces of information from him without the use of harsher techniques,[11] [54] [55] as well as other government officials.[14] [75]

The Iraq War

The U.S. Government used questionable intel from Abu Zubaydah in order to justify the invasion of Iraq. U.S. officials stated that the allegations that Iraq and al-Qaeda were linked in the training of people on chemical weapons came from Abu Zubaydah.[175] [176] [177] The officials noted there was no independent verification of his claims.[175] [176] The U.S. Government included statements made by Abu Zubaydah in regards to al Qaeda's ability to obtain a dirty bomb in its attempts to show a link between Iraq and al Qaeda.[178] According to a Senate Intelligence Committee report Abu Zubaydah also "indicated that he had heard that an important al Qaeda associate, Abu Musab al Zarqawi, and others had good relationships with Iraqi intelligence."[179] However, in June 2003 Abu Zubaydah and Khalid Sheikh Mohammed were reported as stating there was no link between Saddam Hussein and al Qaeda.[180] [181]

In the Senate Armed Services Committee report on the abuses of detainees, it was revealed that pressure from the Bush administration was applied to interrogators to find a link between Iraq and Al-Qaeda.[182] Indeed, Major Paul Burney, a psychiatrist with the United States Army, reported that "while we were [at Guantanamo] a large part of the time we were focused on trying to establish a link between Al Qaeda and Iraq and we were not being successful."[182] [183] He stated that higher-ups for more "frustrated" and applied "more and more pressure to resort to measures that might produce more immediate results."[182] [183] [184]

Additionally, Colonel Lawrence B. Wilkerson, the former chief of staff for former Secretary of State Colin Powell has stated that:

> Likewise, what I have learned is that as the administration authorized harsh interrogation in April and May of 2002—well before the Justice Department had rendered any legal opinion—its principal priority for intelligence was not aimed at pre-empting another terrorist attack on the U.S. but discovering a smoking gun linking Iraq and al-Qa'ida.
>
> So furious was this effort that on one particular detainee, even when the interrogation team had reported to Cheney's office that their detainee "was compliant" (meaning the team recommended no more torture), the VP's office ordered them to continue the enhanced methods. The detainee had not revealed any al-Qa'ida-Baghdad contacts yet. This ceased only after Ibn al-Shaykh al-Libi, under waterboarding in Egypt, "revealed" such contacts. Of course later we learned that al-Libi revealed these contacts only to get the torture to stop.[183]

The Military Commissions Act

President Bush also asked Congress in a speech in September 2006 to formulate special rules in order to try Abu Zubaydah via military commission in Guantanamo Bay.[185] In fact, in late April 2002 less than one month after Abu Zubaydah's capture, Justice Department officials stated Abu Zubaydah "is a near-ideal candidate for a tribunal trial."[186] However, only several months later US Officials acknowledged there was "no rush" to try Abu Zubaydah via military commission.[187] In September 2006, President Bush stated in an interview that if Congress could pass a "good bill" out of the Senate in regards to setting up a military commission system, then Abu Zubaydah "is going to go on trial."[188] The U.S. Government has yet to try Abu Zubaydah by military commission, article 3 court, or in any other capacity.

The Bush Administration's domestic spying program

In addition to justifying the use of presently illegal torture techniques, the Bush administration used Abu Zubaydah's capture as justification to accelerate its domestic spying program to allow quick action on the phone numbers and addresses seized during Abu Zubaydah's capture.[189] Inexplicably the NSA expanded its surveillance beyond the numbers seized during Abu Zubaydah's capture.[190] The spying program would later be revamped in order to make it theoretically legal.[11]

Growing concerns about Abu Zubaydah's indefinite CIA detention

In 2004 media coverage of Abu Zubaydah began listing him as a "disappeared" prisoner claiming he had no access to the International Red Cross.[191]

In February 2005, U.S. officials claim the CIA was growing uncomfortable keeping Abu Zubaydah in indefinite custody.[192] Less than 18 months later Abu Zubaydah and the other high value detainees who had been in secret CIA custody were transferred to Guantanamo Bay.[193]

After his transfer the CIA denied access to Abu Zubaydah even to the Department of Justice's Inspector General who was investigating the United States' treatment of its detainees.[194]

According to U.S. officials quoted in Ron Suskind's book, President Bush allegedly told CIA director George Tenet "I said [Abu Zubaydah] was important. You're not going to let me lose face on this, are you?"[193] Tenet replied "No sir, Mr. President."[193]

U.S. government admits Abu Zubaydah was never a member of al Qaeda

Top officials in the U.S. government refused to believe Abu Zubaydah was not the operative they believed him to be.[14] [74] The May 30, 2005 Department of Justice memo noted that while on-scene interrogators believed Abu Zubaydah no longer had any information to disclose, CIA Headquarters ordered additional waterboarding.[74] The interrogators believed the waterboarding was "unnecessary."[74] Orders for the additional waterboarding likely came from Dick Cheney directly.[183] Additionally, the Bush White House and CIA officials couldn't believe Abu Zubaydah didn't have additional information.[14] One official stated the pressure from upper levels of government was "tremendous," and that "[t]hey couldn't stand the idea that there wasn't anything new."[14] The official said, "[t]hey'd say, 'You aren't working hard enough.' There was both a disbelief in what he was saying and also a desire for retribution – a feeling that 'He's going to talk, and if he doesn't talk, we'll do whatever.'"[14]

However, in September 2009, the United States Government finally admitted, during Abu Zubaydah's habeas corpus petition, that Abu Zubaydah had never been a member of al-Qaeda, nor involved in the attacks on the African embassies in 1998, or the attacks on the United States on September 11, 2001.[195] The motion, filed by the United States Government, states:

> Evidence indicating that Petitioner is not a member of al-Qaida or had ideological differences with al-Qaida is not inconsistent with the factual allegations made in the Government's factual return, because the Government has not contended in this proceeding that Petitioner was a member of al-Qaida or

otherwise formally identified with al-Qaida. Pg. 35, 36

Respondent [The United States Government] does not contend that Petitioner [Abu Zubaydah] was a "member" of al-Qaida in the sense of having sworn a bayat (allegiance) or having otherwise satisfied any formal criteria that either Petitioner [Abu Zubaydah] or al-Qaida may have considered necessary for inclusion in al-Qaida. Nor is the Government detaining Petitioner [Abu Zubaydah] based on any allegation that Petitioner [Abu Zubaydah] views himself as part of al-Qaida as a matter of subjective personal conscience, ideology, or worldview. Pg. 36

The Government has not contended in this proceeding that Petitioner [Abu Zubaydah] had any direct role in or advance knowledge of the terrorist attacks of September 11, 2001. Pg. 82

For example, for purposes of this proceeding the Government has not contended that Petitioner [Abu Zubaydah] had any personal involvement in planning or executing either the 1998 embassy bombings... or the attacks on September 11, 2001. Pg. 34

Charge sheet discrepancies

In 2005, the U.S. Government charged several detainees at Guantanamo Bay and ordered them to stand trial in Military Commission Tribunals.[196] Of the ten detainees charged, four of them, Binyam Mohamed, Ghassan Abdullah al Sharbi, Sufyian Barhoumi, and Jabrad Said bin Al-Qahtani referenced Abu Zubaydah in their charge sheets.[197] [198] [199] [200] Three of the four detainees, Ghassan Abdullah al Sharbi, Sufyian Barhoumi, and Jabran Said bin Al-Qahtani, had identical 2005 charge sheets and referenced Abu Zubaydah in six different paragraphs.[197] [198] [199] The fourth detainee, Binyam Mohamed, referenced Abu Zubaydah five times.[200]

In 2006, before the U.S. Government could hold a trial of the four detainees in a Military Commission Tribunal, the Supreme Court struck down the Tribunals as unconstitutional in Hamdan v. Bush.[201] [202] The U.S. Government responded by amending the tribunals, although they were forced to recharge Binyam Mohamed, Ghassan Abdullah al Sharbi, Sufyian Barhoumi, and Jabran Said bin Al-Qahtani in 2008.[172]

The new 2008 charge sheets contained almost identical charges against each of the four detainees, as were found in their 2005 charge sheets.[203] [204] [205] [206] However, Abu Zubaydah's name was entirely removed from each detainee's charge sheet.[203] [204] [205] [206] On October 21, 2008, all charges were dropped against the four men.[207] [208] Binyam Mohamed was released from Guantanamo Bay on February 20, 2009 despite his alleged connections to Abu Zubaydah, and thus has been removed from the Department of Defense Military Commissions Page.[172] [209]

In addition, the alleged emir of Khalden Camp, Noor Uthman Muhammed, was charged in 2008 with conspiring with Abu Zubaydah against the United States. However, the charges against Noor Uthman Muhammed were also completely dropped on October 21, 2008.[207] [208] The U.S. Government recharged Noor Uthman Muhammed on December 22, 2008.[23] Abu Zubaydah's name is not found in any of the charges even though he was a facilitator for Noor Uthman Muhammed's Khalden Camp.[23]

Abu Zubaydah's name is also not mentioned in any other detainee's charge sheets, not even Khalid Sheikh Mohamed, the admitted mastermind of the 9/11 attacks.[172] [210] The U.S. Government has also failed to charge Abu Zubadyah, even though they have charged other high ranking al-Qaeda members, such as Khalid Sheikh Mohamed, who has been in custody less time than Abu Zubaydah.[172] [211]

In 2009, the U.S. Government recharged two of the men, Ghassan Abdullah al Sharbi and Jabran Said bin Al-Qahtani.[212] [213] Sufyian Barhoumi has not been recharged.[214] The 2009 charge sheet of al Sharbi reincorporates Abu Zubaydah in three paragraphs, while the 2009 charge sheet of Al-Qahtani reincorporated Abu Zubaydah in four paragraphs.[212] [213] Unlike the 2005 charge sheets, which are archived on the Department of Defense website, Ghassan Abdullah al Sharbi's 2008 Charge Sheet has been removed from his page.[215]

International cases involving Abu Zubaydah

Several individuals being held or tried internationally, who have been connected by the U.S. Government to Abu Zubaydah, have had their charges dropped, been released, or received other relief from their handlers.

Abousofian Abdelrazik

Abousofian Abdelrazik was alleged by the State Department to be closely associated with Abu Zubaydah.[216] In 2008, Canada asked the United Nations to remove Abousfian Abdelrazik from its terrorism watch-list.[217] In June 2009, Canada agreed to repatriate Abousfian Abdelrazik from Sudan where he has been stranded since 2008.[218]

Mohamed Harkat

Abu Zubaydah supposedly knew Mohamed Harkat "since the early 90's"[219] and claimed Harkat ran a guest house in Pakistan.[220] Mohamed Harkat's attorney sought access to Abu Zubaydah for testimony relating to Harkat's trial, but the US refused to respond to his requests.[221]

Originally in Harkat's Canadian trial, however, Abu Zubaydah's claims were not part of the charges brought against Harkat.[222] After CIA director Michael Hayden's public admittance of Zubaydah's waterboarding, Canadian officials deleted all references to Zubaydah's statements in its public dossier.[223] A spokesman for the Canadian Security Intelligence Service stated "The CSIS director has stated publicly that torture is morally repugnant and not particularly reliable. CSIS does not knowingly use information which has been obtained through torture."[224] Mohamed Harkat was released on bail by Canadian authorities.[225] [226]

Ahmed Ressam and the Millennium Plot

In April 2001, Ahmed Ressam was convicted of plotting to detonate a bomb at the Los Angeles International Airport.[227] [228] Ahmed Ressam had previously trained at the Khalden camp before coming to the United States to undertake his mission.[15] [227] [228] Abu Zubaydah admits in his CSRT testimony that he recommended to the leader of Khalden Camp that Ahmed Ressam be allowed to train there.[15] Abu Zubaydah further testified that he facilitated Ahmed Ressam's travel to the camp as well as to Algeria once Ressam's training was complete.[15] The U.S. Government alleges, in Abu Zubaydah's summary of evidence, that Ahmed Ressam identified Abu Zubaydah as the leader of the Khalden camp and an associate of Usama Bin Laden "equal to and not subordinate to UBL."[229] The U.S. Government further alleges that Ahmed Ressam stated that Abu Zubaydah had "known of Ahmed Ressam's operation, although not specifically the date and exact target" and that Abu Zubaydah wanted Ressam to acquire "fraudulently-obtained Canadian passports" for himself and five others, in order to facilitate their travel into the United States to "possibly bomb several cities."[229]

Abu Zubaydah admits he attempted to procure Canadian passports for Ahmed Ressam and other trainees, but not for "terrorist-related activities.".[15] Abu Zubaydah denies ever having participated in the planning of the Millennium Plot or encouraging Ahmed Ressam to attack American targets or civilians.[15] The U.S. Government has not produced any further evidence of Abu Zubaydah's alleged role in the Millennium Plot, nor was this connection even mentioned in Ahmed Ressam's criminal trial.[230] In fact, Ressam's confessions were only brought up during his sentencing in an attempt to mitigate the sentence he would receive, and he later recanted them.[230]

Abu Zubaydah's mental health

Some people contest Abu Zubaydah's mental health. Ron Suskind noted in his book, *The One Percent Doctrine*, that Zubaydah turned out to be mentally ill, keeping a diary "in the voice of three people: Hani 1, Hani 2, and Hani 3" – a boy, a young man and a middle-aged alter ego.[17] Abu Zubaydah's diaries spanned ten years and recorded in numbing detail "what he ate, or wore, or trifling things [people] said."[16] Dan Coleman, then the FBI's top al-Qaeda analyst, told a senior bureau official, "This guy is insane, certifiable, split personality."[17] According to Suskind, this judgment was "echoed at the top of CIA and was briefed to the President and Vice President."[17] Dan Coleman, the

FBI's senior expert on al Qaeda, echoed many of Suskind's sentiments in an interview with the Washington Post. Coleman stated Zubaydah was a "safehouse keeper" with mental problems, who "claimed to know more about al-Qaeda and its inner workings than he really did."[16] Abu Zubaydah's co-counsel, Joseph Margulies, wrote in an OpEd in the LA Times that:

> Partly as a result of injuries he suffered while he was fighting the communists in Afghanistan, partly as a result of how those injuries were exacerbated by the CIA and partly as a result of his extended isolation, Abu Zubaydah's mental grasp is slipping away. Today, he suffers blinding headaches and has permanent brain damage. He has an excruciating sensitivity to sounds, hearing what others do not. The slightest noise drives him nearly insane. In the last two years alone, he has experienced about 200 seizures. Already, he cannot picture his mother's face or recall his father's name. Gradually, his past, like his future, eludes him.[231]

Saudi and Pakistani connection allegations

Gerald Posner in his book "Why America Slept: The Failure to Prevent 9/11" claimed that Abu Zubaydah was duped by U.S. interrogators masquerading as Saudis and using painkillers and sodium pentathol, sometimes called "truth serum". Zubaydah, he writes, thought he was in a Saudi prison, when in fact he was in Afghanistan. Posner says Zubaydah, "relieved" to find he was being quizzed by people he thought were Saudis, provided them with phone numbers for a senior member of the Saudi royal family who would "tell you what to do." He says the U.S. interrogators were stunned when the numbers were traced to Prince Ahmed bin Salman bin Abdul Aziz, a nephew of King Fahd.

When the agents accused Zubaydah of lying, he revealed more details of Saudi and Pakistani ties to bin Laden, the author says, and in the space of one week, three of the four persons named by him were dead. Prince Ahmed, 43, died of a heart attack on July 22, 2002. The next day, a car crash killed Saudi Prince Sultan bin Faisal bin Turki al-Saud, 41. A week later, Prince Fahd bin Turki bin Saud al-Kabir, 25, reportedly died "of thirst" while traveling east of Riyadh. Seven months later, a plane crash "in clear weather," Posner says, killed Pakistani Air Marshal Mushaf Ali Mir, his wife and several aides in Pakistan.[232]

See also

- Shaker Aamer

References

[1] "Americas | Profile: Key US terror suspects" (http://news.bbc.co.uk/2/hi/5322694.stm). BBC News. 2008-02-11. . Retrieved 2012-04-04.

[2] "CIA Chief: We Waterboarded - ABC News" (http://abcnews.go.com/Blotter/TheLaw/story?id=4244423&page=1). Abcnews.go.com. 2008-02-05. . Retrieved 2012-04-04.

[3] "Abu Zubaydahâ€™s Torture Diary" (http://www.andyworthington.co.uk/2010/03/15/abu-zubaydahs-torture-diary/). Andy Worthington. . Retrieved 2012-04-04.

[4] "Headlines for March 30, 2009" (http://www.democracynow.org/2009/3/30/headlines). Democracy Now!. 2009-03-30. . Retrieved 2012-04-04.

[5] "Do CIA Cables Show Doctors Monitoring Torture?" (http://www.propublica.org/article/do-cia-cables-show-doctors-monitoring-torture-528). ProPublica. 2009-05-28. . Retrieved 2012-04-04.

[6] "Abu Zubaydah: Tortured for Nothing" (http://www.andyworthington.co.uk/2010/04/06/abu-zubaydah-tortured-for-nothing/). Andy Worthington. . Retrieved 2012-04-04.

[7] Shane, Scott (April 20, 2009). "Waterboarding Used 266 Times on 2 Suspects" (http://www.nytimes.com/2009/04/20/world/20detain.html). *The New York Times*. .

[8] "President Obama's first 100 days: Case studies - Amnesty International Australia" (http://www.amnesty.org.au/hrs/comments/20872/). Amnesty.org.au. 2009-04-28. . Retrieved 2012-04-04.

[9] "Who Authorized The Torture of Abu Zubaydah?" (http://www.andyworthington.co.uk/2009/04/24/who-authorized-the-torture-of-abu-zubaydah/). Andy Worthington. . Retrieved 2012-04-04.

[10] Jason Leopold (2012-05-29). "EXCLUSIVE: From Hopeful Immigrant to FBI Informant - the Inside Story of the Other Abu Zubaidah" (http://truth-out.org/news/item/9344-from-hopeful-immigrant-fbi-informant-the-other-abu-zubaidah). Truthout. . Retrieved 2012-06-04. "One of ten children born to Palestinian parents in Saudi Arabia, Hesham is five years younger than Hani, and was only 11 or 12 years old when his brother left home for good, and he recalls him only as a happy-go-lucky guy, and something of a womanizer. At that time, he insists, there was no hint of religious extremism."

[11] Mayer, Jane, "The Dark Side: The Inside Story of How the War on Terror Turned Into a War on American Ideals", 2008. p. 140

[12] Christopher Dickey *Securing The City: Inside America's Best Counterterror Force—The NYPD* (http://www.npr.org/templates/story/story.php?storyId=100559912) Simon & Schuster February 3, 2009

[13] R. Raman (February 14, 2010). "Pune's Past Jihadi Connections" (http://www.webcitation.org/query?url=http://www.eurasiareview.com/2010/02/31774-punes-past-jihadi-connections.html&date=2010-02-14). Eurasia Review. Archived from the original (http://www.eurasiareview.com/2010/02/31774-punes-past-jihadi-connections.html) on February 14, 2010. . "Pune as a possible centre for jihadi activities came to notice in March 2002, when Abu Zubaidah, the then No.3 [*sic*] to Osama bin Laden, was arrested by the Pakistani authorities acting at the instance of the USA's Federal Bureau of Investigation (FBI), in the house of an activist of the Lashkar-e-Toiba [*sic*] (LET) at Faislabad in Pakistani Punjab and handed over to the FBI. He is now in the Guantanamo Bay Detention Centre in Cuba. Sections of the Pakistani media had reported at that time that Abu Zubaidah, a Palestinian, had studied computer science in Pune before crossing over into Pakistan and joining Al Qaeda."

[14] Peter Finn and Joby Warrick, *Detainee's Harsh Treatment Foiled No Plots* (http://www.washingtonpost.com/wp-dyn/content/article/2009/03/28/AR2009032802066.html) *The Washington Post*, March 29, 2009

[15] "Abu Zubaydah Unclassified Verbatim Combatant Status Review Tribunal Transcript" (http://www.defenselink.mil/news/transcript_ISN10016.pdf). Department of Defense. 2007. .

[16] Dan Eggen and Walter Pincus, *FBI, CIA Debate Significance of Terror Suspect: Agencies Also Disagree On Interrogation Methods* (http://www.washingtonpost.com/wp-dyn/content/article/2007/12/17/AR2007121702151_pf.html). *The Washington Post*, December 18, 2007

[17] Ron Suskind, *The One Percent Doctrine: Deep Inside America's Pursuit of Its Enemies Since 9/11* Simon & Schuster, 2006

[18] "Guantanamo Prisoner - Noor Uthman Muhammed Guantanamo Story" (http://www.esquire.com/features/guantanamo-prisoner-0911?click=main_sr). Esquire. 2011-08-18. . Retrieved 2012-04-04.

[19] Khalid Sulaymanjaydh Al Hubayshi Unclassified Verbatim Combatant Status Review Tribunal Transcript, Pgs. 65–73 Department of Defense (http://www.dod.mil/pubs/foi/detainees/csrt_arb/Set_1_0001-0097.pdf)

[20] Noor Uthamn Muhammed Unclassified Verbatim Combatant Status Review Tribunal (http://www.dod.mil/pubs/foi/detainees/csrt_arb/Set_38_2608-2628.pdf), Pg. 15, Department of Defense]

[21] *9/11 Commission Report: Final Report of the National Commission on Terrorist Attacks Upon the United States* (http://www.gpoaccess.gov/911/pdf/fullreport.pdf) July 22, 2006

[22] Brynjar Lia, *Architect of Global Jihad: The Life of Al-Qaida Strategist Abu Mus'ab al-Suri* pg. 242–243 (http://books.google.com/books?id=WJefraeEKs4C&pg=PA242&lpg=PA242&dq=brynjar+lia+khalden+camp&source=bl&ots=3SqJD2tq9c&sig=dnbTxJ4SbsNKQiY_qosYHLbe1JA&hl=en&ei=c8_5SduyF4_SNMHfpLUE&sa=X&oi=book_result&ct=result&resnum=1), Columbia University Press, 2008

[23] *Noor Uthamn Muhammed Charge Sheets* (http://www.defenselink.mil/news/commissionsNoor.html) Department of Defense Website

[24] "New York Times Guantanamo Docket" (http://projects.nytimes.com/guantanamo). Projects.nytimes.com. . Retrieved 2012-04-04.

[25] Colin Freeze (September 12, 2009). "Jailed Arab details ties to tortured Canadians" (http://www.webcitation.org/query?url=http://www.theglobeandmail.com/news/opinions/jailed-arab-details-ties-to-tortured-canadians/article1285350/#article&date=2009-09-14). *Globe and Mail* (Canada). Archived from the original (http://www.theglobeandmail.com/news/opinions/jailed-arab-details-ties-to-tortured-canadians/article1285350/#article) on September 14, 2009. .

[26] *Pak leadership under Sharif offered to try to capture Osama* The Press Trust of India Ltd. Through Asia Pulse. March 28, 2004 (http://www.accessmylibrary.com/coms2/summary_0286-4648752_ITM) (Access My Library Link, requires free membership)

[27] David A. Vise and Lorraine Adams, *Bin Laden Weakened, Officials Say* (http://www.highbeam.com/doc/1P2-513940.html) *The Washington Post*. March 11, 2000 (Highbeam News Database Link, requires free membership)

[28] *Six Muslim militants sentenced to death for plotting to attack tourists in Jordan* (http://www.independent.co.uk/news/world/middle-east/six-muslim-militants-sentenced-to-death-for-plotting-to-attack-tourists-in-jordan-699684.html) The Independent September 19, 2000

[29] Jamal Halaby, *Arabs of Terror Linked to bin Laden* AP News Online (http://www.highbeam.com/doc/1P1-26114220.html) (Highbeam News Database Link, requires free membership)

[30] *Trainees in Terror* (http://community.seattletimes.nwsource.com/archive/?date=20000305&slug=4008236) *The Seattle Times*, Sunday, March 5, 2000

[31] *Transcript: Bin Laden determined to Strike in US* (http://edition.cnn.com/2004/ALLPOLITICS/04/10/august6.memo/index.html) CNN.com, Saturday April 10, 2004

[32] Elaine Ganley, *Terror Plot Took Typical Path* (http://www.highbeam.com/doc/1G1-80592999.html) AP Online, December 19, 2001 (Highbeam News Database Link, requires free membership)

[33] Andy Worthington *The Guantanamo Files* Pluto Press, 2007

[34] Tim McGirk, *Anatomy of a Raid* (http://www.time.com/time/magazine/article/0,9171,227584,00.html) TIME Magazine, April 8, 2002

[35] John Burns, *A NATION CHALLENGED: THE FUGITIVES, In Pakistan's Interior, A Troubling Victory in Hunt for Al Qaeda* (http://query.nytimes.com/gst/fullpage.html?res=9A00EFDA123CF937A25757C0A9649C8B63). *New York Times*, April 14, 2002
[36] *Anti-terror raids yield bonanza for U.S. intelligence* (http://community.seattletimes.nwsource.com/archive/?date=20020403&slug=zub03). *The Seattle Times*, April 2, 2002
[37] Arun Kumar, *US imposes sanctions on four Lashkar-e-Toiba leaders* (http://www.hindustantimes.com/StoryPage/StoryPage.aspx?id=04859c7f-8276-4944-acf6-cebaaf311614&MatchID1=55&TeamID1=1&TeamID2=7&MatchType1=5&SeriesID1=1&MatchID2=56&TeamID3=5&TeamID4=8&MatchType2=5&SeriesID2=1&PrimaryID=55&Headline=US+imposes+sanctions+on+f) *Hindustan Times*, May 28, 2008
[38] Brian Ross, *CIA- Abu Zubaydah: Interview with John Kiriakou: Transcript* (http://abcnews.go.com/images/Blotter/brianross_kiriakou_transcript1_blotter071210.pdf) ABC News, December 10, 2007
[39] J.J. Green, *Former CIA Officer: Waterboarding is Wrong, but it Worked* (http://www.wtop.com/?sid=1368866&nid=251) WTOPnews.com, March 20, 2008
[40] *President George Bush, Speech* (http://georgewbush-whitehouse.archives.gov/news/releases/2006/09/20060906-3.html) September 6, 2009
[41] Risen, James. "State of War: The Secret History of the CIA and the Bush Administration", 2006
[42] Shane, Scott (June 22, 2008). "Inside a 9/11 Mastermind's Interrogation" (http://www.nytimes.com/2008/06/22/washington/22ksm.html?pagewanted=all). *New York Times*. . Retrieved January 24, 2012.
[43] Soufan, Ali. "The Black Banners".
[44] Dan Froomkin, *Bush's Exhibit A for Torture* (http://www.washingtonpost.com/wp-dyn/content/blog/2007/12/18/BL2007121800862.html) The Washington Post December 18, 2007
[45] Dana Priest, *CIA Holds Terror Suspects in Secret Prisons* (http://www.washingtonpost.com/wp-dyn/content/article/2005/11/01/AR2005110101644.html). *The Washington Post*, November 2, 2005
[46] *Lawmakers to examine claims Indian Ocean island used in secret prison network* (http://www.iht.com/articles/ap/2007/10/19/europe/EU-GEN-Britain-Diego-Garcia.php). *International Herald Tribune*, October 19, 2007
[47] Dick Marty, *Secret detentions and illegal transfers of detainees involving Council of Europe member states: Second report* (http://assembly.coe.int/CommitteeDocs/2007/Emarty_20070608_NoEmbargo.pdf) Council of Europe Parliamentary Assembly, June 7, 2007
[48] Brian Ross and Richard Esposito, *EXCLUSIVE: Sources Tell ABC News Top Al Qaeda Figures Held in Secret CIA Prisons* (http://abcnews.go.com/WNT/Investigation/story?id=1375123) ABC News, December 5, 2005
[49] *CIA Shuffled Prisoners Out of Poland* (http://www.globalpulse.net/archives/security/cia_shuttled_pr_000087.php) Global Pulse December 5, 2005
[50] Jason Burke, *Secret World of U.S. Jails* (http://observer.guardian.co.uk/international/story/0,6903,1237589,00.html) Observer, June 13, 2004
[51] *Enforced Disappearance, Illegal Interstate Transfer, and Other Human Rights Abuses Involving the UK Overseas Territories: Executive Summary* (http://www.reprieve.org.uk/documents/FinalReprieveFASCExecutiveSummary.pdf) Reprieve
[52] Jay Bybee Department of Justice (http://www.usdoj.gov/opa/documents/memorandumpresidentpower03132002.pdf), March 13, 2002
[53] James Rowley (http://www.bloomberg.com/apps/news?pid=20601087&sid=aJchF0ELduak&refer=home) Bloomberg, March 5, 2009
[54] Ali Soufan *My Tortured Decision* (http://www.nytimes.com/2009/04/23/opinion/23soufan.html). *New York Times*, April 22, 2009
[55] Michael Isikoff "We Could Have Done This the Right Way" (http://www.newsweek.com/id/195089) Newsweek, April 25, 2009
[56] *Congressional Testimony of Ali Soufan, May 13, 2009* (http://judiciary.senate.gov/hearings/testimony.cfm?id=3842&wit_id=7906) Congressional Testimony, May 13, 2009
[57] *International Committee of the Red Cross Report on the Treatment of Fourteen "High Value Detainees" in CIA Custody* (http://www.nybooks.com/icrc-report.pdf) International Committee of the Red Cross, February 2007
[58] David Johnston *At a Secret Interrogation, Dispute Flared Over Tactics* (http://www.nytimes.com/2006/09/10/washington/10detain.html?pagewanted=1). *New York Times*, September 10, 2006
[59] Scott Shane *Book Cites Secret Red Cross Report of CIA Torture of Qaeda Captives* (http://www.nytimes.com/2008/07/11/washington/11detain.html?scp=1&sq=zubaydah&st=nyt). *New York Times*, July 11, 2008
[60] Jane Meyer *The Black Sites: A rare look inside CIA's secret interrogation program* (http://www.newyorker.com/reporting/2007/08/13/070813fa_fact_mayer) The New Yorker, August 13, 2007
[61] Scott Horton *Six Questions for Jane Meyer, Author of The Dark Side* (http://harpers.org/archive/2008/07/hbc-90003234) Harper's Magazine, July 14, 2008
[62] Paul Koring *New book slams Bush's black ops* (http://www.webcitation.org/5yFpbGc1P) Globe and Mail, January 4, 2006
[63] "United States' "Disappeared" CIA Long-term "Ghost Detainees"" (http://www.hrw.org/backgrounder/usa/us1004/index.htm) Human Rights Watch, October 2004
[64] *Impunity for the Architects of Illegal Policy* (http://www.hrw.org/reports/2005/us0405/6.htm) Human Rights Watch, 2005
[65] Serrin Turner and Stephen J. Schullhoffer *The Secrecy Problem in Terrorism Trials* (http://brennan.3cdn.net/6a0e5de414927df95e_lbm6iy66c.pdf) Brennan Center for Justice 2005
[66] Eun Young Choi *Veritas, Not Vengeance: An Examination of the Evidentiary Rules for Military Commissions in the War Against Terrorism*] 42 Harvard Civil Rights Civil Liberties Law Review, 2007

[67] Charles H. Brower II *The Lives of Animals, the Lives of Prisoners, and the Revelations of Abu Ghraib*] 37 Vanderbilt Journal of Transnational Law, 2004

[68] A. John Radsan *Symposium on Reexamining the Law of War: The Collision Between Common Article Three and The Central Intelligence Agency*] 56 Catholic University Law Review, 2007

[69] Tommy Harnden *Gloves off, the screws go on 9/11 suspect* (http://www.theage.com.au/cgi-bin/common/popupPrintArticle.pl?path=/articles/2003/03/05/1046826435361.html) The Age, March 6, 2003

[70] Raymond Bonner, Don Van Natta Jr, and Amy Waldman *THREATS AND RESPONSES: INTERROGATIONS; Questioning Terror Suspects In a Dark and Surreal World* (http://query.nytimes.com/gst/fullpage.html?res=9C01E4DC1F3FF93AA35750C0A9659C8B63). *New York Times*, March 9, 2003

[71] Walter Pincus, *Waterboarding Historically Controversial* (http://www.washingtonpost.com/wp-dyn/content/article/2006/10/04/AR2006100402005.html). *The Washington Post*, October 5, 2006

[72] Set_11_1145-1178.pdf#33 *Ibrahim Zeidan Summarized transcripts* (http://www.dod.mil/pubs/foi/operation_and_plans/Detainee/csrt_arb/) Department of Defense]

[73] *Full ICRC Report Further Underscores Extent of Torture And Abuse By U.S. Officials* (http://www.aclu.org/safefree/torture/39296prs20090407.html) ACLU, April 4, 2007

[74] Steven Bradbury *Memoradnum for John A. Rizzo Senior Deputy General Counsel, Central Intelligence Agency Re: Application of United States Obligations Under Article 16 of the Convention Against Torture to Certain Techniques that May Be Used in the Interrogation of High Value al Qaeda Detainees* (http://luxmedia.vo.llnwd.net/o10/clients/aclu/olc_05302005_bradbury.pdf) Department of Justice, Office of Legal Counsel, May 30, 2005

[75] *Abu Zubaydah* (http://topics.nytimes.com/top/reference/timestopics/people/z/abu_zubaydah/index.html). *New York Times*, April 20, 2009

[76] Tanya Nolan *Former CIA Agent Reveals Use of 'Waterboarding' in Interrogations* (http://www.abc.net.au/worldtoday/content/2007/s2116703.htm) The World Today, December 12, 2007

[77] Ray McGovern *Destruction of CIA Tapes Can't Hide Barbaric U.S. Torture Methods* (http://www.alternet.org/story/70490/) AlterNet, December 13, 2007

[78] *Lauer Stirs Scandal Over How Water Boarding Saved Lives* (http://www.mediaresearch.org/cyberalerts/2007/cyb20071212.asp#1) Media Research Center, December 12, 2007

[79] *Ex-CIA Agent: Waterboarding 'Saved Lives'* (http://www.cnn.com/2007/POLITICS/12/11/agent.tapes/#cnnSTCText) CNNPolitics.com, December 11, 2007

[80] Liam Stack *Is waterboarding effective? CIA did it 266 times on two prisoners* (http://www.csmonitor.com/2009/0420/p99s01-duts.html) Christian Science Monitor, April 20, 2009

[81] Richard Esposito & Brian Ross *Coming in From the Cold: CIA Spy Calls Waterboarding Necessary But Torture* (http://abcnews.go.com/Blotter/Story?id=3978231) ABC News, December 10, 2007

[82] Brian Stetler *How '07 Interview Tilted Torture Debate* (http://www.nytimes.com/2009/04/28/business/media/28abc.html?ref=business). *The New York Times*, April 27, 2009

[83] *C.I.A. Interrogations* (http://topics.nytimes.com/top/reference/timestopics/organizations/c/central_intelligence_agency/cia_interrogations/index.html?inline=nyt-classifier). *New York Times*, April 28, 2009

[84] Michael Scherer and Bobby Ghosh *How Waterboarding Got Out of Control* (http://www.time.com/time/nation/article/0,8599,1892708,00.html) *Time Magazine*, April 20, 2009

[85] Scott Shane *Waterboarding Used 266 Times on 2 Suspects* (http://www.nytimes.com/2009/04/20/world/20detain.html). *New York Times*, April 19, 2009

[86] *Taking on Torture* (http://www.latimes.com/news/opinion/editorials/la-ed-torture27-2009apr27,0,6092738.story) *Los Angeles Times*, April 27, 2009

[87] Jeff Stein (January 26, 2010). "CIA Man Retracts Claim on Waterboarding; A study in "enhanced reporting techniques" (http://www.foreignpolicy.com/articles/2010/01/26/cia_man_retracts_claim_on_waterboarding?print=yes&hidecomments=yes&page=full). *Foreign Policy*. .

[88] Steven Bradbury *Memorandum for John A. Rizzo Senior Deputy General Counsel, Central Intelligence Agency, Office of Legal Counsel Re: Application of 18 U.S.C. Sections 2340-2340A to Certain Techniques That May Be Used in the Interrogation of a High Value al Qaeda Detainee* (http://luxmedia.vo.llnwd.net/o10/clients/aclu/olc_05102005_bradbury46pg.pdf) Department of Justice, Office of Legal Counsel, May 10, 2005

[89] Joseph Abrams *Despite Reports, Khalid Sheikh Mohammed Was Not Waterboarded 183 Times* (http://www.foxnews.com/politics/2009/04/28/despite-reports-khalid-sheikh-mohammed-waterboarded-times) Fox News, April 28, 2009

[90] Josh White *Waterboarding Is Torture, Says Ex-Navy Instructor* (http://www.washingtonpost.com/wp-dyn/content/article/2007/11/08/AR2007110802150.html). *The Washington Post*, November 9, 2007

[91] Richard E. Mezo *Why It Was Called 'Water Torture'* (http://www.washingtonpost.com/wp-dyn/content/article/2008/02/08/AR2008020803156.html?sub-AR). *The Washington Post*, February 10, 2008

[92] Alfred McCoy *The U.S. has a History of Using Torture* (http://hnn.us/articles/32497.html) History News Network, December 4, 2006

[93] Rachel Weiner *Mancow Waterboarded (VIDEO): Conservative Radio Host Say It's Torture* (http://www.huffingtonpost.com/2009/05/22/mancow-waterboarded-video_n_206906.html) *Huffington Post*, May 22, 2009

[94] Amy Goodman *The Dark Side: Jane Mayer on the Inside Story of How the War on Terror Turned Into a War on American Ideals, Interview Transcript* (http://www.democracynow.org/2008/7/18/the_dark_side_jane_mayer_on) Democracy Now, July 18, 2008

[95] Mark Mazzetti *Bush Aides Linked to Talks on Interrogations* (http://www.nytimes.com/2008/09/25/washington/25detain.html). *New York Times*, September 24, 2008

[96] *Bush aware of advisers' interrogation talks* (http://abclocal.go.com/kfsn/story?section=news/national_world&id=6076727) ABC News, April 11, 2008

[97] Joby Warrick and Dan Eggen *Hill Briefed on Waterboarding in 2002* (http://www.washingtonpost.com/wp-dyn/content/article/2007/12/08/AR2007120801664.html?hpid=topnews). *The Washington Post*, December 9, 2007

[98] *As Bush Adviser, Rice Gave OK to Waterboard* (http://www.foxnews.com/politics/2009/04/22/timeline-released-senate-shows-condoleezza-rice-okd-waterboarding/) Fox News, April 22, 2009

[99] *Senate Report: Rice, Cheney OK'd CIA use of waterboarding*] CNN, April 23, 2009

[100] Jason Leopold *Cheney Admits He 'Signed Off' on Waterboarding of Three Guantanamo Prisoners* (http://www.atlanticfreepress.com/news/1/6983--cheney-admits-he-signed-off-on-waterboarding-of-three-guantanamo-prisoners.html) Atlantic Free Press, December 29, 2008

[101] Ewen MacAskill *Obama releases Bush torture memos* (http://www.guardian.co.uk/world/2009/apr/16/torture-memos-bush-administration) The Guardian, April 16, 2009

[102] Mark Mazetti and Scott Shane *Interrogation Memos Detail Harsh Tactics by the C.I.A.* (http://www.nytimes.com/2009/04/17/us/politics/17detain.html). *New York Times*, April 16, 2009

[103] Jon Swaine *Barack Obama releases torture memos: details of techniques used by CIA* (http://www.telegraph.co.uk/news/worldnews/northamerica/usa/barackobama/5169211/Barack-Obama-releases-torture-memos-details-of-techniques-used-by-CIA.html) The Telegraph, April 17, 2009

[104] James Comey *James Comey Internal CIA E-mails* (http://www.truthout.org/files/Comey-emails-2005.pdf)

[105] *Rachel Maddow Interview of Philip Zelikow, Transcript* (http://www.msnbc.msn.com/id/30369127/) MSNBC, April 23, 2009

[106] Philip Sherwell and Leonard Doyle *CIA ignored warnings from US soldiers that torture and extreme stress would not work* (http://www.telegraph.co.uk/news/worldnews/northamerica/usa/5221190/CIA-ignored-warnings-from-US-soldiers-that-torture-and-extreme-stress-would-not-work.html) *The Daily Telegraph*, April 25, 2009

[107] *Report: Top Members of Congress Were OK With Waterboarding in 2002* (http://www.foxnews.com/story/0,2933,316256,00.html) Fox News, December 9, 2007

[108] Jane Mayer *C.I.A. Former Inspector General Pushed For Disclosure on Interrogations* (http://www.newyorker.com/online/blogs/newsdesk/2009/06/cia-interrogations-disclosure.html) *The New Yorker*, June 17, 2009

[109] *Paul Gigot Interviews* (http://online.wsj.com/article/SB124681141574196601.html?mod=googlenews_wsj) Wall Street Journal, July 6, 2009

[110] Pam Benson *Senator: Cheney and alleged secret CIA program 'a problem'* (http://www.cnn.com/2009/POLITICS/07/11/cheney.surveillance/) CNN, July 12, 2009

[111] Scott Shane *Cheney Is Linked to Concealment of C.I.A. Project* (http://www.nytimes.com/2009/07/12/us/politics/12intel.html?hp). *The New York Times*, July 11, 2009

[112] *Previously Secret Torture Memo Released* (http://www.cnn.com/2008/POLITICS/07/24/cia.torture/index.html) CNN.com, July 24, 2008

[113] Joby Warrick and Peter Finn *Harsh Tactics Readied Before Their Approval* (http://www.washingtonpost.com/wp-dyn/content/article/2009/04/21/AR2009042104055_2.html?sid=ST2009042101921). *The Washington Post*, April 22, 2009

[114] Alex Koppelman *Ashcroft suggests CIA sought legal approval after torture began* (http://www.salon.com/politics/war_room/2008/07/17/ashcroft/index.html?source=refresh) Salon, July 17, 2008

[115] David Johnston and James Risen *The Reach Of War: The Interrogations; Aides Say Memo Backed Coercion Already In Use* (http://query.nytimes.com/gst/fullpage.html?res=9904E5DA1738F934A15755C0A9629C8B63). *New York Times*, June 27, 2004

[116] Jason Leopold *DOJ Report Says Yoo's Torture Memo Failed To Cite Supreme Court Case* (http://www.pubrecord.org/torture/701-doj-report-blasts-yoo-for-not-citing-court-case-in-torture-memo.html) The Public Record, February 22, 2009

[117] Hilary Andersson *Did America break its torture law?* (http://news.bbc.co.uk/2/hi/americas/8148202.stm) BBC Panorama, July 13, 2009

[118] *US 'waterboarding' row rekindled* (http://news.bbc.co.uk/2/hi/americas/8144625.stm) BBC, July 13, 2009

[119] Zachary Coile *Gonzales unflappable in 6 hours of testimony Bush nominee says torture will not be tolerated* (http://www.sfgate.com/cgi-bin/article/article?f=/c/a/2005/01/07/MNG41AMQ8R1.DTL) San Francisco Chronicle, January 7, 2005

[120] Jay Bybee and John Yoo *Memorandum for John Rizzo Acting General Counsel of the Central Intelligence Agency: Interrogation of al Qaeda Operative* (http://luxmedia.vo.llnwd.net/o10/clients/aclu/olc_08012002_bybee.pdf) U.S. Department of Justice, Office of Legal Counsel, August 1, 2002

[121] Jeffrey Rosen *Conscience of a Conservative* (http://www.nytimes.com/2007/09/09/magazine/09rosen.html?pagewanted=all). *New York Times*, September 9, 2007

[122] Michael Isikoff *A Torture Report Could Spell Big Trouble For Bush Lawyers* (http://www.newsweek.com/id/184801) *Newsweek*, February 14, 2009,

[123] Bobby Ghosh *Partisan Passions Dominate Interrogation Hearings* (http://www.time.com/time/nation/article/0,8599,1898125,00.html) *Time*, May 13, 2009

[124] Steven Bradbury *Memorandum For John A. Rizzo Senior Deputy General Counsel, Central Intelligence Agency Re: Application of 18 U.S.C. §§ 2340-2340A to the Combined Use of Certain Techniques in the Interrogation of High Value al Qaeda Detainees* (http://luxmedia.vo.llnwd.net/o10/clients/aclu/olc_05102005_bradbury20pg.pdf) Department of Justice, Office of Legal Counsel, May 10, 2005

[125] Mark Mazzetti and David Johnston *Inquiry Begins Into Destruction of Tapes* (http://www.nytimes.com/2007/12/09/washington/09zubaydah.html?_r=1&ref=world&oref=slogin). *New York Times*, December 9, 2007

[126] Mark Benjamin *Soufan: CIA torture actually hindered our intelligence gathering* (http://www.salon.com/news/feature/2009/05/14/torture/) Salon, May 14, 2009

[127] Scott Shane *Interrogations' Effectiveness May Prove Elusive* (http://www.nytimes.com/2009/04/23/us/politics/23detain.html?_r=1&th&emc=th). *New York Times*, April 22, 2009

[128] *Cheney Calls for Release of Memos Showing Results of Interrogation Efforts* (http://www.foxnews.com/politics/2009/04/20/cheney-calls-release-memos-showing-results-interrogation-efforts-1862515294/) April 20, 2009

[129] *Transcript for THE SITUATION ROOM with Wolf Blitzer* (http://transcripts.cnn.com/TRANSCRIPTS/0606/20/sitroom.02.html) CNN, June 20, 2006

[130] George Tenet *At the Center of the Storm: My years at the CIA* HarperCollins, April 30, 2007

[131] *Alleged 9-11 Mastermind Nabbed* (http://www.cbsnews.com/stories/2003/03/02/terror/main542459.shtml) CBS News, March 1, 2003

[132] *NEWS SUMMARY: Arrest May Deter Attacks* (http://query.nytimes.com/gst/fullpage.html?res=9F0DE5DB1E3AF937A35757C0A9649C8B63). *New York Times*, April 4, 2002

[133] *World: United States* (http://www.highbeam.com/doc/1G1-84361428.html) Western Mail, April 3, 2002

[134] Donald Rumsfeld *News Transcript* (http://www.defenselink.mil/transcripts/transcript.aspx?transcriptid=3390) Department of Defense, April 3, 2002

[135] Scott Shane, *Divisions Arose on Rough Tactics for Qaead Figure* (http://www.nytimes.com/2009/04/18/world/middleeast/18zubaydah.html?ref=global-home). *The New York Times*, April 17, 2009

[136] *Transcript of Representative John Conyers Jr. Hearing on C.I.A.'s Destruction of Tapes* (http://www.accessmylibrary.com/coms2/summary_0286-33608853_ITM) Political/Congressional Transcript Wire, December 20, 2007 (Access My Library Link, requires free membership)

[137] Peter Finn and Julie Tate *CIA Mistaken on 'High-Value' Detainee, Document Shows* (http://www.washingtonpost.com/wp-dyn/content/article/2009/06/15/AR2009061503045.html?hpid=moreheadlines). *The Washington Post*, June 16, 2009

[138] *Report: Insider May Testify On Zubaydah* (http://www.highbeam.com/doc/1P1-51764739.html) April 2, 2002 (Highbeam News Database Link, requires free membership)

[139] *United States Helsinki Commission Briefing Transcript* (http://www.accessmylibrary.com/coms2/summary_0286-31338935_ITM) Political/Congressional Transcript Wire, June 22, 2007 (Access My Library Link, requires free membership)

[140] *Transcript of Video News Story on Guantanamo Bay with Kelli Arena Reporting* (http://www.accessmylibrary.com/coms2/summary_0286-18437742_ITM) CNN, September 24, 2006 (Access My Library Link, requires free membership)

[141] *Department of Defense News Briefing* (http://www.accessmylibrary.com/coms2/summary_0286-25210130_ITM) April 2, 2002 (Access My Library Link, requires free membership)

[142] Gerry Gilmore *Rumsfeld Confirms Capture of Senior Al Qaeda Leader* (http://www.defenselink.mil/news/newsarticle.aspx?id=44203) Department of Defense, April 2, 2002

[143] *Profile: Abu Zubaydah* (http://news.bbc.co.uk/2/hi/south_asia/1907462.stm) BBC News, April 2, 2002

[144] *Remarks by the President at Connecticut Republican Committee Luncheon* (http://georgewbush-whitehouse.archives.gov/news/releases/2002/04/20020409-8.html) White House website, April 9, 2002

[145] George W. Bush's Remarks at the Virginia Military Institute, April 17, 2002

[146] George W. Bush (June 6, 2002). "Remarks by the President in Address to the Nation" (http://georgewbush-whitehouse.archives.gov/news/releases/2002/06/20020606-8.html) (Press release). White House. .

[147] George W. Bush *Remarks by the President at Thaddeus McCotter for Congress Dinner* (http://georgewbush-whitehouse.archives.gov/news/releases/2002/10/20021014-3.html) White House website, October 14, 2002

[148] Jeff Bliss and Tony Capaccio, *Iraq Group May Attack Outside Nation, McConnell Says* Bloomberg.com, February 5, 2008

[149] Richard Esposito and Jason Ryan *CIA Chief: We Waterboarded"* (http://abcnews.go.com/Blotter/TheLaw/story?id=4244423&page=1). ABC News, February 5, 2008

[150] *The White House's Perverse Argument: Opinion Watch* (http://www.washingtonpost.com/wp-dyn/content/blog/2008/02/07/BL2008020701681_pf.html) *The Washington Post*, February 7, 2008

[151] Philip Shenon *MIDEAST TURMOIL: INTELLIGENCE; Officials Say Qaeda Suspect Has Given Useful Information* (http://query.nytimes.com/gst/fullpage.html?res=9401EFDD123EF935A15757C0A9649C8B63). *New York Times*, April 26, 2002

[152] Amanda Ripley *A look at the evolution of a terrorist-attack warning: one part truth, one part rumor and one part fear* (http://archives.cnn.com/2002/ALLPOLITICS/05/27/time.decoding/index.html) CNN.com, May 27, 2002

[153] *Terrorism Notebook. More attacks have been prevented, officials say* (http://community.seattletimes.nwsource.com/archive/?date=20030111&slug=notebook11). *The Seattle Times*, January 11, 2003

[154] Massimo Calabresi and Romesh Ratnesar, *Can we stop the next attack?* (http://www.cnn.com/ALLPOLITICS/time/2002/03/11/next.html) CNN News, March 4, 2002

[155] *Who's Who in al-Qaeda?* (http://news.bbc.co.uk/nolpda/ukfs_news/hi/newsid_2780000/2780525.stm) BBC News

[156] Nick Schou *One Degree of Separation* (http://www.ocweekly.com/2001-10-11/news/one-degree-of-separation/) Orange County Weekly, October 11, 2001

[157] Marlise Simons *A NATION CHALLENGED: FRANCE; Ninth Man Held in Suspected Plot Against Paris Embassy* (http://www.nytimes.com/2001/10/04/world/nation-challenged-france-ninth-man-held-suspected-plot-against-paris-embassy.html). *New York Times*, October 4, 2001

[158] *AP Top News at 8 p.m. EDT* (http://www.highbeam.com/doc/1P1-47362355.html) Associated Press, October 5, 2001

[159] *Bin Laden Videos Suggest Location* (http://www.highbeam.com/doc/1G1-80592999.html) The Cincinnati Post, December 5, 2001

[160] Eric Schmitt and Erik Eckholm *A NATION CHALLENGED: THE HUNTED; U.S. Takes Custody of a Qaeda Trainer Seized by Pakistan* (http://query.nytimes.com/gst/fullpage.html?res=9C03E3D71639F935A35752C0A9649C8B63). *New York Times*, January 6, 2002

[161] Philip Shenon *A NATION CHALLENGED: BALKAN TRAIL; U.S. Labels an Arab Captive a Planner of Qaeda Attacks* (http://query.nytimes.com/gst/fullpage.html?res=9F01EEDD103BF930A15752C0A9649C8B63). *New York Times*, January 23, 2002

[162] *Global Impact News Alert. U.S. Seeks New Head of Al Qaida Anti-American Operations* (http://www.highbeam.com/doc/1G1-82887691.html) United Press International, February 15, 2002] (Highbeam News Database Link, requires free membership)

[163] *BANK TERROR ATTACK FEAR; Warning issued after interview* (http://www.highbeam.com/doc/1G1-84981557.html) Birmingham Evening Mail, April 20, 2002

[164] *Al-Qaeda claims 'dirty bomb' know-how* (http://news.bbc.co.uk/2/hi/americas/1945765.stm) BBC News, April 23, 2002

[165] John J. Lumpkin *Al-Qaida Captive Talks Of Terror* (http://www.highbeam.com/doc/1P1-52527138.html) AP News, April 24, 2002

[166] *NY terror alert came from bin Laden aide* (http://www.usatoday.com/news/nation/2002/05/23/ny-alert.htm) USA Today, May 23, 2002

[167] *Bush Faces Dissent on European Trip* (http://edition.cnn.com/TRANSCRIPTS/0205/23/mlld.00.html) CNN News Transcripts, May 23, 2002

[168] *Bush: 'No war plans on my desk' for Iraq* (http://archives.cnn.com/2002/US/05/23/gen.war.on.terror/index.html) CNN.com, May 23, 2002

[169] John J. Lumpkin *Many Al-Qaida Chiefs Remain Unreachable* (http://www.highbeam.com/doc/1G1-95918094.html) The Cincinnati Post, December 27, 2002

[170] *CNN American Morning With Paula Zahn* (http://premium.edition.cnn.com/TRANSCRIPTS/0207/01/ltm.02.html) CNN.com, July 1, 2002

[171] *Travel Watch* (http://tvnz.co.nz/view/tvnz_portable_story_skin/112640) ONE News, July 1, 2002

[172] Department of Defense Military Commissions Cases Website (http://www.defenselink.mil/news/commissions.html), April 30, 2009

[173] *George W. Bush September 2006 Speech* (http://www.nytimes.com/2006/09/06/washington/06bush_transcript.html?pagewanted=all) September 2006

[174] *Bush Concedes CIA Held Suspects in Secret Prisons* (http://www.npr.org/templates/story/story.php?storyId=5776107) NPR, September 6, 2006

[175] Ron Hutcheson and James Kuhnhenn *Iraq deal with Congress nears, Bush says* (http://www2.jsonline.com/news/nat/sep02/83149.asp) Milwaukee Journal Sentinel, September 26, 2002

[176] *Bush Says He and Congress Will Band Together on Iraq; Capitol Hill Still Sour* (http://www.accessmylibrary.com/coms2/summary_0286-7430560_ITM) Knight Ridder/Tribune Business News, September 27, 2002] (Access My Library Link, requires free membership)

[177] Andrew Sullivan *One tortured lie: that's all it took for war* (http://www.timesonline.co.uk/tol/comment/columnists/andrew_sullivan/article6168270.ece) The Sunday Times, April 26, 2009

[178] Fritz Umbach *Bush's bogus document dump* (http://www.salon.com/opinion/feature/2006/04/13/document_dump/) The Salon, April 13, 2006

[179] Stephen F. Hayes *The Rice Stuff?* (http://www.weeklystandard.com/Content/Public/Articles/000/000/004/803czhfn.asp?pg=2) The Daily Standard, October 20, 2004

[180] John Diamond and Bill Nichols *CIA in spotlight over reports leading to war* (http://www.usatoday.com/news/washington/2003-06-08-cia-usat_x.htm) June 8, 2003

[181] James Risen *THREATS AND RESPONSES: C.I.A.; Captives Deny Qaeda Worked With Baghdad* (http://query.nytimes.com/gst/fullpage.html?res=9D0DE4D61539F93AA35755C0A9659C8B63). *The New York Times*, June 9, 2003

[182] Frank Rich *The Banality of Bush White House Evil* (http://www.nytimes.com/2009/04/26/opinion/26rich.html). *The New York Times*, April 25, 2009

[183] Senate Armed Services Committee *Inquiry Into the Treatment of Detainees in U.S. Custody* (http://armed-services.senate.gov/Publications/Detainee Report Final_April 22 2009.pdf) Senate Armed Services Committee, November 20, 2008

[184] Ed Brayton *The Lies on Torture Just Keep Coming* (http://scienceblogs.com/dispatches/2009/05/the_lies_on_torture_just_keep.php) Science Blogs, May 4, 2009

[185] Warren Richey *'Alternative' CIA tactics complicate Padilla case* (http://www.csmonitor.com/2006/0915/p01s02-usju.html?s=widep) Christian Science Monitor, September 15, 2006

[186] Neail A. Lewis *A NATION CHALLENGED: THE DETAINEES; U.S. Is Seeking Basis to Charge War Detainees* (http://query.nytimes.com/gst/fullpage.html?res=9D04E2DE103FF932A15757C0A9649C8B63). *New York Times*, April 21, 2002

[187] Frank Davies *U.S. readies tribunals for terrorism trials* (http://www.accessmylibrary.com/coms2/summary_0286-8932707_ITM) The Miami Herald, December 26, 2002

[188] Wolf Blitzer *Interview of George W. Bush* (http://www.accessmylibrary.com/coms2/summary_0286-18409075_ITM) CNN, September 20, 2006

[189] James Risen and Eric Lichtblau *Bush altered rules on spying* (http://www.iht.com/articles/2005/12/16/news/spy.php?page=2). *International Herald Tribune*, December 17, 2005

[190] Aziz Huq *Who's watching whom?* (http://www.accessmylibrary.com/coms2/summary_0286-26038899_ITM) League of Women Voters, October 1, 2006

[191] Reed Brody *Prisoners who disappear* (http://www.iht.com/articles/2004/10/11/opinion/edbrody.php). *International Herald Tribune*, October 12, 2004

[192] Douglas Jehl, David Johnston, and Neil A. Lewis *CIA called uncomfortable with role on war prisoners Holding terror suspects raises concerns* (http://www.highbeam.com/doc/1P1-105455175.html). *International Herald Tribune*, February 16, 2005

[193] Barton Gellman *The Shadow War, In a Surprising New Light* (http://www.washingtonpost.com/wp-dyn/content/article/2006/06/19/AR2006061901211.html). *Washington Post*, June 20, 2006

[194] Emma Schwartz *Justice Dept. Inspector General Claims CIA Hampered its Investigation* (http://www.usnews.com/articles/news/2008/05/20/justice-dept-inspector-general-claims-cia-hampered-its-investigation.html) US News & World Report, May 20, 2008

[195] *Zayn al Abidin Muhammad Husayn v. Robert Gates*, Respondents Memorandum of Points and Authorities in Opposition to Petitioner's Motion for Discover and Petitioner's Motion for Sanctions. Civil Action No. 08-cv-1360 (RWR) (http://www.truth-out.org/files/memorandum.pdf) (September 2009).

[196] *Department of Defense Website: Archived Charge Sheets* (http://www.defenselink.mil/news/commissionsarchives.html) Department of Defense

[197] *Jabran Said bin Al-Qahtani 2005 Charge Sheet* (http://www.defenselink.mil/news/Nov2005/d20051104qahtani.pdf) Department of Defense Website

[198] *Ghassan Abdullah al Sharbi 2005 Charge Sheet* (http://www.defenselink.mil/news/Nov2005/d20051104sharbi.pdf) Department of Defense Website

[199] *Sufyian Barhoumi 2005 Charge Sheet* (http://www.defenselink.mil/news/Nov2005/d20051104barhoumi.pdf) Department of Defense Website

[200] *Binyam Mohamed 2005 Charge Sheet* (http://www.defenselink.mil/news/Nov2005/d20051104muhammad.pdf) Department of Defense Website

[201] Joseph Margulies *Guantanamo and the Abuse of Presidential Power* Simon & Schuster, 2007

[202] Suzanne Goldenberg *Supreme Court Rejects Bush Terror Powers* (http://www.guardian.co.uk/world/2006/jun/30/usa.guantanamo) The Guardian, June 30, 2006

[203] *Jabran Said bin Al-Qahtani 2008 Sworn Charges* (http://www.defenselink.mil/news/d20080529Qahtani.pdf) Department of Defense Website

[204] *Ghassan Abdullah al Sharbi's 2008 Sworn Charges* (http://www.defenselink.mil/news/d20080529Sharbi.pdf) Department of Defense Website

[205] *Sufyian Barhoumi's 2008 Sworn Charges* (http://www.defenselink.mil/news/d20080529Sufyian.pdf) Department of Defense Website

[206] *Binyam Mohamed 2008 Sworn Charges* (http://www.defenselink.mil/news/Mohamed - sworn0603.pdf) Department of Defense Website

[207] Jane Sutton *U.S. drops charges against 5 Guantanamo captives* (http://www.reuters.com/article/topNews/idUSTRE49K65120081021?sp=true) Reuters, October 21, 2008

[208] Carol J. Williams *Guantanamo judge defies Obama* (http://articles.latimes.com/2009/jan/30/nation/na-guantanamo-judge30?pg=1) *Los Angeles Times*, January 30, 2009

[209] *Profile: Binyam Mohamed* (http://news.bbc.co.uk/2/hi/uk_news/7906381.stm) BBC News, February 23, 2009

[210] Ed Pilkington, "I was responsible for 9/11, from A to Z' – a confession from Guantánamo Bay" (http://www.guardian.co.uk/world/2007/mar/15/alqaida.september11), The Guardian, March 15, 2007

[211] *Khalid Sheikh Mohammed Charge Sheet* (http://www.defenselink.mil/news/d20080509Mohammed.pdf) Department of Defense Website

[212] *Ghassan Abdullah al Sharbi 2009 Charge Sheet* (http://www.defenselink.mil/news/alSharbi.pdf) Department of Defense Website

[213] *Jabran Said bin Al-Qahtani 2009 Charge Sheet* (http://www.defenselink.mil/news/alQahtanisworn.pdf) Department of Defense Website

[214] *Sufyian Barhoumi Military Commissions Page* (http://www.defenselink.mil/news/commissionsBarhoumi.html) Department of Defense Website

[215] *Ghassan Abdullah al Sharbi Military Commissions Page* (http://www.defenselink.mil/news/commissionsSharbi.html) Department of Defense Website

[216] Brad Foss *State Department moves against Sudanese-Canadian for alleged terror ties* (http://www.accessmylibrary.com/coms2/summary_0286-16119183_ITM) The America's Intelligence Wire, July 26, 2006

[217] Paul Koring *Ottawa wants terrorism suspect taken off watch list* (http://www.theglobeandmail.com/servlet/Page/document/v5/content/subscribe?user_URL=http://www.theglobeandmail.com/servlet/story/LAC.20080502.SUDAN02/TPStory/Front&

ord=123654973&brand=theglobeandmail&force_login=true) Globe and Mail, February 5, 2008

[218] Michael Comte *Canada will repatriate man said linked to Al-Qaeda* (http://www.google.com/hostednews/afp/article/ALeqM5hUaf3TZSzmnZORHAtaID8uB1DNnw) AFP, June 18, 2009

[219] Peter Cheney *Bin Laden lieutenant identified suspect in Canada* (http://www.theglobeandmail.com/servlet/ArticleNews/front/RTGAM/20021218/wxalge1218/Front/homeBN/breakingnews) The Globe and Mail, December 18, 2002

[220] *Court upholds Harkat's security certificate* CTV News, March 22, 2005

[221] *Lawyer for accused terrorist seeks access to Guantanamo detainee* (http://www.edmontonsun.com/News/Canada/2007/08/26/4448108.html) Edmonton Sun, August 26, 2007

[222] Colin Freeze, Tu Thanh Ha, and Omar El Akkad *Secret files against terror suspects revealed* (http://www.theglobeandmail.com/servlet/story/RTGAM.20080222.wcertificate0222/BNStory/National/?cid=al_gam_mostdiscuss) Globe and Mail, February 23, 2008

[223] The Bryant Park Project *Canada Takes a Stand Against Alleged U.S. Torture* (http://www.npr.org/templates/story/story.php?storyId=87975164) National Public Radio, March 7, 2008

[224] *Canada Won't Use "Waterboard" Evidence* (http://www.npr.org/blogs/news/2008/03/canada_wont_use_waterboard_evi.html) The NPR News Blog, March 6, 2008

[225] *Canada court OKs release of al Qaeda Suspect* (http://www.swissinfo.org/eng/international/ticker/detail/Canadian_court_OKs_release_of_al_Qaeda_suspect.html?siteSect=143&sid=6798381&cKey=1149878259000) Reuters, Swiss Info, June 9, 2006

[226] *Terrorism suspect released on bail* Dose Canada, May 24, 2006

[227] ""Amhed Ressam's Millennium Plot" PBS" (http://www.pbs.org/wgbh/pages/frontline/shows/trail/inside/cron.html). Pbs.org. . Retrieved 2012-04-04.

[228] ""Ahmed Ressam's Testimony at Mokhtar Haouari's Trial" PBS" (http://www.pbs.org/wgbh/pages/frontline/shows/trail/inside/testimony.html). Pbs.org. 1999-12-14. . Retrieved 2012-04-04.

[229] *Abu Zubaydah Summary of Evidence* (http://www.dod.mil/pubs/foi/detainees/csrt_arb/000784-000819.pdf) Department of Defense]

[230] Ressam Case Documents, Accessed via PACER, December 2008

[231] Joseph Margulies *Abu Zubaydah's suffering* (http://www.latimes.com/news/opinion/la-oe-margulies30-2009apr30,0,3309097.story) *Los Angeles Times*, April 30, 2009

[232] Pyle, Richard. *Did Saudis know of 9–11 plot?*.

External links

- The Final 9/11 Commission Report (http://www.gpoaccess.gov/911/index.html)
- Approving Torture and Destroying Documents: More Notes on the "Zelikow Memo" (http://nsarchive.wordpress.com/2012/04/04/approving-torture-and-destroying-documents-more-notes-on-the-zelikow-memo/)
- Committee on Armed Services United States Senate: Inquiry Into the Treatment of Detainees in U.S. Custody (http://armed-services.senate.gov/Publications/Detainee Report Final_April 22 2009.pdf)
- Jay Bybee and John Yoo *Memorandum for John Rizzo Acting General Counsel of the Central Intelligence Agency: Interrogation of al Qaeda Operative* U.S. Department of Justice, Office of Legal Counsel, August 1, 2002 (http://luxmedia.vo.llnwd.net/o10/clients/aclu/olc_08012002_bybee.pdf)
- Steven Bradbury *Memorandum for John A. Rizzo Senior Deputy General Counsel, Central Intelligence Agency, Office of Legal Counsel Re: Application of 18 U.S.C. Sections 2340-2340A to Certain Techniques That May Be Used in the Interrogation of a High Value al Qaeda Detainee* Department of Justice, Office of Legal Counsel, May 10, 2005 (http://luxmedia.vo.llnwd.net/o10/clients/aclu/olc_05102005_bradbury46pg.pdf)
- Steven Bradbury *Memorandum For John A. Rizzo Senior Deputy General Counsel, Central Intelligence Agency Re: Application of 18 U.S.C. §§ 2340-2340A to the Combined Use of Certain Techniques in the Interrogation of High Value al Qaeda Detainees* Department of Justice, Office of Legal Counsel, May 10, 2005 (http://luxmedia.vo.llnwd.net/o10/clients/aclu/olc_05102005_bradbury20pg.pdf)
- Steven Bradbury *Memoradnum for John A. Rizzo Senior Deputy General Counsel, Central Intelligence Agency Re: Application of United States Obligations Under Article 16 of the Convention Against Torture to Certain Techniques that May Be Used in the Interrogation of High Value al Qaeda Detainees* Department of Justice, Office of Legal Counsel, May 30, 2005 (http://luxmedia.vo.llnwd.net/o10/clients/aclu/olc_05302005_bradbury.pdf)
- *U.S. Department of Justice, Office of Inspector General: A Review of the FBI's Involvement in and Observations of Detainee Interrogations in Guantanamo Bay, Afghanistan, and Iraq* Department of Justice, Office of Inspector

General, May 2008 (http://www.usdoj.gov/oig/special/s0805/final.pdf)

- Human Rights First; Tortured Justice: Using Coerced Evidence to Prosecute Terrorist Suspects (2008) (http://www.humanrightsfirst.info/pdf/08307-etn-tortured-justice-web.pdf)
- Human Rights First; Undue Process: An Examination of Detention and Trials of Bagram Detainees in Afghanistan in April 2009 (2009) (http://www.humanrightsfirst.info/pdf/HRF-Undue-Process-Afghanistan-web.pdf)
- Former CIA "Ghost Prisoner" Abu Zubaydah Recognized as "Victim" in Polish Probe of Secret Prison (http://www.andyworthington.co.uk/2011/01/20/former-cia-ghost-prisoner-abu-zubaydah-recognized-as-victim-in-polish-probe-of-secret-prison/) Andy Worthington
- Abu Zubaydah (http://topics.nytimes.com/top/reference/timestopics/people/z/abu_zubaydah/index.html) collected news and commentary at *The New York Times*
- Works by or about Abu Zubaydah (http://worldcat.org/identities/np-zubaydah, abu) in libraries (WorldCat catalog)

Ahmed_al-Nami

Ahmed al-Nami	

Born	Ahmed al-Nami (in Arabic: أحمد النامي)December 7, 1977'Asir Province, Saudi Arabia
Died	September 11, 2001 (aged 23)Shanksville, Pennsylvania, U.S.

A former law student and muezzin, **Ahmed bin Abdullah al-Nami** (Arabic: أحمد بن عبد الله النامي, *Aḥmad bin 'Abdullāh an-Nāmī*; also transliterated as **Alnami**) (December 7, 1977 – September 11, 2001) was one of four hijackers of United Airlines Flight 93 as part of the September 11 attacks.

Born in Saudi Arabia, Nami had served as a muezzin and was a college student. He left his family in 2000 to complete the Hajj, but later went to Afghanistan bound for an al-Qaeda training camp where he befriended other future hijackers and would soon be chosen to participate in the attacks.

He arrived in the United States in May 2001, on a tourist visa, where he would settle in Florida up until the attacks. On September 11, 2001, Nami boarded United 93 and assisted in the hijacking of the plane, which crashed into a field in rural Shanksville, Pennsylvania, after a passenger uprising, due to the passengers receiving information from their families of the 3 other hijacked planes that hit the World Trade Center and the Pentagon.

History

Nami, much like Abdulaziz al-Omari, Wail al-Shehri, Waleed al-Shehri, and Mohand al-Shehri was born in the 'Asir Province in Saudi Arabia. Born to the Quraish tribe of Saudi Arabia,[1] Nami served as a muezzin at the Seqeley mosque after having reportedly become very religious sometime in early 1999. That autumn he enrolled in the King Khaled University at Abha to study Sharia, he left his family home in Khamis Mushayt in the summer of 2000 to complete the Hajj, but never returned – instead travelling to the Al Farouq training camp in Afghanistan where he met and befriended Waleed and Wail al-Shehri, two brothers from Khamis Mushayt, and Saeed al-Ghamdi. The four reportedly pledged themselves to Jihad in the Spring of 2000, in a ceremony presided over by Wail – who had dubbed himself *Abu Mossaeb al-Janubi* after one of Muhammad's companions.[2] Dubbed "Abu Hashim", Nami was considered "gentle in manner" by his colleagues, and reported that he had a dream in which he rode a mare along with Muhammad, and that the prophet told him to dismount and fight his enemies to liberate his land.[1]

During his time at al-Farooq, there is a curious mention under Mushabib al-Hamlan's details that Nami had recently had laser eye surgery, an uncited fact that does not reappear.

By October he had taken a prospective hijacker Mushabib al-Hamlan from Afghanistan to Saudi Arabia where they both procured B-1/B-2 tourist/business visas on October 28 – but Hamlan then decided not to proceed and is thought to have returned to his family. Nami's Visa application has since been reviewed, and while he mentioned that Mushabib will be travelling with him, he listed his occupation as *student* but failed to provide an address for his school, and listed his intended address in the United States merely as Los Angeles – in the end he never used this Visa to enter the United States, and reported his passport (C115007, which showed evidence of travel to

Afghanistan) as "lost", and procured a new one from Jeddah (C505363). He used the new passport to acquire a new B-1/B-2 visa in Jeddah on April 23, again recopying his answers from previously although crossing out the lines regarding Mushabib and previous attempts to acquire a visa.[3] He was interviewed by a consular officer, who again approved his application. Records at the time only recorded past failures to procure a visa, so the officer had no way of realising that Nami had successfully received an earlier visa.

In mid-November 2000, the 9/11 Commission believes that Nami, Wail and Waleed al-Shehri, all of whom had obtained their U.S. visas in late October, traveled in a group from Saudi Arabia to Beirut and then onward to Iran where they could travel through to Afghanistan without getting their passports stamped. This probably followed their return to Saudi Arabia to get "clean" passports. An associate of a senior Hezbollah operative is thought to have been on the same flight, although this may have been a coincidence.

While in the United Arab Emirates, Nami purchased traveler's cheques presumed to have been paid for by Mustafa al-Hawsawi. Five other hijackers also passed through the UAE and purchased travellers cheques, including Majed Moqed, Saeed al-Ghamdi, Hamza al-Ghamdi, Ahmed al-Haznawi and Wail al-Shehri.

2001

In March 2001, Ahmed al-Nami appeared in an al-Qaeda farewell video showing 13 of the *muscle hijackers* before they left their training centre in Kandahar; while he does not speak, he is seen studying maps and flight manuals.

On April 23, Nami was recorded obtaining a new US Visa.[4]

On May 28, Nami arrived in the United States from Dubai with fellow-hijackers Mohand al-Shehri and Hamza al-Ghamdi. By early June, Nami was living in apartment 1504 at the Delray Racquet Club condominiums with Saeed al-Ghamdi in Delray Beach, Florida. He telephoned his family in 'Asir shortly after arriving in the country.

In June, he phoned his family for the last time.[5]

He was one of 9 hijackers to open a SunTrust bank account with a cash deposit around June 2001, and on June 29 received either a Florida State Identification Card or Drivers License.[6]

He may have been one of three hijackers that listed the Naval Air Station in Pensacola, Florida as their permanent address on drivers' licenses, though other sources claim he listed the Delray condominium.

On August 28, Nami and Ahmed al-Haznawi reportedly bothered a Delray Beach resident, Maria Siscar Simpson, to let them through her apartment to retrieve a towel that had fallen off their balcony onto hers.[7]

On September 5, Nami and Saeed al-Ghamdi purchased tickets for a September 7 flight to Newark at Mile High Travel on Commercial Boulevard—paying cash for their tickets. Ziad Jarrah and Ahmed al-Haznawi also purchased tickets for the same flight from Passage Tours.

On September 7, all four Flight 93 hijackers flew from Fort Lauderdale to Newark International Airport aboard Spirit Airlines.

Attacks

On September 11, 2001, Nami arrived in Newark to board United Airlines Flight 93 along with Saeed al-Ghamdi, Ahmed al-Haznawi and Ziad Jarrah. Some reports suggest Haznawi was pulled aside for screening while others claim there is no record of whether any of the four were screened; the lack of CCTV cameras at the time has compounded the problem. Nami boarded the plane between 7:39 am and 7:48 am; seated in First Class 3C, next to Saeed al-Ghamdi.

Due to the flight's routine delay, the pilot and crew were notified of the previous hijackings and were told to be on the alert, though within two minutes Jarrah had stormed the cockpit leaving the pilots dead or injured.

At least two of the cellphone calls made by passengers indicate that all the hijackers they saw were wearing red bandanas, which some have questioned may have signified an allegiance to the Egyptian Islamic Jihad. The calls also indicated that one of the men had tied a box around his torso, and claimed there was a bomb inside; it is not known which hijacker this was.

Passengers on the plane heard through phone calls the fates of the other hijacked planes, and organized a brief assault to retake the cockpit. The hijackers crashed the plane into the Pennsylvania farmland rather than cede control of the plane. All aboard died.

Aftermath

After the attacks, an employee of Saudi Arabian Airlines named Ahmed al-Nami came forward to say that he feared his identity had been stolen, although he had never lost his passport.

He has been portrayed by Jamie Harding in the film United 93 and Asim Wali in the film Flight 93

See also

- PENTTBOM

References

[1] Videotape of recorded will of Abdulaziz al-Omari and others
[2] Boston Globe, Driving a Wedge: Part I (http://www.boston.com/news/packages/underattack/news/driving_a_wedge/part1_side.shtml)
[3] "Staff Monograph on 9/11 and Terrorist Travel" (http://www.9-11commission.gov/staff_statements/911_TerrTrav_Monograph.pdf) (PDF). 9/11 Commission. 2004. .
[4] "Page 542" (http://www.faqs.org/docs/911/911Report-542.html). Faqs.org. . Retrieved May 3, 2010.
[5] Murphy, Caryle; Ottaway, David B. (September 25, 2001). "Some Light Shed On Saudi Suspects" (http://www.washingtonpost.com/ac2/wp-dyn?pagename=article&node=&contentId=A19549-2001Sep24&notFound=true). *The Washington Post*. . Retrieved April 23, 2010.
[6] "Chapter 2- Chronology" (http://www.9-11commission.gov/staff_statements/911_TerrTrav_Ch2.pdf) (PDF). . Retrieved May 3, 2010.
[7] "911: Florida: terror's launching pad" (http://www.sptimes.com/2002/09/01/911/Florida__terror_s_lau.shtml). Sptimes.com. . Retrieved May 3, 2010.

External links

- The Final 9/11 Commission Report (http://www.gpoaccess.gov/911/index.html)

Ahmed_al-Haznawi

Ahmed al-Haznawi	

Born	Ahmed al-Haznawi (in Arabic: احمد الحزناوي)October 11, 1980Hazna, al Bahah Province, Saudi Arabia
Died	September 11, 2001 (aged 20)Shanksville, Pennsylvania, U.S.

Ahmed Ibrahim al-Haznawi (Arabic: احمد ابراهيم الحزناوي, *Aḥmad Ibrāhīm al-Ḥaznāwī*) (October 11, 1980 – September 11, 2001) was one of four hijackers of United Airlines Flight 93 as part of the September 11 attacks.

A Saudi, Haznawi had trained in Afghanistan after leaving his family to fight in Chechnya in 2000. He arrived in the United States in June 2001 under the direction of Al-Qaeda for terrorist attacks, on a tourist visa. Once he was in the U.S., he settled in Florida and helped plan out how the attacks would take place.

On September 11, 2001, Haznawi boarded United Airlines Flight 93 and assisted in the hijacking of the plane, which crashed into a field in Shanksville, Pennsylvania, after a passenger uprising.

Early life

Ahmed al-Haznawi was the son of a Saudi imam from the Al-Bahah province, a province in the south west of Saudi Arabia. It is the capital of Al Bahah Province nestled between the resorts of Mecca and Abha, Al Bahah is one of the Kingdom's prime tourist attractions. Haznawi grew up in the village of Hazna, where his father was a cleric at the mosque in the central marketplace section of the village. Haznawi belonged to a family that was part of the larger, al-Ghamdi tribe, sharing the same tribal affiliation with fellow hijackers Saeed al-Ghamdi, Hamza al-Ghamdi, and Ahmed al-Ghamdi.[1] He memorised the Quran, giving him the title Hafiz.[2]

This group is noted as being some of the more religiously observant of the hijackers, and they are thought to have met each other some time in 1999.

History

1999–2000

Haznawi announced he was leaving his family in 1999 to help fight in Chechnya, although his father forbade him.[3] His father and brother, Abdul Rahman al-Haznawi, reportedly last heard from him in late 2000, after he made references to training in Afghanistan.

On November 12, 2000, Haznawi applied for and received a two-year US B-1/B-2 (tourist/business) visa in Jeddah, Saudi Arabia.[4]

From November 27, 2000 through December 27, 2000, Haznawi was in Saudi Arabia for Ramadan. It is theorized that during this trip, he may have initially told Saeed and Hamza al-Ghamdi about the operation.

Some time late in 2000, Haznawi traveled to the United Arab Emirates, where he purchased traveler's cheques presumed to have been paid for by Mustafa al-Hawsawi. Five other hijackers also passed through the UAE and purchased travellers cheques, including Majed Moqed, Saeed al-Ghamdi, Hamza al-Ghamdi, Wail al-Shehri and Ahmed al-Nami.

2001

He was one of four hijackers believed to be staying at a Kandahar guest house in March 2001, where they were seen by Mohammed Jabarah. Jabarah remembered Haznawi specifically, saying that he was "very devout and could recite the entire Koran from memory".[5]

Haznawi appeared in an al-Qaeda video, pledging to give his life to martyrdom.

On June 8 he arrived in Miami, Florida with fellow hijacker Wail al-Shehri. He was one of 9 hijackers to open a SunTrust bank account with a cash deposit around June 2001. He is believed to have moved in with Ziad Jarrah, who got a new apartment on Bougainvilla Dr. in Lauderdale-by-the-Sea, after both men gave the landlord photocopies of their German passports, which he later turned over to the FBI.[6]

On June 25, Jarrah took Haznawi to Holy Cross Hospital in Fort Lauderdale on advice of his landlord Charles Lisa. Haznawi was treated by Dr. Christos Tsonas, who gave him antibiotics for a small gash on his left calf. While he told staff that he had bumped into a suitcase,[7] the media briefly reported it as a sign of cutaneous anthrax and a possible link to the 2001 anthrax attacks, although FBI later addressed the rumors stating that "Exhaustive testing did not support that anthrax was present anywhere the hijackers had been."[8]

On July 10 Haznawi obtained a Florida driver's license, later obtaining another copy on September 7, 2001 by filling out a change-of-address form. Five other suspected hijackers also received duplicate Florida licenses in 2001, and others had licenses from different states. Some have speculated that this was to allow multiple persons to use the same identity.[9]

Jarrah and Haznawi both received their one-way tickets for United Airlines Flight 93, on September 5.[10] On September 7, all four Flight 93 hijackers flew from Fort Lauderdale to Newark International Airport aboard Spirit Airlines.

Attacks

On September 11, 2001, Haznawi arrived at Newark International Airport to board Flight 93. Although he was selected for additional security by CAPPS and screened, he was able to board the flight without incident, with only his checked bags requiring extra screening for explosives.[11]

Due to the flight's delay, the pilot and crew were notified of the previous hijackings that day and were told to be on the alert. Within minutes, Flight 93 was hijacked as well.

At least two of the cellphone calls made by passengers indicate that all the hijackers they saw were wearing red bandanas, which some believe may have signified an allegiance to the Egyptian Islamic Jihad. The calls also indicated that one of the men had tied a box around his torso, and claimed there was a bomb inside - it is not known which hijacker this was. Some passengers expressed doubt that the bomb was real.

Passengers on the plane heard through phone calls the fates of the other hijacked planes. A passenger uprising soon took place. Hijacker-pilot, Ziad Jarrah, crashed the plane into an empty field near Shanksville, Pennsylvania in order to prevent the passengers from gaining control of the plane. The crash killed everyone on board.

Aftermath

After the attacks, before the release of the FBI pictures of the hijackers, Arab News reported that Haznawi's brother Abdul Rahman had told al-Madinah newspaper that a photograph published by local newspapers bore no resemblance to his brother.[12]

A videotape released with The Wills of the New York and Washington Battle Martyrs, was aired on Al Jazeera on April 16, 2002. While the name beneath the speaker read *al-Ghamdi*, the image is of Haznawi speaking.[3] Officials suggested that the name was merely a reference to his tribal affiliation. The film was thought to have been made in March 2001. In it, he talked about his plans to bring the "bloodied message" to America. In September 2002, a similar tape made by Abdulaziz al-Omari appeared.

He has been portrayed by actors Omar Berdouni in United 93, and Zak Santiago in Flight 93.

See also

- PENTTBOM

References

[1] "Driving a Wedge - Bin Laden, the US and Saudi Arabia (Part 1)". The Boston Globe. 2002-03-03.

[2] Bell, Stewart. "The Martyr's Oath", 2005.

[3] Borger, Julian (April 16, 2002). "Chilling, defiant: the video suicide message of a September 11 killer" (http://www.guardian.co.uk/international/story/0,3604,685102,00.html). *The Guardian* (London). . Retrieved April 23, 2010.

[4] "Staff Monograph on 9/11 and Terrorist Travel" (http://www.9-11commission.gov/staff_statements/911_TerrTrav_Monograph.pdf) (PDF). 9/11 Commission. 2004. .

[5] Ressa, Maria. "Seeds of Terror", 2003. p. 165

[6] "Miami Herald: Hijack plotters used S. Florida as a cradle for conspiracy" (http://web.archive.org/web/20011019090410/www.miami.com/herald/special/news/worldtrade/digdocs/046328.htm). Web.archive.org. 2001-10-19. . Retrieved 2012-01-06.

[7] Memo on Florida Case Roils Anthrax Probe (http://www.ph.ucla.edu/epi/bioter/memoroilsanthraxprobe.html) Washington Post, March 29, 2002

[8] "Report raises question of anthrax, hijacker link" (http://www.cnn.com/2002/US/03/23/anthrax/?related). *CNN*. March 23, 2002. . Retrieved April 23, 2010.

[9] "Multiple identities of hijack suspects confound FBI: South Florida Sun-Sentinel" (http://web.archive.org/web/20040604015841/http://www.sun-sentinel.com/news/local/southflorida/sfl-worldtrade-suspectprofiles.story). Web.archive.org. 2004-06-04. . Retrieved 2012-01-06.

[10] "Statement of Robert S. Mueller: Joint Investigation Into September 11: (published September 26, 2002)" (http://www.fas.org/irp/congress/2002_hr/092602mueller.html). Fas.org. . Retrieved 2012-01-06.

[11] "The Aviation Security System and the 9/11 Attacks - Staff Statement No. 3" (http://www.9-11commission.gov/staff_statements/staff_statement_3.pdf) (PDF). 9/11 Commission. .

[12] Hanjour family denies Hani's involvement in crime (http://www.arabnews.com/?page=1&section=0&article=9490&d=1&m=10&y=2001) Arab News, October 1, 2001

External links

- The Final 9/11 Commission Report (http://www.gpoaccess.gov/911/index.html)
- Bin Laden tape with Haznawi (http://www.cnn.com/2002/WORLD/meast/04/15/terror.tape/)

Article Sources and Contributors

Peter_Brownback *Source*: http://en.wikipedia.org/w/index.php?title=Peter_Brownback *Contributors*: Dhartung, Drive 55, Geo Swan, Kafziel, Klemen Kocjancic, Kumioko (renamed), Looper5920, Nobunaga24, Rich Farmbrough, Rjwilmsi, Sherurcij, Startstop123, Ulric1313, Vortexjk, Waacstats, 2 anonymous edits

Ramzi_bin_al-Shibh *Source*: http://en.wikipedia.org/w/index.php?title=Ramzi_bin_al-Shibh *Contributors*: Alberuni, All Hallow's Wraith, Astanhope, Aude, Basawala, Bender235, Bonadea, Bratsche, Chris the speller, Cla68, Clerks, Colonies Chris, CommonsDelinker, Courcelles, Cravendekere, D6, Daniel C. Boyer, Darouet, DasIstRicht, Daveman 84, Delirium, DerAmerikaner, Duiarmia, EamonnPKeane, EverettMcGill, Falcon9x5, Fanatix, Fluppy, Fragglet, Gabbe, Gaius Cornelius, Geo Swan, Geoffrey Gibson, Good Olfactory, HIGHSCHOOLFAILURE, Hooperbloob, II MusLiM HyBRiD II, Iqinn, Jamesmullarkeyy, Jeronimo, John, JohnCengiz77, Jonathan ryan, Josiah Rowe, Jprg1966, Katarighe, Khanada, Kingturtle, Klemen Kocjancic, Kukini, L Kensington, LDH, Lllo44, Lord Voldemort, Magioladitis, Maximus Rex, Mild Bill Hiccup, Mleonard85032, Mnnlaxer, Nadav1, Patrick, Paul E Ester, Paul Pot, Peripatetic, Quadell, RedWolf, Rich Farmbrough, Rjwilmsi, Robofish, SamuelTheGhost, Sardanaphalus, Sherurcij, Sisyph, Sophos II, Stefanomione, Steven Russell, Thriley, Toby Bartels, Treybien, Ulric1313, VeryVerily, WhisperToMe, WikiFlier, Will Beback Auto, WojPob, Zahid Abdassabur, Zer0faults, Zipotur, Zvar, 52 anonymous edits

Unlawful_combatant *Source*: http://en.wikipedia.org/w/index.php?title=Unlawful_combatant *Contributors*: A D Monroe III, ANONYMOUS COWARD0xC0DE, Abunakhli, Agge1000, Alai, Alansohn, Anders.Warga, Andjam, Angr, ArielGlenn, Arjayay, Assawyer, Avengerx, BYF, Badagnani, Bart133, Ben Ben, BenAveling, Bender235, Betacommand, Bgwhite, Billinghurst, Binary TSO, Bryan Derksen, Canada Jack, Cassowary, Cecropia, Chadloder, Chain27, Colonies Chris, Copperchair, Crystallina, Cstalberg, Cybbe, Cyferx, Dabbler, Daniel Quinlan, Date delinker, David Newton, DavidCary, Dbachmann, Declare, DrMajestico, Draeco, Dreaded Walrus, Ed Poor, Edwy, EnviroGranny, Epeefleche, Esaborio, Esprit15d, Evil Monkey, Ferkelparade, Fingers-of-Pyrex, Fluzwup, FrankWilliams, GPS Pilot, Gabbe, Gaimhreadhan, Gargantrithor, Geo Swan, George Burgess, Get-back-world-respect, Grafen, GreatGatsby, Gretchen, Ground Zero, Guanaco, Gul Dukat, Hairy Dude, Happy Sapper, Harold f, Harriv, Hephaestos, Ideal gas equation, Iqinn, JHunterJ, JJCaesar, JPD, James086, Jan van Male, Jburt1, Jcobb, Jiang, Jmarchn, Johnnyw, JonHarder, Jordi Burguet Castell, Jpbrenna, Jpers36, Jrockley, Jschnur, Jyril, Kaliz, KarlFrei, Katana0182, Ken Gallager, Kevin Baas, Kingboyk, Kingturtle, Klemen Kocjancic, KnowledgeOfSelf, Kosebamse, LK, Lan Di, Lapaz, Larkusix, LeeHunter, Liftarn, Ligulem, LilHelpa, Lir, Madder, Magioladitis, Marvan Hogan, Mast3r andrew, Matthewi, Maurice Carbonaro, Maurreen, Mdchachi, Melab-1, Metamagician3000, Mets501, Mhenneberry, Michael Hardy, Mikue, Millahnna, Motor, NSR, NeilTarrant, NekoDaemon, Nescio, Neutrality, Nickj, Nobs01, Noosphere, PBS, PZFUN, PeaceThroughStrength, Pearle, PennState21, Peter Greenwell, Petri Krohn, Phe, Postdlf, Quackslikeaduck, Rama, Random account 47, Randy2063, Ray Radlein, Rchamberlain, Reagle, Reddi, Rich Farmbrough, Richard Keatinge, Richardmilgate, Rjwilmsi, RobPetrone, Roger Davies, SPECTRE, Sam Etler, Sam Hocevar, Sam8, Sbump, Sexyeamo, Shadowjams, Shakescene, Shangrilaista, Sheehan, Simetrical, Sir Anon, Sloane, Small black sun, Snoyes, Sommers, Sorenr, Sosa1998y, Srich32977, Staxringold, Steven Andrew Miller, Stevertigo, Stingray86, Stor stark7, Swatjester, TShilo12, Tabletop, Tazmaniacs, The Epopt, The Founders Intent, TomStar81, Toytoy, Triona, Uriber, Vegaswikian, Ventolin, Viajero, Westoe, Wfaxon, Wik, WikiDao, Will Beback, Wing Nut, Wknight94, Wookipedian, Wrolf, YUL89YYZ, Zer0faults, Zero0000, Zotel, 179 anonymous edits

Khalid_Sheikh_Mohammed *Source*: http://en.wikipedia.org/w/index.php?title=Khalid_Sheikh_Mohammed *Contributors*: 33marcy, 4twenty42o, A4arjunwiki, A8UDI, ATLian, Adam sk, AdjustShift, Adluolas, Adoniscik, Aec is away, Afranelli, Ahoerstemeier, Alan Pierce, Alasdair, Alexscho, Alpha Quadrant, Amer-aln7l, Andreasmperu, Andrewlp1991, Andropod, Anomalocaris, Antandrus, Aquila89, Arab-, Aranea Mortem, Arigoldberg, Arjun G. Menon, Arno, Arwel Parry, Arzel, Ashwinsbhat, Aude, Avnjay, Azmurath, B1atv, B33R, BACbKA, Badagnani, Bairam, BazookaJoe, Bazza 7, Bender235, Bil1, Bill Shillito, BillC, Bitoffish, Bkonrad, BlackTarHeroin, Blaske, BlastOButter42, Bleeding Blue, Bletch, Blue Tie, Bluefist, Bobblewik, BobbyHedgehog, Boomshadow, BorgQueen, BowlerFowl, Bratsche, Brenont, Brianski, Bucksburg, Clreland, Callanecc, Calliopejen1, Can't sleep, clown will eat me, Causa sui, Cbarlow3, Cdc, Charles Stewart, CharlieTC, Chissh, Chocolateluvr88, Chris Capoccia, ChrisO, ChristianGL, City of Tragedy, CityOfSilver, Cla68, Cleared as filed, Cloj, ClovisPt, ColenFace, Collins432, CommonsDelinker, Connelly, Corwin8, Courcelles, CowboySpartan, Crazycoolman, Cresix, Cs32en, Css, Cuba pete, D6, DMCer, DanKeshet, Daniel J. Leivick, Danny, Dantt777, Daren420c, Darkwarriorblake, DasIstRicht, Dave Farquhar, Dave souza, Davidhorman, Davidlawrence, DeadEyeArrow, Decltype, Derek R Bullamore, Deselliers, Dhaluza, Djallal, Dogru144, Donreed, Doomcrawler, Doradus, Dori, Doristanghue, Dorshil, Dpeters11, Dr. Blofeld, Drewmm, Dsp13, Dthomsen8, E Wing, Egeymi, Ego White Tray, Eleland, Enemenemu, EnzoRoyale, Epbr123, Epeefleche, Equilibriumgirl, Ericleb01, Erik9, Esemono, EverettMcGill, Fahd101, Faithlessthewonderboy, Fanatix, Farosdaughter, Farry, Fayenatic london, Filter1987, Fireice, Fluppy, Fraggle81, Francisx, François Pichette, Friedo, Fsotrain09, FunkMonk, GCarty, GHcool, Gabbe, Garing, Garion96, GaryKeorkunian, GateKeeper, Gazpacho, Geo Swan, Gidonb, Gighli, Gimboid13, Glane23, Gobonobo, GoingBatty, Good Olfactory, Goooter, Graft, Greenguy1090, Grenavitar, Ground Zero, Grsz11, Gtg204y, Gunnman07, Halda, Hamiltondaniel, Haramzadi, Harry Potter's Hugest Fan, Hbdragon88, Hecht, HenkvD, Henry Flower, Hephaestos, HighTouch, Hippie Metalhead, Hmains, Hmich176, Holocron, Holon67, Hooperbloob, Hotlorp, Hux, Hypnosadist, Icestorm815, Imblack1234, Indio, Insommia, Intelligentsium, InverseHypercube, Ioannes Pragensis, Iqinn, Irish Pearl, IshmaelMarcos, It Is Me Here, JFreeman, JPG-GR, Jakedaniel, Jambalaya, Jamonbo, Jared, Jasonchesworth, Jaybee's psychiatrist, Jcforge, Jeff G., Jeffrey Smith, Jersyko, Jevansen, Jleon, Jnestorius, Joefromrandb, Joelkubickijr, John, John Bentley, John of Reading, JohnCengiz77, Johnex1, Jon4033, Joseph Solis in Australia, Joshyco, Josiah Rowe, Jredmond, Julesd, Jwisenter, Jxraynor, Kaisershatner, Kalmia, Katarighe, Katieh5584, Kbdank71, Kennvido, Kermitbuns, Kesal, KimChee, Kingal86, Kingturtle, Knucmo2, Kpalion, KrakatoaKatie, Kribbeh, Ksmplusfive, LDH, LachlanA, Lahiru k, Laurinavicius, Lawyer2b, Leutha, Logical Cowboy, Loisel, Lyk4, MER-C, Magioladitis, Makeemlighter, Malo, Mannerheim, Mar4d, MarXidad, Marek69, Mark5677, MarkOller, Marokwitz, Materialscientist, Mato, Mattyjdub, Max Thayer, Maximus Rex, Mdmcginn, Meco, Melsaran, Metstotop333, Mhsbored, Minimac, Mirwais Hotak, Mnalpha, Mocirne, Mohsenali93, Moncrief, Moonriddengirl, Morninggloryseed, Moses Horwitz, Moshe Constantine Hassan Al-Silverburg, Mr. Billion, Mruss1975, Mystichumwipe, NEMT, NWOTruther, NYCJosh, Nakon, Napoleon Dynamite42, Nappyrootslistener, Naqui786, Natalie Erin, Neilgravir, Nellis, Neutrality, Nexus Goof, Niac7, Nightscream, Nightstallion, Nihil, Nil Einne, Ninjarrr, Noor Aalam, Notreallydavid, Nphs2010, Nyttend, Odysseyandoracle, Oleg-ch, Olegwiki, Omnipaedista, Ottre, PFHLai, PaZuZu, Pakismine, Pat1792A, Patchouli, Patrick, Paul E Ester, Pcb21, Pepsidrinka, Philip Trueman, Phlar, PhnomPencil, Pie Man 360, Pigsonthewing, Pilotguy, Plasticup, Pokipsy76, PopUpPirate, ProveIt, Psychonaut, Puffin, Quadell, Quasipalm, Quatloo, R'n'B, R. fiend, RFabian, Rama, Ramesh Ramaiah, RayAYang, Raymondwinn, Rbraunwa, Reinhardheydt, RenniePet, Retiono Virginian, RexNL, Rhombus, Rick Block, Rjwilmsi, Rmhermen, Robcotton, Ronark, SG, SGGH, SINFUL OCTOPUS, SJP, SailorAlphaCentauri, Salamurai, Sam Blacketer, Sammybhushan, SarajevoSprite, Sardanaphalus, Satre21, Scharferimage, Sciurinæ, SeanMD80, Seav, Seb az86556, Semperf, Seth Ilys, Shanes, Sherurcij, Shoefly, Siddiqui, Sinisterscrawl, Sirsamizzle, Skimnc, Sloane, Slon02, Smyth, Sparrowgoose, Spirit Of Truth, Squeemu, Srikeit, Starlemusique, Steel, Stefanomione, Steve099, Steven Russell, Storkk, Stubblyhead, Sundarnut, SuperHero2111, Supertouch, Suprah, Susan Mason, Szhaider, Tad Lincoln, Tadpole256, Tariqabjotu, Tbsdy lives, Tdadamemd, Tegwarrior, Tel555, Terence, Tgeairn, The Original Wildbear, The Thing That Should Not Be, The wub, Thegoodsirspud, ThinkEnemies, Tidus the BlitzStar, TimothyHorrigan, Tinton5, TogetherinParis, Tom harrison, Tommy2010, Tony Myers, Tpbradbury, Treybien, TurtleShroom, Twas Now, Tyrenius, Ud0ntn0m3, Ulric1313, UnexpectedBanana, UnitedStatesian, Usadid911, V7-sport, Vacuumloops, Vanobamo, VeryVerily, Vexorg, VincentG, ViriiK, Visite fortuitement prolongée, Visnoskiify, Visor, Vlad, Vzbs34, WLRoss, Walloon, Waqashacker, Warmfuzzygrrl, Warren Dew, Wayne Miller, Werdna, What was this guy eating?, Whatthree16, WhisperToMe, Whouk, Wikimike, Wikireader41, Wil137, Wknight94, Woohookitty, Worobiew, Writegeist, X686, Xenophrenic, Xoloz, Yottie, Zeppelin42, Zer0fighta, Zoe, Zoicon5, Zulu127, Zvar, Zzuuzz, Zzyzx11, Тивереполник, דפרסה, خالد شيخ محمد 679 anonymous edits

Riduan_Isamuddin *Source*: http://en.wikipedia.org/w/index.php?title=Riduan_Isamuddin *Contributors*: *drew, AI, Ahoerstemeier, Anas Salloum, AxelBoldt, Baseballbaker23, Behun, Bhadani, BhaiSaab, Bobblewik, BorgQueen, Bratsche, Chris the speller, Comandante, D6, Edward, FT2, Flauto Dolce, Fooishbar, Fowler Pierre, Gazkthul, Geo Swan, Good Olfactory, Goodcapcay, HandThatFeeds, Heri Zaelani, Hmains, Hypnosadist, Inbetweener, Indon, Iqinn, Jedonnelley, Joseph Solis in Australia, KrakatoaKatie, LDH, Longhair, Lupo, Magioladitis, Maximus Rex, Merbabu, Monegasque, Mushroom, Nikai, Patrick, Permafrost, Prester John, Quadell, Quatloo, Rettetast, Rich Farmbrough, Rjwilmsi, Robofish, Sherurcij, Sutrannu, The wub, VeryVerily, Vsion, WhisperToMe, Wizardman, Zipotur, Zvar, 41 anonymous edits

Walid_bin_Attash *Source*: http://en.wikipedia.org/w/index.php?title=Walid_bin_Attash *Contributors*: Abberley2, Anas Salloum, BD2412, Bobblewik, Bratsche, CommonsDelinker, DBigXray, David Gerard, Duiarmia, EamonnPKeane, Egeymi, Fjmustak, Geo Swan, GoingBatty, Hooperbloob, Hugo999, Iqinn, Iridescent, Italldepends, John of Reading, JohnCengiz77, Josiah Rowe, Kfrogers, Longhair, MK2, Magioladitis, Mar4d, Michael2-, Mnnlaxer, Neutrality, Open2universe, Otto4711, Parsecboy, Phatom87, Philip Stevens, Quadell, Rich Farmbrough, Rjwilmsi, Robofish, Sardanaphalus, Sherurcij, Swpb, Vegaswikian, VeryVerily, WMSBRBA, WRK, WhisperToMe, Zipotur, Zvar, Δ, דפרסה, 10 anonymous edits

Ahmed_Ghailani *Source*: http://en.wikipedia.org/w/index.php?title=Ahmed_Ghailani *Contributors*: 2001:558:6026:97:44AF:E2BF:9B1E:3DD9, AAAAA, AI, Abductive, Andrewlp1991, Anomalocaris, Bratsche, Bryan Derksen, BryanG, D6, Damacyboy, Denichmsk, Deville, Doc Accurate, Droll, EamonnPKeane, Falcorian, Gavia immer, Geo Swan, Good Olfactory, Ground Zero, Hbdragon88, Hmains, Hypnosadist, Iqinn, Iridescent, JohnCengiz77, Josephf, Jossi, Katarighe, KeptSouth, Korossyl, Kumioko (renamed), Kww, LDH, Lambiam, Lanternix, Leshonai, MDEVER802, Magioladitis, Marie01, Marokwitz, Marrkkymarc, Memorex1987, Mulad, Omoo, Onceonthisisland, Pauly04, Pcb21, Pikpik, Quadell, Redhanker, Redthoreau, Rjwilmsi, Rmhermen, Robofish, Roy McCoy, Scartol, Sherurcij, Steven Russell, Tstrobaugh, Twalls, Ulric1313, VeryVerily, Victor falk, Welsh, WhisperToMe, YUL89YYZ, Zhoban, 65 anonymous edits

American_Airlines_Flight_77 *Source*: http://en.wikipedia.org/w/index.php?title=American_Airlines_Flight_77 *Contributors*: -xwingsx-, 2T, 66.92.166.xxx, A Quest For Knowledge, AOCJedi, AVM, Acroterion, Adashiel, AdjustShift, Ageekgal, Agorski, Airline, Airodyssey, Alan Kurtz, Alansohn, Alaric the Goth, Aliwalla, American champagne, Anastrophe, Andonic, Antandrus, Anyeverybody, ArXivist, Arag0rn, Arbero, Arno, Art LaPella, Arthur Rubin, Astropata, Atomist, Aude, Awardgive, Ayebretwalda, Ayudante, BD2412, Badagnani, BaseballDetective, Bender235, Biffpincus, Bjb29407, Bjmarfito, Black Kite, Blake Burba, BlckKnght, Blerg1, Bobby122, Bobo192, Boing! said Zebedee, Bongwarrior, Booksworm, Boomshadow, Bov, Brewhaha, Brian78046, Brothejr, Bryan Derksen, BryanG, Bueller 007, CJK, CMG, CT Cooper, CWii, CanisRufus, Carbonite, Caribbean H.Q., CeeGee, Ched, Cheznouss, ChicXulub, Chris Roy, CiTrusD, Circeus, Cla68, Cleared as filed, Clipper471, Cnelson, CommonsDelinker, Complete Truth, Conversion script, Courcelles, Cptnono, Crum375, Cs32en, Culnacreann, DARTH SIDIOUS 2, DReifGalaxyM31, Dale Arnett, Damien Shiest, Dan Gluck, Daniel, Daniel C. Boyer, Dantheman4297, Darrell Greenwood, DarthKieduss, Dave, Dave6, Daveschroeder, Dchall1, Decora, Deleting Unnecessary Words, Der Eberswalder, Descendall, Dkendr, DobbyD, Dobie80, Donpayette, Dontlookforme, DrKiernan, Draicone, Drakew, Duiarmia, Durin, Dysprosia, ENeville, ESkog, Ed Poor, Edcolins, Edward, Elcounte, Elipongo, Ellmist, Emerson7, Emote, Epbr123, Ergative rlt, EricSerge, Ericoides, Erik9, Espoo, Eurosong, Facilitatetruth, Fanatix, Felipestrong, Fences and windows, Fighting for Justice, FourPtRoll, Fra5gt, Fraggle81, Frank Freeman, Fresheneesz, Freunde, Fuhghettaboutit, FunkMonk, G Clark, GD, Gabbe, Gaius Cornelius, Gazpacho, Ghost of starman, Gibbyboy15, Glen, Golbez, Graeme Bartlett, Graham87, GregU, Guaka, Gurch, Hadal, Halgin, Haus, Hbdragon88, Headbomb, Hirudo, Hmains, Hugo999, Hut 8.5,

Hydrargyrum, Hydrogen Iodide, I Help When I Can, Ian Pitchford, Irbisgreif, IronGargoyle, Ixfd64, J JMesserly, JCO312, JCaesar, JForget, JHM345, Jake Wartenberg, Jake sanchard, JamesBWatson, Jaranda, Jeff G., Jefferson Whig, Jeffr, JimmyBallgame, Jimwilson, Jklamo, Joebengo, John, JohnCengiz77, Jonathan ryan, Jonathunder, Joseph Solis in Australia, Joyous!, Jpfagerback, Jpgordon, Jrkarp, Juliancolton, JzG, KJK::Hyperion, Kaldari, Kbdank71, Keegbarn, Knucmo2, KoshVorlon, KyuuA4, Lamrock, LarryMorseDCOhio, Laser brain, Lemmey, Lightmouse, Lotje, Lucas Duke, MONGO, Magioladitis, Mailbox101, Mainstreetmark, Malo, Mandsford, Mannyjr95, MartinHarper, Math Champion, Maurreen, Mav, Mentifisto, Merovingian, Meursault2004, Mfoch, MickMacNee, Mieciu K, Mike Denicola, MindstormsKid, Miquonranger03, Misterrick, Mlindroo, Modster, Modular, Moni3, Morriswa, Morton devonshire, Moxy, Mr.Z-man, Myskoxen, N328KF, N419BH, NTK, NYDCSP, NatalieKushner, NawlinWiki, NekoDaemon, Neutrality, Nightscream, Nixdorf, Noahcs, Noclip, Notreallydavid, Nsaa, NuclearWarfare, Nwt, OOODDD, Oaktree b, Ombudsman, Orthogonal, Ospalh, Ottershrew, Oxymoron83, PS., Pablothegreat85, ParaGreen13, Paradiver, Paranoid, Parkwells, Patrick, Paul Benjamin Austin, PaulHanson, PhilipO, Piledhigheranddeeper, Pilotguy, Postdlf, President Rhapsody, Pride Affair, Prodego, Promethean, Pronoun, Proofreader, Proofreader77, ProudIrishAspie, ProudPrimate, Public awareness, Puffdragon18, Pvsheridan, Pygar2, Quadell, Quantumentanglement, RFBailey, RL0919, Raemie, Ratfox, Redrose64, Renamed user 1752, Reywas92, Rhobite, Riccardov, Rich Farmbrough, RichardMills65, Richb1111, Rivertorch, Rjwilmsi, RobertG, Roguenine2000, Roke, Rominandreu, RoyBoy, RxS, RyanGerbil10, SEWilco, SNIyer1, SNIyer12, SQGibbon, Saberwolf116, Sacxpert, Salon Essahj, SandyGeorgia, Sardanaphalus, Savidan, SchuminWeb, Scjiwillbe, Scrappy.doo, ScreaminEagle, Sdpurdy, Seabhcan, Sellyme, Sepa, Sertrel, Sforgue, Shorele17, Siebrand, Slashme, Sloane, Soetermans, Sp33dyphil, SqueakBox, Starionwolf, Stature791, Stephenf1113, Steven Russell, Stevenwarran, StuHarris, Sulfis, Sumsum2010, Sun Creator, Suomi Finland 2009, Tbhotch, TeaDrinker, Tex, Tgeairn, The Bushranger, The Cunctator, The Original Wildbear, The Rambling Man, TheGoodBadWorst, Thorpe, Tim!, Tim1965, TimShell, Time Becton, Tom, Tom harrison, Tomservo3000, Tony1, Tpbradbury, Trevor MacInnis, Trilobite, TripleH1976, Truthanado, Twas Now, Ultra megatron, Undescribed, Ungeniusman, Vaoverland, Vegaswikian, VegitaU, VeryVerily, Viriditas, Visium, Vorellanaj, Wackywace, WereSpielChequers, WhisperToMe, Whosasking, Whosyourjudas, Wiher, Wikidan829, Wikievil666, Wikimedes, Wpktsfs, XalD, Xiutwel, Y2kcrazyjoker4, YSSYguy, Yettaw, Zipotur, 731 anonymous edits

al-Qaeda *Source*: http://en.wikipedia.org/w/index.php?title=Al-Qaeda *Contributors*: *drew, -Majestic-, 0o64eva, 172, 1Chunk, 208.60.199.xxx, 212.67.100.xxx, 23prootie, 23teotwawki, 3dom, A Fantasy, A bit iffy, A vision of it, A.Khalil, A.Z., AAAAA, AC+79 3888, ACSE, AI, ALL YOUR WASTED SOUL ARE BELONG TO US, ALargeElk, AOCJedi, Aaronfreeman, Abdullais4u, Abonazzi, Accurizer, Achlochan, Acroterion, Adam McMaster, Adamrce, Adcor13, Adilou1981, AdjustShift, Adnan bogi, AdultSwim, Aecis, Aesculapius75, Aeusoes1, Afranelli, Againme, Agathoclea, Ahmad2099, Ahoerstemeier, Aitias, Akadruid, Akamad, Akerans, Akronpow, Alai, Alan Zatkow, Alberdi, Alberuni, Alcuin, Aldie, Aleksi Peltola, Alex '05, Alex LaPointe, AlexHammer, AlexLevyOne, AlexiusHoratius, Alexjohnc3, Ali masharli, Alison, All5Horizons, Allansteel, Allstar86, Alphachimp, Alrasheedan, Amcbride, Americanprogress, Aminullah, Amire80, Amoruso, Anarchist42, Anaselshamy, Andicat, Andonic, Andre Engels, Andrejj, Andrensath, Andrew Gray, Andrew Levine, Andrewpmk, Android79, Andys627, Andyshan, Anentiresleeve, Angelpeream, Angielaj, Angstmann, Anime Editor, Animeelf, Animum, Anniika, Annoynmous, AnonEMouse, Anonymous editor, Anshelm, Antandrus, Anthony, Anthonywaldock12, Antiuser, AntonioMartin, ApocalypseNow115, Appleboy, Appraiser, Arab editor 9212, Aranel, Arda Xi, Ardric47, AreJay, Ariconte, Arindam Biswas, Arjun01, Arjun024, Ark30inf, Armyable, Armyrifle9, Arnabdas, Arno, ArqMage, Arshad3m, Art LaPella, Artaxiad, Arthur Rubin, Arzel, Asbestos, Ashmoo, Ashthecat3, Asix, AssistantX, Asterion, Astronaut, Atlant, Atp997, AuburnPilot, Aude, Audry2, Auntof6, Avjoska, Avono, Awesomepastorjoebob, Ayrton Prost, B4hand, BGOATDoughnut, Bachcell, Bad Night, Badmoon36, Bakhsm0a, Baldhur, Ballplay3, Bandit333, Barbara Shack, Bart133, Bartok77, Basalisk, Baseballrocks538, Batmanand, Bcorr, Bdell555, Bdj, Beardsley, Beh-nam, Beland, Bellerophon5685, BenB4, BenFrantzDale, Benbest, Benjamin9832, Benvon, Benwildeboer, Benwing, BeringStrait, Bertilvidet, Bgeer, Biblbroks, Big Axe, Big Brother 1984, Bigmacd24, Bill, Billbrock, Billim1, BinaryTed, Bishop^, Bjarki S, Bjornyvan, Bkessler, Black Falcon, BlackGT, Blackcats, Blah3, BlastOButter42, Blau, BleedingSky84, Bless sins, Blindman shady, Bluedustmite, Bnguyen, Bob4449, Bobblewik, Bobisbob, Bobmack89x, Bobo192, Bobrayner, Boinger, Bongwarrior, Bonked116, BoogaLouie, Borisknezevic, Bowlhover, BrOnXbOmBr21, Bratsche, BrekekekexKoaxKoax, BrentDanzig, Brentdax, Brewcrewer, Brian Kendig, Brianboulder, Brianfredieu, Brianga, Briantw, BrightStarSky, Brighterorange, Brmwk, Broomwick, Bruce1ee, Bruinfan13, Bryan Derksen, Bssc, Buchanan-Hermit, Budhen, BuickCenturyDriver, Bull Market, Bullsfan7210, Butros, Byeee, Bzero, C.Fred, CARRRRRRRRRR!, CHIMERA, CJGB, CLW, CMacMillan, CSWarren, Caffelice, Caltas, Calton, Calvinung2000, Cam, CambridgeBayWeather, Camembert, Can't sleep, clown will eat me, Canadian-Bacon, CanadianCaesar, Candent shlimazel, Canderson7, Cantus, Caper13, Captain Smartass, Carachi, Careless hx, Carlon, Carlson288, CarterHays, Cassowary, Catdude, Causa sui, Cdc, Cdogsimmons, Ce garcon, Cedrus-Libani, Cenarium, Centonup, Cerebral Warrior, Cgoakley, Chachilongbow, Chaosfeary, Chendy, Chiton magnificus, Chkiss, Chmod007, Chong shao wei, Chowbok, Chris Roy, ChrisCork, Chrisch, Chrislk02, Christian List, Christopher Parham, Chriswiki, Chuckiesdad, Chuckkos, Chudogg, Cielomobile, Cjarbo2, Clamster5, ClaudiaTR80, Clawson, Clayoquot, Clearcrash1, Cloaked Romulan, Cmmmm, Cocoacow911, Coconuteire, CombatRock, Cometstyles, Commodore Sloat, CommonGround, CommonsDelinker, ComputerJA, Comrade42, Comte0, Conversion script, CoolFox, CoolKid1993, Cooldude64, Coolguyforshore, Cornischong, Corusant, Corwinstephen, Cprompt, Crabula, Craigster00, Crazyeddie, Cremepuff222, Crystallina, Cs32en, CultureTrainer, Cunado19, Curps, Cybercobra, Cyrius, CzarB, DJ Tricky86, DMcanada, DO'Neil, DVD R W, Daa89563, Dabloodz, Dabomb87, Dainomite, Dally 17, DanKeshet, Dancter, Dangerousnerd, Danianjan, Daniel, Daniel Case, Daniel J. Leivick, Danieldis47, Dankezman, Danlev, Danthemankhan, Daren420c, DarkAdonis255, DarkXWeatherX, Darkness Shines, Darkohead, Darkshadow24, Darthyorktown, Dasha14, Date delinker, Daved46, David.Mestel, Davidmccairley, Davidmiller2192, Davsch2, Dawdler, Dawkeye, Dawsonj, Dbachmann, Dcandeto, Dcaron, Dcflyer, Dchall1, De Administrando Imperio, DeadEyeArrow, Deadcorpse, Deadflagblues, DearPrudence, Deathhatefear, DeeKenn, Deflective, Delirium, Delldot, Demec2, Dentren, Deodar, DerHexer, Dermo69, DevOhm, Dianaliu, Dicttrshp, Diderot, Dirty sys, Discospinster, Dismas, Distinguisher, Diving2010, Diyako, Djlayton4, Dmyersturnbull, DocWatson42, Dojarca, Donimus6969, Dori, Dowse, Dpen2000, Dpm12, Dpol, Dr. Blofeld, Dr.alf, DrDoley, DragonsBloodAxe, Dreadnaught, Dreadstar, Dreambeaver, Drumguy8800, DrunkSkunk, Dtbain, Duagloth, Dus7, Dynablaster, E0N, EMENIE, ESkog, Eagleamn, Earthsound, Earthworm Makarov, East Nakfa, EdRicardo, Edward, Edward321, Ejstarchuk, El C, ElKevbo, Eleland, Elenseel, Elfguy, EllsworthSK, Elmarco, Emadd, EmmDee, Emory W. Smith, Emperor Genius, Emperor001, Enemenemu, Energybone, Enigma059, Enrico Dirac, Epbr123, Epeefleche, Equilibrial, Eraserhead1, Eric Sandholm, Erik Williamson, Erxnmedia, Esanchez7587, EscapingLife, Escientist, Esowteric, Esteffect, Eszett, EternalSoul, Eubulides, Eurleif, Euryalus, Evanh2008, Evanreyes, Evenfiel, Evenmadderjon, Evercat, Everton, Everyking, Evlekis, EvocativeIntrigue, Evrenos, Ewlong3, Exiledone, Exilian, Eyas, Eyreland, F13nd, FF2010, Fabiform, Falcon9x5, Falcorian, Farside6, FayssalF, Federal Street, Feedmecereal, Feezo, Fennessy, Fezz, Figureground, FilipeS, Finlander, Fish and karate, FisherQueen, Fitch, Flatterworld, Flockmeal, Florkle, FloydRTurbo, Fluri, Flyguy649, FokkerTISM, Francs2000, Frank Freeman, Freakofnurture, Frecklefoot, Fredrik, Freestylefrappe, Frymaster, Fsotrain09, Fudoreaper, Funandtrvl, Fuzheado, Fvw, GABaker, GCarty, GDonato, Gabi S., Gadfium, Gaius Cornelius, Galloisian, Galmicmi, Garkbit, Garzo, Gavin Compton, Gazpacho, GazzTaylor, Gbinal, Gdo01, Geeta, Gene Nygaard, Geo Swan, Gerbrant, Gergis, Geshxd, Ghettoboy911, GhostPirate, Gidonb, Gilliam, Gimboid13, Gimmetrow, Gira2be, Giraffedata, Glacialfox, Glen, Glenjenvey, Glich, Godofalamut, Gogo Dodo, Golbez, GoldenXuniversity, GomiTaroGeorge42, GoneAwayNowAndRetired, Gonzalo84, Gorgonzilla, Gorgutz, Gr8opinionater, Gracefool, GraemeL, Graft, Graham87, Gramaic, Gramscis cousin, Grandy Grandy, Grant.Alpaugh, Green Beret 625, Green caterpillar, Gregb, Gregbard, Grey Fox-9589, Grika, Grm wnr, Ground Zero, Grsz11, Grunon, Grunt, Gscshoyru, Gtg204y, Gtru e, Guaka, Guanaco, Gun Powder Ma, Gunsfornuns, Gurch, Gurchzilla, Guy M, Gw2005, Gwen-chan, Gyrofrog, HCPUNXKID, HJ32, Hadal, Haemo, Haha169, Haham hanuka, Halma10, Halo Storm, Hamiltondaniel, Hammersoft, HangingCurve, Hans Zarkov, HanzoHattori, HappyCamper, Harakashi, Harej, Harksaw, Hashar, Hatewiki, Havohej, Hazard-SJ, Hbdragon88, Hdt83, Heckhgs, Heimstern, Helohe, Hemanshu, Heron, Herri18, Hhsc, Historypre, Histrion, Hkelkar, Hmains, Hodgdon's secret garden, Hodja Nasreddin, Honette, Hontogaichiban, Hookedupmusi, Hopelesscross, Horsten, Hourick, HoyaProff, Htaccess, Hu!tz!l0p0chtl!, Hu12, Huckamike, Huckfinne, Hugracollective, Hulkster1996, Humus sapiens, Huntster, Hut 8.5, Hvz886, Hypnosadist, I am Ward Cunningham and I am Mad!, ICAPTCHA, IMNOTAVANDALK6, INkubusse, IReceivedDeathThreats, IZAK, Iamajpeg, Iamunknown, Ian Pitchford, Iapetus, Ibn Kofi, Id447, Idlpeach, Igor82, Ikanreed, Illegitimate Barrister, Iloveandrea, ImmortalYawn, Imorthodox23, Indon, Infocidal, Infrogmation, Insanity Incarnate, Insommia, Intangible, InverseHypercube, IoneWells1994, Iqinn, IronDuke, IronGargoyle, Isahries, IshmaelMarcos, Islami, Ismailmountassir, Isria, It Is Me Here, Ivan Ilir, Ixfd64, J Cricket, J Di, J Milburn, J'raxis, J.A.McCoy, JCRules, JEB90, JForget, JFreeman, JLGD, JRSpriggs, JW1805, Jack248, Jackismybestfriend, Jacobmal, Jacobolus, Jacoplane, Jag123, Jagged 85, Jak879, Jake Nelson, Jake381, Jamesheadspeath2006, Jamezcd, Jamosite, Jamott, Jamyskis, JanDeFietser, Jandalhandler, Jaranda, Jarble, JarlaxleArtemis, JaseFace, Jason from nyc, Jauhienij, Javizzle21, Jayjg, Jbull, Jcarroll, Jcoffland, Jdlyall, Jeancey, Jeepday, Jeff Silvers, Jeffrey S, Jembay, JeremyA, Jerseypolitics, JesseRafe, Jet57, JetLover, Jfdwolff, Jfiling, Jim Douglas, Jkaki00, Jklamo, Jklin, Jmoney007, Jni, Jnothman, Joan-of-arc, Joao, Jobbus McKnockey, JoeBlogsDord, JoeM, JoetheMoe25, Joffeloff, Jogershok, Johann Wolfgang, Johanna-Hypatia, John, John Fader, John Vandenberg, John-107, John-1107, JohnC, JohnCengiz77, JohnI, JohnInDC, JohnSawyer, JohnWittle, Johnfos, Johnny McCracken, Jojhutton, JokerXtreme, Jon C., Jonathan F, JonathanDP81, Jonny-mt, JonnyLate, Jons63, Joowwww, JordeeBec, Joseph Solis in Australia, Josephtate, JoshDuffMan, Joshdboz, Jossi, Joyous!, Jpgordon, Jrevill, Jrevill79, Jusjih, JustW353, Jwissick, Jwray, Jwrosenzweig, JzG, K0hlrabi, K1Bond007, KI, Kahmed, Kaihsu, Kalathalan, Karada, Karam.Anthony.K, Karelj, Karimarie, Karkachev, KarlHallowell, Karlstar, Karlthegreat, Kassjab, Katarighe, Kathanar, Kathyedits, KazakhPol, Kazvorpal, Kbdank71, Kdar, Keenan Pepper, Kelly Martin, Kendrick7, Kent Witham, Kenyon, Kermanshahi, Kesaloma, Kesla, Khalid Mahmood, Khillerbhudda, Khizarsaud, Khoikhoi, Khutuck, Kick the cat, Killa123, Kilo-Lima, King Isaac 3, Kingal86, Kingturtle, Kintetsubuffalo, Kirin4, Kiteinthewind, Kitesailor, Kitrus, Kiwimhm, Kizor, Kkeurope, Kkm010, Klemen Kocjancic, Klingoncowboy4, Klinnmirt, Kmkn00, KnowledgeOfSelf, Knucmo2, Koavf, Kotjze, Kozuch, KrakatoaKatie, Krich, Krisnabest, Krsont, Krun, Ksanexx, Ktwombley, Kubigula, Kukini, Kungfuadam, Kuronue, Kurtis, Kurtle, Kuru, Kurykh, Kwamikagami, Kwantus, Kwekubo, Kwertii, Kymacpherson, Kzzl, L Kensington, LGreen, LM1973, Ladnav666, Lamrock, Lance6Wins, Lancemurdoch, LardoBalsamico, Larsro, LatinoMuslim, Latka, Ldmerriam, Leandrod, Lectonar, Ledenierhomme, LeedsKing, Lemuel Gulliver, Leroy65X, Leutha, Lexicon, Lexivore, Lft6771, Liberalcynic, Liftarn, Lifthz, Lightdarkness, Lightpositive, Lights, Ligulem, Lihaas, LilHelpa, Lilac Soul, Lilstarfish, Ling.Nut, Lipton sale, Little birdy, Livna-Maor, Lo2u, Localzuk, Lord British, Lothar von Richthofen, Lotje, Louche7, Louis Waweru, Loyalist Cannons, LtNOWIS, Luatha, Lucas9000, Lucias21, Lucobrat, LuftWaffle0, Luigifan, Luk, Lukelastic, Lukey9706, Lulz234567, Lumos3, Lutherlayman, M. E. Smith, M.O.X, MA1988, MBisanz, MER-C, MONGO, MXL52, MacStep, Mackensen, Maebmij, Maelnuneb, MagicTom, Magzy T, Mahakaya, Maheshkumaryadav, Mahmudmasri, Mailer diablo, Malcolm Farmer, Malcomdavidkelly, Malik Shabazz, Mani1, Manishearth, Manormadman, Manuel007, Manwhatsup, Mapletip, Mar4d, MarcK, Marcika, Marcus Qwertyus, Marek69, Marianocecowski, Marijuanarchy, Mario Profaca, Mark, Mark Micallef, Mark Millard, Mark Richards, MarkSutton, Markeilz, Marokwitz, Martarius, Martial75, Master of Puppets, Master son, MasterRadius, Mastoo, Matanariel, Matchups, Mattfoetus, Matthew Fennell, Mattworld, Mav, Max Naylor, Max rspct, MaxSem, MaxVeers, Maxim, Mb1000, McSnatherson, Mcenroeucsb, Mclay1, Mechanothist, Meco, Meekywiki, Meisterkoch, Merovingian, Metallurgist, Mets501, Metstotop333, Michael Shields, Michael Snow, MichaelBillington, Michaelzeng7, Mick gold, Micler, Midwesternerforlife, Mike Rosoft, MikeLieman, Miketheburrito, Mild Bill Hiccup, Millionsandbillions, Milo99, Minesweeper, Mini-Geek, Minority Report, Mirandaaa, Mirv, Mirwais Hotak, Miskwito, MisterHand, Mistress Selina Kyle, Mitrebox, Mjb1981, Mmortal03, Mnnlaxer, Mo ainm, Modulatum, Mogism, Mohamed Ouda, Mohammadkidwai, Mohonu, Monkeyjunky, Monni95, Monowiki, Monty845, Mooo, MoraSique, Moreau36, Moreschi, Morning star, Mortense, Morton devonshire, Mortus Est, Moshirah, Mountlebanon, Mov, Mpatel, Mporter, Mqi.media, Mr Adequate, Mr. Billion, Mr.Careless, Mr.Z-man, Mrbojanglescj, Mrchadsexington, Mrpink2605, Mrprez2016, Msin10, Mufka, Mughalnz, Mulp, Munci, Murderbike, Musa as-Salafi, Mustafaa, Muéro, Myk60640, N'Dreash, N512ma, NByz, NEMT, NYCJosh, Nabelon, Nascar1996, Nasz, Nathan Johnson, NationalSecurityForums, NawlinWiki, Necromancer44, Neilbeach, NeoFreak, Neodammerung, NerdyScienceDude, Nertzy, Neutrality, NewName, NiTessine, Nick.mon, NickBush24, Nickrds09, Nightghoul, Nightstallion, Nightwatch24, Nikkimaria, Ninjutsu, Nivix, Nixdorf, Nlaporte, No Guru, Noclador, Node ue, NodnarbLlad, Nolatime, Noleander, Nomenphile, Noorkhanuk85, Noroton, Nr9, Nsayer, Nubsnubbed, NuclearUmpf, Nunamiut, Nuno Raimundo, Nuno Tavares, Nurg, Nv8200p, OCNative, Obamiac, Oberst, ObsidianOrder, Occlasty, OdeSSa, Ohconfucius, Ohnoitsjamie, OlEnglish, OldakQuill, Oleg-ch, Olivier, Ollie,

Omicronpersei8, Onefourfive, Oneiros, Onthegogo, Opelio, Opie, Oppolord, Optimist on the run, OrangesRyellow, Oren neu dag, Oren.tal, Orlyyahalom, Orthuberra, Ortonmc, Osli73, Ostinato2, Otto Knell, Ottre, Ouedbirdwatcher, OverSS, Owen, OwenX, Oxymoron83, Ozdog, P. S. Burton, PAK Man, PBP, PBS, PCHS-NJROTC, PDH, PDanger, PFHLai, Pachaaspaisa11, Packcigs, Pahari Sahib, Pajz, Paki.tv, Palang hernan, Paolo Tagliano, Parsecboy, Pass a Method, Patchouli, Paul Willocx, PaulHanson, PaulinSaudi, Paulmaddon, Paulyche, Pax:Vobiscum, Paxosmotic, Pbroks13, Peepeedia, Peeteeree, PerfectStorm, Perfectblue97, Persian Poet Gal, Perspicacite, Peter, Peter Isotalo, Peter McConaughey, Petercorless, Pexise, Pfalstad, Pgk, Pgreenfinch, Phakedc, Phil Boswell, PhilKnight, Philippe, Phirazo, Phoenix2, Phoenix521, PiRSquared17, Picaroon, Picus viridis, Pinktulip, Pir, Plasticup, Plattopus, Plot Spoiler, Plrk, Pmusser, Poindexter Propellerhead, Polyhister, Pomte, Poorleno, Populus, Postdlf, Prasi90, Prester John, Prestonmarkstone, Primalchaos, Professorleah, Pshermo564, Public awareness, Publicus, Pucao, Pudeo, Punungwe, Pwjb, Pwrm, Qaddosh, Quadell, Quendus, QuentinUK, Quercusrobur, Quest for Truth, Qwyrxian, Qxz, R'n'B, REfreakk55, RachelEhrenfeld2, Racooon, Radon210, Raed 5, Ragib, Rajeevmass, Rakanqat, Ral315, Rama, Rammso 22, Ramsmobile, RanEagle, Random account 47, Randor1980, Randroide, Rangeley, Ranveig, Raprchju, Raven4x4x, RayAYang, RazorICE, Rbarreira, Rdavi404, Rdsmith4, Rebroad, Reconfirmer, Recurring dreams, Redaktoer, Reddi, Redquark, Redthoreau, Reesespieces86, Relmeligy, Remiel, RenniePet, Retiono Virginian, Retired username, Rettetast, RexNL, Rfl, Rhobite, Rhsimard, Rich Farmbrough, RickK, Riddley, River10, Rjd0060, Rjm656s, Rjwilmsi, Rmhermen, Rmt2m, Roadrunner, Robbinato, Robneild, Robsinden, Rodneyorpheus, Rogerb67, RonCram, Ronark, Rooty Tooty Fruity, Ross Burgess, Rowantoad, RoyBoy, Royalguard11, Rrburke, Rrius, Rui Gabriel Correia, Rursus, Ruuta 25, Ruy Lopez, RxS, Ryan Vesey, RyanGerbil10, RyanWesleySmith, Ryudo, Ryulong, SEWilco, SGGH, SJP, SNEST, SNIyer1, SU Linguist, Saeed Jahed, Saketh, SallyForth123, Salmankhanpisces, Sam Korn, Sam Spade, Sam42, SamuelTheGhost, Sarcasticninja, Sasha best, Savidan, Saypotatoes, Sbwiddi, Scanlyze, Scarpy, Sceptre, Schultmc, ScienceApe, Scope creep, Sct72, Scythre, Sears4life, Sebastian scha., SecretAgentMan00, SeiteNichtGefunden, Senshi32, Sephodwyrm, Serpent-A, Settersr, Setu, Shadowin, Shady1012, Shamir1, Shanel, Shanes, Shanesterman, Shangrilaista, Shariq, Sharveet, Sheavsey33, Shenme, Sherurcij, Shii, Shoaler, Shoeofdeath, Shoy, Shreshth91, Shrigley, Shrike, Shroomyi3h, ShurizenVenra, Sigma 7, Signalhead, SilentOne17, SilhouetteSaloon, Silly rabbit, Silverback, Simon Shek, SimpleJYouTube, Sinenomine, SingingSenator, Sirkablaam, Sjb5093, Sjbcen, Sjc, Sjontow, Skagedal, Skarioffszky, Skinnyweed, Skobelief, SkyWalker, Slakr, Slarson, Slearn2, Sleigh, SlimVirgin, Slimface, SlipperyN, Slipperybeans, Slon02, Smee, SmilesALot, Smilingsuzy, SmthManly, Smurfy, Smyth, Snowolfd4, Sohail2506, Solargroovy, SoledadKabocha, Somedifferentstuff, Someone65, Sophie means wisdom, Soporaeternus, Spacepotato, Spangineer, Sparks1911, Special-T, Specialkaye33, Spellmaster, Spencer1212, SpidErxD, Spikethehawk, SpuriousQ, Srich32977, Srm07, Ssmith619, St.Krekeler, Stanley011, Starquake, StaticGull, Stephen G. Brown, Stephen Gilbert, Steve3849, SteveW, Steven Russell, Steven X, Stevertigo, Stevietheman, Stevob2006, StillWalking, Str1977, StradivariusTV, Stroppolo, Stubalove69, Studerby, Stwalkerster, Subdolous, Subman758, SudoGhost, Suimaru, Sumdumguy66, Sumedh gawai, Sumnjim, Sundberg, SupaStarGirl, Super-Magician, Super-top-bre, Superbeecat, Superm401, Supertouch, Surlyduff50, Svetlana Miljkovic, Swalot, Switzpaw, Swpb, Sycthos, Symphy, Synchrite, TBadger, TDC, TOttenville8, TRosenbaum, Tadpole256, Tainter, Tapered, Tapir Terrific, Tarc, Tariqabjotu, Taw, Taxman, Tazmaniacs, Tbasherizer, Tbhotch, Tcrowl18, Teamga, Tekleni, Telex, Tequendamia, TexasAndroid, Texture, ThaddeusB, Thameen, That Guy, From That Show!, The Anome, The Dragon of Bosnia, The Evil Spartan, The Hybrid, The Last Angry Man, The Lizard Wizard, The Man in Question, The Minister of War, The Rambling Man, The Random Element, The Red Hat of Pat Ferrick, The Rock And Roll Pirate, The Storm Surfer, The Thing That Should Not Be, The Ungovernable Force, The ace12003, The machine512, The wub, TheCheeseManCan, TheTimesAreAChanging, TheWilliamson, TheYmode, Thecoolmusic, Theguyfromupthere, Themightyquill, TheoloJ, Therius, Theshard, Thewayforward, Thewonderwaffle, Thryduulf, Thucydides of Thrace, Thumperward, Tiddly Tom, Tim1357, Timc, Timeshifter, Timwi, TinaSparkle, TitaniumCarbide, Titoxd, Tkephart, Tnkr111, Tobby72, Tobycat, Tom, Tom harrison, Tompagenet, TopGun, Topbanana, Toryblue, Toya, Tpbradbury, Tree Hugger, Tregoweth, Trengarasu, Trey Stone, Trilobite, TrogdorPolitiks, Trojan traveler, Trusilver, Tuncrypt, Tv316, Twyford, TylerSci, Tylop, Tzaquiel, Túrelio, U5K0, UberEmo, UberMan5000, Ucanlookitup, Ugen64, Ultimus, Unnatural20, Unreal128, Unreal7, Uriah923, Uriber, Utcursch, Ute in DC, Utrechtse, Uusitunnus, V7-sport, VCV16712, Vague Rant, Van helsing, Vanished user 39948282, Vaquero100, Vashti, Vcj, VelairWight, VeryVerily, Vexorg, Vice regent, Victoria2007, Vindheim, Violask81976, Viperes, Visite fortuitement prolongée, Viva43, VolatileChemical, Vovkav, Vranak, Vssun, Vzbs34, W guice, WLRoss, WacoJacko, Wahdeh, Wai Hong, Walloon, Walterego, Waltersimons, Wareh, Wasell, Wavelength, Wayward, Wb1111, Weaponbb7, Welsh, Weregerbil, West Brom 4ever, Westlundwu, Weswilson4, Wewelsburg, Weyes, What was this guy eating?, Whatsyourexcuse, WhisperToMe, White Wolf, Whitechocolate84, Why Not A Duck, Wik, WikHead, Wiki alf, WikiLaurent, Wikifan12345, Wikimanso, Wikipedian64, Wikireader41, Wikishia, Wikiwhat?, Wikiwikifast, WillOakland, William M. Connolley, Williamb, Willowx, Wimt, Windward1, Winston365, Wipe, Wiz9999, Wjemather, Wknight94, Wolfcm, Wolfkeeper, Woodsylass, Woohookitty, Worldofwarcraft69, Woycgpbbr, Woyjenkins, Wrp103, X-factor, XMaster4000, Xabian40409, Xasf, Xiaopo, Xiutwel, Xixtas, Xmike920, Xompanthy, Yahel Guhan, Yamaguchi, Yamamoto Ichiro, Yamla, Yandman, YellowMonkey, Yellowdesk, Yerpo, Yidisheryid, Yintan, Yom, Yonatan, Yono, Yotna, Yousaf465, Yrtneg, Ysangkok, Ytrck, Yummifruitbat, Yuval a, Zain engineer, Zakhebeone, ZapThunderstrike, Zeerover, Zeeshanhasan, Zeimusu, Zephram Stark, Zer0faults, Zhoban, Zigger, Zik-Zak, Zikzak23, Zimzimzallabim, Zntrip, Zoe, Zondi, Zouavman Le Zouave, Zundark, Zzuuzz, Zzyzx11, Методије, , , , 2261 anonymous edits

Abdulaziz_al-Omari *Source*: http://en.wikipedia.org/w/index.php?title=Abdulaziz_al-Omari *Contributors*: 900lb Gorilla, Addshore, AdjustShift, Ahoerstemeier, Andycjp, Angel74dm, Anthony, Arx Fortis, Aude, AxelBoldt, Bobblewik, Bookandcoffee, Burntsauce, Chzeteck, Cleared as filed, Closedmouth, Conversion script, Courcelles, Crosbiesmith, D6, DRK, Darwinek, Derek R Bullamore, Dismas, Dover, Dysprosia, EamonnPKeane, Edward, Epbr123, Fanatix, Fjmustak, GRider, Gabbe, Gilliam, Good Olfactory, Graham87, GregorB, HIGHSCHOOLFAILURE, Hailey C. Shannon, Hmains, Hooperbloob, Huysman, Iqinn, JBC3, John Fader, JohnCengiz77, Kbdank71, LOL, Lightmouse, Lir, Lotje, Maracana, Marek69, Martpol, Mav, Maximus Rex, Moe Aboulkheir, Morhange, Mr. Billion, Mrdthree, N5iln, NEPats, Nappyrootslistener, Neutrality, NickBush24, OSU80, Onthegogo, Oswego52, P4k, Quadell, Ramesh Ramaiah, RayAYang, Rich Farmbrough, Rjwilmsi, Rmhermen, RobNS, Sannita, Sardanaphalus, Scriberius, Serein (renamed because of SUL), Seresin, Shadowjams, Sherurcij, Simon12, Sir Arthur Williams, StephenGonzalo, Sublium, Sverdrup, Tel555, The Brain, Tubedogg, Ucucha, VeryVerily, Washburnmav, WhisperToMe, Wtmitchell, Zipotur, Zsero, 78 anonymous edits

Abu_Zubaydah *Source*: http://en.wikipedia.org/w/index.php?title=Abu_Zubaydah *Contributors*: 1804solstice, Abductive, Adamdaley, Akhristov, Ammar shaker, Anthonymendoza, ArthurThomas24, Ashley Y, AuburnPilot, B33R, Badagnani, Beefyt, Big Bob the Finder, Biltritewave, Blue Tie, BlueTruth, Bratsche, Brigittboat, Cc162, Ce155576, Chameleons84, ChrisGualtieri, Cla68, Commodore Sloat, CommonsDelinker, Conversion script, D6, DHeyward, Da5id403, DasIstRicht, Delirium, DeltaQuad, Dhaluza, Dirwulf, Dysprosia, EamonnPKeane, Edward, EricEnfermero, EverettMcGill, Flatterworld, Fluppy, Fshafique, Fsotrain09, GWU-NSA, Geo Swan, George Ho, Giraffedata, Gnuish, Gojiberry1, Good Olfactory, Hammersoft, Hmains, Hooperbloob, Iqinn, Jack O'Lantern, Jahiegel, JamesAM, JasonMay1997, Jeff3000, Jeffcardille, Jlcoving, John Broughton, KConWiki, Katarighe, Kbdank71, Kingboyk, Klemen Kocjancic, Kwiki, Lambiam, Legomyeggo252, Lotje, Macaddct1984, Magioladitis, Mandarax, Mar4d, Marklar95, Marokwitz, Maximus Rex, Menaus, Mifter, Mike Selinker, Mild Bill Hiccup, Mnnlaxer, NameIsRon, Nick Number, Nightkey, NinaDownstreet, Nurg, Ohconfucius, OlEnglish, Olegwiki, OrangeDog, Papercrab, Prometheus.pyrphoros, Quadell, Raphael1, Raul654, RavenThePackIsBack, Rdavi404, Rebecpat, Reconfirmer, RenniePet, Rich Farmbrough, Richard Arthur Norton (1958-), Rjwilmsi, Rl, Robofish, Romanm, Rursus, Scharferimage, Secretlondon, Seetel, Seventruckloads, ShawnLee, Sherurcij, Sleeper99999, Snori, Soczyczi, SpyMagician, Sue Anne, TMLutas, Tadhirsch, Tantaris, Tfine80, TheAznSensation, Thismightbezach, Tinton5, Tom Pippens, Tregoweth, Ulric1313, Underpants, V7-sport, Vchimpanzee, VeryVerily, Viriditas, Waacstats, Webmgr, WhisperToMe, WhiteCrane, Wiki13, Woohookitty, XsHoa10pSogRcAVgjAxsVMs1C1BbYUspRh, Yakushima, Yojakanaaman, Yonmei, Zvar, 119 anonymous edits

Ahmed_al-Nami *Source*: http://en.wikipedia.org/w/index.php?title=Ahmed_al-Nami *Contributors*: 169.197.11.xxx, A.V., Adambro, AdjustShift, Ageekgal, Angr, Antandrus, Arno, Arundhati bakshi, Arx Fortis, Atmospherica, Bart133, BetaCentauri, Bobblewik, Brion VIBBER, Bryan Derksen, CIS, Conversion script, D6, DabMachine, DangApricot, Darwinek, Docu, Dodo bird, Dolovis, EamonnPKeane, El C, Epeefleche, Facts707, Fakealnamidata, Geo Swan, Good Olfactory, HIGHSCHOOLFAILURE, Hooperbloob, Impala2009, Iqinn, JJLeahy, JohnCengiz77, Joseph Solis in Australia, Kbdank71, Kernow, Klemen Kocjancic, Koavf, Kumioko (renamed), LOL, Le Anh-Huy, Lotje, MarkSweep, Merovingian, Moondyne, Morel, Morhange, Neutrality, Nightscream, Noahcs, Osarius, Ottawahitech, Owen, Paxse, Quadell, Rjwilmsi, ST47, Salsa Shark, Sango123, Sardanaphalus, Seetel, Seloloving, Sharveet, Sherurcij, Sligocki, StephenGonzalo, Steve8675309, Synergy, Tinss, Tinton5, Tonywalton, Tubedogg, VeryVerily, WhisperToMe, Zhoban, Zipotur, Σ, 58 anonymous edits

Ahmed_al-Haznawi *Source*: http://en.wikipedia.org/w/index.php?title=Ahmed_al-Haznawi *Contributors*: AI, AdjustShift, Ageekgal, Antandrus, Arx Fortis, Atmospherica, Aude, Boscobiscotti, CIS, Chzeteck, Closedmouth, Cmcnicoll, Conversion script, Cyan, D6, Dantheman4297, Darwinek, Docu, Facts707, Fluppy, Fortdj33, Good Olfactory, Hallows AG, Hooperbloob, Huysman, Iqinn, Jamesmullarkeyy, Jamespatrickphillip, Jamonbo, JohnCengiz77, Jonathan ryan, Kbdank71, Kumioko (renamed), Ledenierhomme, Lightmouse, Lotje, Loyd415, MarkSweep, MartinHarper, Maximus Rex, Merovingian, Mohammed0052548, Morhange, Nightscream, Noahcs, Quadell, Quarl, Rjwilmsi, Sardanaphalus, ScottS, Seloloving, Sherurcij, Sloane, Tinton5, Tom, Tubedogg, VeryVerily, WhisperToMe, Wikipelli, Woohookitty, Zipotur, 31 anonymous edits

Image Sources, Licenses and Contributors

File:Ramzibinalshibh2.jpg *Source*: http://en.wikipedia.org/w/index.php?title=File:Ramzibinalshibh2.jpg *License*: unknown *Contributors*: Duiarmia, Leoboudv

File:Ramzi bin Alshibh at Guantanamo.jpg *Source*: http://en.wikipedia.org/w/index.php?title=File:Ramzi_bin_Alshibh_at_Guantanamo.jpg *License*: unknown *Contributors*: Geo Swan, JohnCengiz77, Katarighe

File:100franctireur.jpg *Source*: http://en.wikipedia.org/w/index.php?title=File:100franctireur.jpg *License*: unknown *Contributors*: Beyond My Ken, Kintetsubuffalo, Rama, Ras67

File:Camp x-ray detainees.jpg *Source*: http://en.wikipedia.org/w/index.php?title=File:Camp_x-ray_detainees.jpg *License*: unknown *Contributors*: Shane T. McCoy, U.S. Navy

File:Sheikh july2009.jpg *Source*: http://en.wikipedia.org/w/index.php?title=File:Sheikh_july2009.jpg *License*: unknown *Contributors*: Geo Swan, Mystichumwipe, SuperMarioMan, 2 anonymous edits

File:Khalid Shaikh Mohammed after capture.jpg *Source*: http://en.wikipedia.org/w/index.php?title=File:Khalid_Shaikh_Mohammed_after_capture.jpg *License*: unknown *Contributors*: Taken by U.S. forces when KSM was captured.

File:Riduan Isamuddin.jpg *Source*: http://en.wikipedia.org/w/index.php?title=File:Riduan_Isamuddin.jpg *License*: unknown *Contributors*: Pmolloy291

File:BinAttash.jpg *Source*: http://en.wikipedia.org/w/index.php?title=File:BinAttash.jpg *License*: unknown *Contributors*: Geo Swan, WMSBRBA, 1 anonymous edits

File:Trailer where CSR Tribunals were held.jpg *Source*: http://en.wikipedia.org/w/index.php?title=File:Trailer_where_CSR_Tribunals_were_held.jpg *License*: unknown *Contributors*: Deadstar, Doldrums, Geo Swan, Shell Kinney, Túrelio, 1 anonymous edits

File:Ghailani.jpg *Source*: http://en.wikipedia.org/w/index.php?title=File:Ghailani.jpg *License*: unknown *Contributors*: Nachcommonsverschieber, Szczebrzeszynski, Themightyquill, Túrelio,

File:48-tentcity.standalone.prod affiliate.56.JPG *Source*: http://en.wikipedia.org/w/index.php?title=File:48-tentcity.standalone.prod_affiliate.56.JPG *License*: unknown *Contributors*: Department of Defense

File:MCC New York jeh.JPG *Source*: http://en.wikipedia.org/w/index.php?title=File:MCC_New_York_jeh.JPG *License*: unknown *Contributors*: User:Jim.henderson

File:Supermax prison, Florence Colorado.jpg *Source*: http://en.wikipedia.org/w/index.php?title=File:Supermax_prison,_Florence_Colorado.jpg *License*: unknown *Contributors*: US Bureau of Prisons

Image:Flightpath-AA77.gif *Source*: http://en.wikipedia.org/w/index.php?title=File:Flightpath-AA77.gif *License*: unknown *Contributors*: User:Yumetriz

File:Aa757.jpg *Source*: http://en.wikipedia.org/w/index.php?title=File:Aa757.jpg *License*: unknown *Contributors*: Ardfern, Denniss, Dmitry G, MB-one, Zscout370

File:Pentagon Security Camera 1.ogv *Source*: http://en.wikipedia.org/w/index.php?title=File:Pentagon_Security_Camera_1.ogv *License*: unknown *Contributors*: Unmanned security camera owned by the US government.

File:Flight 77 wreckage at Pentagon.jpg *Source*: http://en.wikipedia.org/w/index.php?title=File:Flight_77_wreckage_at_Pentagon.jpg *License*: unknown *Contributors*: U.S. Navy Photo by Journalist 1st Class Mark D. Faram. (RELEASED)

File:Pentagon fire AFIP.jpg *Source*: http://en.wikipedia.org/w/index.php?title=File:Pentagon_fire_AFIP.jpg *License*: unknown *Contributors*: U.S. Air Force Tech. Sgt. Louis Briscese , Forensic Photography, Office of the Armed Forces Medical Examiner, AFIP.

File:DM-SD-02-03886.JPEG *Source*: http://en.wikipedia.org/w/index.php?title=File:DM-SD-02-03886.JPEG *License*: unknown *Contributors*: DoD photo by: CPL JASON INGERSOLL, USMC

File:Aerial view of the Pentagon during rescue operations post-September 11 attack.JPEG *Source*: http://en.wikipedia.org/w/index.php?title=File:Aerial_view_of_the_Pentagon_during_rescue_operations_post-September_11_attack.JPEG *License*: unknown *Contributors*: TSGT CEDRIC H. RUDISILL, USAF

File:DN-SD-04-12744.JPEG *Source*: http://en.wikipedia.org/w/index.php?title=File:DN-SD-04-12744.JPEG *License*: unknown *Contributors*: JO1 MARK D. FARAM, USN

File:FirstFloor Pentagon Bodies.png *Source*: http://en.wikipedia.org/w/index.php?title=File:FirstFloor_Pentagon_Bodies.png *License*: unknown *Contributors*: U.S. Government

File:Flight 77 CVR.jpg *Source*: http://en.wikipedia.org/w/index.php?title=File:Flight_77_CVR.jpg *License*: unknown *Contributors*: US government during 9/11 investigation

File:Damaged Pentagon being reconstructed.jpg *Source*: http://en.wikipedia.org/w/index.php?title=File:Damaged_Pentagon_being_reconstructed.jpg *License*: unknown *Contributors*: User:Tomf688

File:Pentagon Security Camera 2.ogv *Source*: http://en.wikipedia.org/w/index.php?title=File:Pentagon_Security_Camera_2.ogv *License*: unknown *Contributors*: Unmanned security camera owned by the US government.

File:12.6.11TimothyMaudePanelS-74ByLuigiNovi2.jpg *Source*: http://en.wikipedia.org/w/index.php?title=File:12.6.11TimothyMaudePanelS-74ByLuigiNovi2.jpg *License*: unknown *Contributors*: User:Nightscream

File:Flag of the United States.svg *Source*: http://en.wikipedia.org/w/index.php?title=File:Flag_of_the_United_States.svg *License*: unknown *Contributors*: Anomie

File:Flag of the United Kingdom.svg *Source*: http://en.wikipedia.org/w/index.php?title=File:Flag_of_the_United_Kingdom.svg *License*: unknown *Contributors*: Anomie, Good Olfactory, Mifter

File:Flag of Europe.svg *Source*: http://en.wikipedia.org/w/index.php?title=File:Flag_of_Europe.svg *License*: unknown *Contributors*: User:-xfi-, User:Dbenbenn, User:Funakoshi, User:Jeltz, User:Nightstallion, User:Paddu, User:Verdy p, User:Zscout370

File:Flag of India.svg *Source*: http://en.wikipedia.org/w/index.php?title=File:Flag_of_India.svg *License*: unknown *Contributors*: Anomie, Mifter

File:Hamid Mir interviewing Osama bin Laden.jpg *Source*: http://en.wikipedia.org/w/index.php?title=File:Hamid_Mir_interviewing_Osama_bin_Laden.jpg *License*: unknown *Contributors*: Abdul Rahman bin Laden (son of Osama bin Laden) took the photo and released it to Hamid Mir, a Pakistani news reporter at the time.

File:Afghan Muja crossing from Saohol Sar pass in Durand border region of Pakistan, August 1985.png *Source*: http://en.wikipedia.org/w/index.php?title=File:Afghan_Muja_crossing_from_Saohol_Sar_pass_in_Durand_border_region_of_Pakistan,_August_1985.png *License*: unknown *Contributors*: User:Erwin Franzen

File:Omar Abdel-Rahman.jpg *Source*: http://en.wikipedia.org/w/index.php?title=File:Omar_Abdel-Rahman.jpg *License*: unknown *Contributors*: Original uploader was Paul E Ester at en.wikipedia

File:AnschalgInZahran1996 KhobarTower.jpg *Source*: http://en.wikipedia.org/w/index.php?title=File:AnschalgInZahran1996_KhobarTower.jpg *License*: unknown *Contributors*: D1ma5ad, Hohum, Kilom691, Maksim, Paulbe, Pmsyyz, Túrelio, Ultrogothe

File:Anwar al-Awlaki sitting on couch, lightened.jpg *Source*: http://en.wikipedia.org/w/index.php?title=File:Anwar_al-Awlaki_sitting_on_couch,_lightened.jpg *License*: unknown *Contributors*: User:Greg A L

File:Osama bin Laden compound1.jpg *Source*: http://en.wikipedia.org/w/index.php?title=File:Osama_bin_Laden_compound1.jpg *License*: unknown *Contributors*: Sajjad Ali Qureshi

File:Flag of Turkey.svg *Source*: http://en.wikipedia.org/w/index.php?title=File:Flag_of_Turkey.svg *License*: unknown *Contributors*: User:Dbenbenn

File:Flag of Yemen.svg *Source*: http://en.wikipedia.org/w/index.php?title=File:Flag_of_Yemen.svg *License*: unknown *Contributors*: -

File:Flag of Kenya.svg *Source*: http://en.wikipedia.org/w/index.php?title=File:Flag_of_Kenya.svg *License*: unknown *Contributors*: User:Pumbaa80

File:Flag of Tanzania.svg *Source*: http://en.wikipedia.org/w/index.php?title=File:Flag_of_Tanzania.svg *License*: unknown *Contributors*: User:Alkari, User:Madden, User:SKopp

File:RecentAlQaedaAttacks.svg *Source*: http://en.wikipedia.org/w/index.php?title=File:RecentAlQaedaAttacks.svg *License*: unknown *Contributors*: Uploaded and created by NikNaks93.

File:Ramzi Yousef.gif *Source*: http://en.wikipedia.org/w/index.php?title=File:Ramzi_Yousef.gif *License*: unknown *Contributors*: Leoboudv, Man vyi, Nixón, Officer, Steven Andrew Miller, Themightyquill, 1 anonymous edits

File:Kenya bombing 1.jpg *Source*: http://en.wikipedia.org/w/index.php?title=File:Kenya_bombing_1.jpg *License*: unknown *Contributors*: FBI

File:FEMA - 4235 - Photograph by Andrea Booher taken on 09-28-2001 in New York.jpg *Source*: http://en.wikipedia.org/w/index.php?title=File:FEMA_-_4235_-_Photograph_by_Andrea_Booher_taken_on_09-28-2001_in_New_York.jpg *License*: unknown *Contributors*: Multichill

File:Anaconda-helicopter.jpg *Source*: http://en.wikipedia.org/w/index.php?title=File:Anaconda-helicopter.jpg *License*: unknown *Contributors*: FSII, Orlovic, Zaccarias, 1 anonymous edits

File:The Guardian al-Qaeda recruitment.jpg *Source*: http://en.wikipedia.org/w/index.php?title=File:The_Guardian_al-Qaeda_recruitment.jpg *License*: unknown *Contributors*: Good Olfactory, Ottre

File:INTEL-COGNITIVE-Cole.jpg *Source*: http://en.wikipedia.org/w/index.php?title=File:INTEL-COGNITIVE-Cole.jpg *License*: unknown *Contributors*: Docu, Geo Swan, Hcberkowitz, Tangopaso

File:Abu Ayyub al-Masri 1.jpg *Source*: http://en.wikipedia.org/w/index.php?title=File:Abu_Ayyub_al-Masri_1.jpg *License*: unknown *Contributors*: United States Department of State

File:An Al Qaida recruit dreams about Osama bin Laden.jpg *Source*: http://en.wikipedia.org/w/index.php?title=File:An_Al_Qaida_recruit_dreams_about_Osama_bin_Laden.jpg *License*: unknown *Contributors*: Beria, Geo Swan, Jacklee, McZusatz, ZxxZxxZ

File:Abdulaziz al-Omari.png *Source*: http://en.wikipedia.org/w/index.php?title=File:Abdulaziz_al-Omari.png *License*: unknown *Contributors*: NJR ZA, The Evil Spartan, Themightyquill

Image:Atta in airport.jpg *Source*: http://en.wikipedia.org/w/index.php?title=File:Atta_in_airport.jpg *License*: unknown *Contributors*: security camera

Image:Abdulaziz al-omari alive.jpg *Source*: http://en.wikipedia.org/w/index.php?title=File:Abdulaziz_al-omari_alive.jpg *License*: unknown *Contributors*: User:OrphanBot

File:Abu Zubaydah.jpg *Source*: http://en.wikipedia.org/w/index.php?title=File:Abu_Zubaydah.jpg *License*: unknown *Contributors*: Department of Defense

File:Ahmed al-Nami.png *Source*: http://en.wikipedia.org/w/index.php?title=File:Ahmed_al-Nami.png *License*: unknown *Contributors*: Fietsbel, Lowdown, Meno25, Túrelio, 1 anonymous edits

File:AAlhaznawi.JPG *Source*: http://en.wikipedia.org/w/index.php?title=File:AAlhaznawi.JPG *License*: unknown *Contributors*: AnRo0002, Bukk, Fietsbel, Lowdown, 2 anonymous edits

File:Alhaznawi video aljazeera.jpg *Source*: http://en.wikipedia.org/w/index.php?title=File:Alhaznawi_video_aljazeera.jpg *License*: unknown *Contributors*: Quadell, Sherurcij

GNU Free Documentation License Version 1.2, November 2002 Copyright (C) 2000,2001,2002 Free Software Foundation, Inc. 59 Temple Place, Suite 330, Boston, MA 02111-1307 USA Everyone is permitted to copy and distribute verbatim copies of this license document, but changing it is not allowed.

0. PREAMBLE

The purpose of this License is to make a manual, textbook, or other functional and useful document "free" in the sense of freedom: to assure everyone the effective freedom to copy and redistribute it, with or without modifying it, either commercially or noncommercially. Secondarily, this License preserves for the author and publisher a way to get credit for their work, while not being considered responsible for modifications made by others. This License is a kind of "copyleft", which means that derivative works of the document must themselves be free in the same sense. It complements the GNU General Public License, which is a copyleft license designed for free software. We have designed this License in order to use it for manuals for free software, because free software needs free documentation: a free program should come with manuals providing the same freedoms that the software does. But this License is not limited to software manuals; it can be used for any textual work, regardless of subject matter or whether it is published as a printed book. We recommend this License principally for works whose purpose is instruction or reference.

1. APPLICABILITY AND DEFINITIONS

This License applies to any manual or other work, in any medium, that contains a notice placed by the copyright holder saying it can be distributed under the terms of this License. Such a notice grants a world-wide, royalty-free license, unlimited in duration, to use that work under the conditions stated herein. The "Document", below, refers to any such manual or work. Any member of the public is a licensee, and is addressed as "you". You accept the license if you copy, modify or distribute the work in a way requiring permission under copyright law. A "Modified Version" of the Document means any work containing the Document or a portion of it, either copied verbatim, or with modifications and/or translated into another language. A "Secondary Section" is a named appendix or a front-matter section of the Document that deals exclusively with the relationship of the publishers or authors of the Document to the Document's overall subject (or to related matters) and contains nothing that could fall directly within that overall subject. (Thus, if the Document is in part a textbook of mathematics, a Secondary Section may not explain any mathematics.) The relationship could be a matter of historical connection with the subject or with related matters, or of legal, commercial, philosophical, ethical or political position regarding them. The "Invariant Sections" are certain Secondary Sections whose titles are designated, as being those of Invariant Sections, in the notice that says that the Document is released under this License. If a section does not fit the above definition of Secondary then it is not allowed to be designated as Invariant. The Document may contain zero Invariant Sections. If the Document does not identify any Invariant Sections then there are none. The "Cover Texts" are certain short passages of text that are listed, as Front-Cover Texts or Back-Cover Texts, in the notice that says that the Document is released under this License. A Front-Cover Text may be at most 5 words, and a Back-Cover Text may be at most 25 words. A "Transparent" copy of the Document means a machine-readable copy, represented in a format whose specification is available to the general public, that is suitable for revising the document straightforwardly with generic text editors or (for images composed of pixels) generic paint programs or (for drawings) some widely available drawing editor, and that is suitable for input to text formatters or for automatic translation to a variety of formats suitable for input to text formatters. A copy made in an otherwise Transparent file format whose markup, or absence of markup, has been arranged to thwart or discourage subsequent modification by readers is not Transparent. An image format is not Transparent if used for any substantial amount of text. A copy that is not "Transparent" is called "Opaque". Examples of suitable formats for Transparent copies include plain ASCII without markup, Texinfo input format, LaTeX input format, SGML or XML using a publicly available DTD, and standard-conforming simple HTML, PostScript or PDF designed for human modification. Examples of transparent image formats include PNG, XCF and JPG. Opaque formats include proprietary formats that can be read and edited only by proprietary word processors, SGML or XML for which the DTD and/or processing tools are not generally available, and the machine-generated HTML, PostScript or PDF produced by some word processors for output purposes only. The "Title Page" means, for a printed book, the title page itself, plus such following pages as are needed to hold, legibly, the material this License requires to appear in the title page. For works in formats which do not have any title page as such, "Title Page" means the text near the most prominent appearance of the work's title, preceding the beginning of the body of the text. A section "Entitled XYZ" means a named subunit of the Document whose title either is precisely XYZ or contains XYZ in parentheses following text that translates XYZ in another language. (Here XYZ stands for a specific section name mentioned below, such as "Acknowledgements", "Dedications", "Endorsements", or "History".) To "Preserve the Title" of such a section when you modify the Document means that it remains a section "Entitled XYZ" according to this definition. The Document may include Warranty Disclaimers next to the notice which states that this License applies to the Document. These Warranty Disclaimers are considered to be included by reference in this License, but only as regards disclaiming warranties: any other implication that these Warranty Disclaimers may have is void and has no effect on the meaning of this License.

2. VERBATIM COPYING

You may copy and distribute the Document in any medium, either commercially or noncommercially, provided that this License, the copyright notices, and the license notice saying this License applies to the Document are reproduced in all copies, and that you add no other conditions whatsoever to those of this License. You may not use technical measures to obstruct or control the reading or further copying of the copies you make or distribute. However, you may accept compensation in exchange for copies. If you distribute a large enough number of copies you must also follow the conditions in section 3. You may also lend copies, under the same conditions stated above, and you may publicly display copies.

3. COPYING IN QUANTITY

If you publish printed copies (or copies in media that commonly have printed covers) of the Document, numbering more than 100, and the Document's license notice requires Cover Texts, you must enclose the copies in covers that carry, clearly and legibly, all these Cover Texts: Front-Cover Texts on the front cover, and Back-Cover Texts on the back cover. Both covers must also clearly and legibly identify you as the publisher of these copies. The front cover must present the full title with all words of the title equally prominent and visible. You may add other material on the covers in addition. Copying with changes limited to the covers, as long as they preserve the title of the Document and satisfy these conditions, can be treated as verbatim copying in other respects. If the required texts for either cover are too voluminous to fit legibly, you should put the first ones listed (as many as fit reasonably) on the actual cover, and continue the rest onto adjacent pages. If you publish or distribute Opaque copies of the Document numbering more than 100, you must either include a machine-readable Transparent copy along with each Opaque copy, or state in or with each Opaque copy a computer-network location from which the general network-using public has access to download using public-standard network protocols a complete Transparent copy of the Document, free of added material. If you use the latter option, you must take reasonably prudent steps, when you begin distribution of Opaque copies in quantity, to ensure that this Transparent copy will remain thus accessible at the stated location until at least one year after the last time you distribute an Opaque copy (directly or through your agents or retailers) of that edition to the public. It is requested, but not required, that you contact the authors of the Document well before redistributing any large number of copies, to give them a chance to provide you with an updated version of the Document.

4. MODIFICATIONS

You may copy and distribute a Modified Version of the Document under the conditions of sections 2 and 3 above, provided that you release the Modified Version under precisely this License, with the Modified Version filling the role of the Document, thus licensing distribution and modification of the Modified Version to whoever possesses a copy of it. In addition, you must do these things in the Modified Version: A. Use in the Title Page (and on the covers, if any) a title distinct from that of the Document, and from those of previous versions (which should, if there were any, be listed in the History section of the Document). You may use the same title as a previous version if the original publisher of that version gives permission. B. List on the Title Page, as authors, one or more persons or entities responsible for authorship of the modifications in the Modified Version, together with at least five of the principal authors of the Document (all of its principal authors, if it has fewer than five), unless they release you from this requirement. C. State on the Title page the name of the publisher of the Modified Version, as the publisher. D. Preserve all the copyright notices of the Document. E. Add an appropriate copyright notice for your modifications adjacent to the other copyright notices. F. Include, immediately after the copyright notices, a license notice giving the public permission to use the Modified Version under the terms of this License, in the form shown in the Addendum below. G. Preserve in that license notice the full lists of Invariant Sections and required Cover Texts given in the Document's license notice. H. Include an unaltered copy of this License. I. Preserve the section Entitled "History", Preserve its Title, and add to it an item stating at least the title, year, new authors, and publisher of the Modified Version as given on the Title Page. If there is no section Entitled "History" in the Document, create one stating the title, year, authors, and publisher of the Document as given on its Title Page, then add an item describing the Modified Version as stated in the previous sentence. J. Preserve the network location, if any, given in the Document for public access to a Transparent copy of the Document, and likewise the network locations given in the Document for previous versions it was based on. These may be placed in the "History" section. You may omit a network location for a work that was published at least four years before the Document itself, or if the original publisher of the version it refers to gives permission. K. For any section Entitled "Acknowledgements" or "Dedications", Preserve the Title of the section, and preserve in the section all the substance and tone of each of the contributor acknowledgements and/or dedications given therein. L. Preserve all the Invariant Sections of the Document, unaltered in their text and in their titles. Section numbers or the equivalent are not considered part of the section titles. M. Delete any section Entitled "Endorsements". Such a section may not be included in the Modified Version. N. Do not retitle any existing section to be Entitled "Endorsements" or to conflict in title with any Invariant Section. O. Preserve any Warranty Disclaimers. If the Modified Version includes new front-matter sections or appendices that qualify as Secondary Sections and contain no material copied from the Document, you may at your option designate some or all of these sections as invariant. To do this, add their titles to the list of Invariant Sections in the Modified Version's license notice. These titles must be distinct from any other section titles. You may add a section Entitled "Endorsements", provided it contains nothing but endorsements of your Modified Version by various parties--for example, statements of peer review or that the text has been approved by an organization as the authoritative definition of a standard. You may add a passage of up to five words as a Front-Cover Text, and a passage of up to 25 words as a Back-Cover Text, to the end of the list of Cover Texts in the Modified Version. Only one passage of Front-Cover Text and one of Back-Cover Text may be added by (or through arrangements made by) any one entity. If the Document already includes a cover text for the same cover, previously added by you or by arrangement made by the same entity you are acting on behalf of, you may not add another; but you may replace the old one, on explicit permission from the previous publisher that added the old one. The author(s) and publisher(s) of the Document do not by this License give permission to use their names for publicity for or to assert or imply endorsement of any Modified Version.

5. COMBINING DOCUMENTS

You may combine the Document with other documents released under this License, under the terms defined in section 4 above for modified versions, provided that you include in the combination all of the Invariant Sections of all of the original documents, unmodified, and list them all as Invariant Sections of your combined work in its license notice, and that you preserve all their Warranty Disclaimers. The combined work need only contain one copy of this License, and multiple identical Invariant Sections may be replaced with a single copy. If there are multiple Invariant Sections with the same name but different contents, make the title of each such section unique by adding at the end of it, in parentheses, the name of the original author or publisher of that section if known, or else a unique number. Make the same adjustment to the section titles in the list of Invariant Sections in the license notice of the combined work. In the combination, you must combine any sections Entitled "History" in the various original documents, forming one section Entitled "History"; likewise combine any sections Entitled "Acknowledgements", and any sections Entitled "Dedications". You must delete all sections Entitled "Endorsements".

6. COLLECTIONS OF DOCUMENTS

You may make a collection consisting of the Document and other documents released under this License, and replace the individual copies of this License in the various documents with a single copy that is included in the collection, provided that you follow the rules of this License for verbatim copying of each of the documents in all other respects. You may extract a single document from such a collection, and distribute it individually under this License, provided you insert a copy of this License into the extracted document, and follow this License in all other respects regarding verbatim copying of that document.

7. AGGREGATION WITH INDEPENDENT WORKS

A compilation of the Document or its derivatives with other separate and independent documents or works, in or on a volume of a storage or distribution medium, is called an "aggregate" if the copyright resulting from the compilation is not used to limit the legal rights of the compilation's users beyond what the individual works permit. When the Document is included in an aggregate, this License does not apply to the other works in the aggregate which are not themselves derivative works of the Document. If the Cover Text requirement of section 3 is applicable to these copies of the Document, then if the Document is less than one half of the entire aggregate, the Document's Cover Texts may be placed on covers that bracket the Document within the aggregate, or the electronic equivalent of covers if the Document is in electronic form. Otherwise they must appear on printed covers that bracket the whole aggregate.

8. TRANSLATION

Translation is considered a kind of modification, so you may distribute translations of the Document under the terms of section 4. Replacing Invariant Sections with translations requires special permission from their copyright holders, but you may include translations of some or all Invariant Sections in addition to the original versions of these Invariant Sections. You may include a translation of this License, and all the license notices in the Document, and any Warranty Disclaimers, provided that you also include the original English version of this License and the original versions of those notices and disclaimers. In case of a disagreement between the translation and the original version of this License or a notice or disclaimer, the original version will prevail. If a section in the Document is Entitled "Acknowledgements", "Dedications", or "History", the requirement (section 4) to Preserve its Title (section 1) will typically require changing the actual title.

9. TERMINATION

You may not copy, modify, sublicense, or distribute the Document except as expressly provided for under this License. Any other attempt to copy, modify, sublicense or distribute the Document is void, and will automatically terminate your rights under this License. However, parties who have received copies, or rights, from you under this License will not have their licenses terminated so long as such parties remain in full compliance.

10. FUTURE REVISIONS OF THIS LICENSE

The Free Software Foundation may publish new, revised versions of the GNU Free Documentation License from time to time. Such new versions will be similar in spirit to the present version, but may differ in detail to address new problems or concerns. See http://www.gnu.org/copyleft/. Each version of the License is given a distinguishing version number. If the Document specifies that a particular numbered version of this License "or any later version" applies to it, you have the option of following the terms and conditions either of that specified version or of any later version that has been published (not as a draft) by the Free Software Foundation. If the Document does not specify a version number of this License, you may choose any version ever published (not as a draft) by the Free Software Foundation. ADDENDUM: How to use this License for your documents To use this License in a document you have written, include a copy of the License in the document and put the following copyright and license notices just after the title page: Copyright (c) YEAR YOUR NAME. Permission is granted to copy, distribute and/or modify this document under the terms of the GNU Free Documentation License, Version 1.2 or any later version published by the Free Software Foundation; with no Invariant Sections, no Front-Cover Texts, and no Back-Cover Texts. A copy of the license is included in the section entitled "GNU Free Documentation License". If you have Invariant Sections, Front-Cover Texts and Back-Cover Texts, replace the "with...Texts." line with this: with the Invariant Sections being LIST THEIR TITLES, with the Front-Cover Texts being LIST, and with the Back-Cover Texts being LIST. If you have Invariant Sections without Cover Texts, or some other combination of the three, merge those two alternatives to suit the situation. If your document contains nontrivial examples of program code, we recommend releasing these examples in parallel under your choice of free software license, such as the GNU General Public License, to permit their use in free software.

MIX
Papier aus verantwortungsvollen Quellen
Paper from responsible sources
FSC® C105338

Printed by Books on Demand GmbH, Norderstedt / Germany